Biology of the Vestfold Hills, Antarctica

Developments in Hydrobiology 34

Series editor
H.J. Dumont

Biology of the Vestfold Hills, Antarctica

Proceedings of the symposium, Hobart, August 1984

Edited by
J. M. Ferris, H. R. Burton, G. W. Johnstone & I. A. E. Bayly

Reprinted from Hydrobiologia, vol. 165

1988 **KLUWER ACADEMIC PUBLISHERS**
DORDRECHT / BOSTON / LONDON

Biology of the Vestfold Hills, Antarctica.

(Developments in hydrobiology ; 34)
Papers presented at the Symposium on Biological Research in the Vestfold Hills.
"Reprinted from Hydrobiologia."
Includes index.
1. Natural history--Antarctic regions--Vestfold Hills--Congresses. 2. Biology--Antarctic regions--Vestfold Hills--Congresses. 3. Vestfold Hills (Antarctic regions)--Congresses. I. Ferris, J. M. II. Symposium on Biological Research in the Vestfold Hills (1984 : Hobart, Tas.) III. Series.
QH84.2.B56 1988 574.998'9 86-18557

ISBN 90-6193-616-0
ISBN 90-6193-751-5 (series)

Published by Kluwer Academic Publishers
P.O. Box 17, 3300 AA Dordrecht, The Netherlands

Kluwer Academic Publishers incorporates the publishing programmes of D. Reidel, Martinus Nijhoff, Dr W. Junk and MTP Press.

Sold and distributed in the U.S.A. and Canada
by Kluwer Academic Publishers,
101 Philip Drive, Norwell, MA 02061, U.S.A.

In all other countries, sold and distributed
by Kluwer Academic Publishers Group,
P.O. Box 322, 3300 AH Dordrecht, The Netherlands

Printed in The Netherlands.

Dust jacket:
An oblique aerial view to the west over Deep Lake with Lakes Stinear and Dingle beyond. Offshore islands, icebergs and the open sea can be seen in the distance.
(Photo: H. R. Burton, February 1984.)

Tendrils of filamentous algae in a shallow meltstream near Weddell Arm of Long Fjord.
8 January 1980. G. W. Johnstone.

Preface

The present volume was conceived as a companion to 'Antarctic Oasis: Terrestrial environments and history of the Vestfold Hills' edited by J. Pickard and published in 1986 by Academic Press, Sydney. Pickard's book contains accounts of the Vestfold Hills' climate (N. A. Streten) and recent geomorphological history (D. A. Adamson & J. Pickard) which provide a valuable context for understanding their present day biology. Pickard also gives a history of human discovery and occupation of the Vestfold Hills. There is some overlap in the coverage, to the extent that both this volume and Pickard's book describe the terrestrial flora and fauna. The reader specifically interested in the terrestrial ecosystems of the Vestfold Hills should draw from both sources. Together, these works present a broad and descriptive account of the largest truly coastal antarctic oasis: a region that holds a unique variety of opportunities for future scientific investigation.

There are several tasks I wish to accomplish here, apart from expressing my sincere thanks to the many people who have contributed to the completion of this volume. I wish to briefly introduce the Vestfolds and to list some of the features that, in my opinion, make them biologically varied, and unique in the context of other coastal ice-free areas. I wish to describe the phases of biological research in this region, including the directions that have been pursued since the 1984 symposium and to comment upon the future of the Vestfold Hills.

Perhaps the most eloquent introduction to the Vestfold Hills is contained in the photographic essay (Figs. 1–6) that accompanies this preface. Maps of the Vestfold Hills are presented in many of the papers in this volume, and the paper by Miller *et al.* shows many of the named locations within the hills. An excellent map, which includes biological information such as Adélie penguin breeding sites and Elephant seal wallows, is available from the Surveying and Land Information Group, Department of Administrative Services, Canberra (Australian Antarctic Territory, Princess Elizabeth Land, Vestfold Hills – 1:50000, Second Edition, September 1982).

From the air in summer, the Vestfold Hills are an area of brown boulder-strewn ground striped by near-black dykes, dotted with lakes and partitioned into three peninsulas by fjords. The dark hills form a stark contrast to the brilliant, blue-white antarctic ice-cap. The lakes that occupy valleys and depressions are a striking and varied feature of the landscape, some are white with moated ice-cover, others are blue, brown or green (Dust-cover & Fig. 4). They vary from deep freshwater basins like Crooked Lake (157 m deep) to shallow hypersaline and meromictic lakes like Organic Lake. This range of salinities distinguishes the Vestfold Hills from other nearby oases. Both the Bunger and Larsemann Hills are dotted with lakes but almost all are less saline than seawater; at least one hypersaline lake occurs on Cachalot Island in the Bunger Hills. Lake formation in the Vestfold Hills reflects the combination of ice retreat from a low lying area which was then partially flooded by the sea, and subsequent isostatic readjustment which has isolated sea water basins to further develop according to the balance of evaporation and snowmelt inflow. Lakes of such different salinities show a variety of density stratification cycles. The saltiest remain ice-free throughout the year and have classic monomictic thermal stratification patterns but with winter water temperatures down to −17 °C. Less salty lakes and the fjords are ice-covered for much of the year and in some of these bottom layers of anoxic water have developed and are maintained over many years. These anoxic basins trap sediments and hold nearly undisturbed microfossil and organic geochemical records of the recent past. Within this small area (c. 400 km^2), there are about 15 meromictic lakes and at least five stratified basins in marine-connected embayments and fjords. Some lakes are at intermediate stages of separation from the sea, being flooded seasonally or at especially high tides, while the small meromictic basin at the head of Ellis Fjord is isolated for about four months by the formation of sea-ice over the 1 m deep sill at its western side. The Vestfold Hills saline lakes do not have very diverse phytoplankton communities. Ace Lake, for example, has only four species of flagellates. However, the diversity of aquatic environments in the hills is the basis for a remarkable diversity of biological responses in terms of community structure, seasonal cycles and production. The extremes of planktonic primary production are exemplified in Deep Lake, with its sparse plankton comprised of one alga and two strains of a bacterium, on the one hand and Ellis Fjord, in which chlorophyll can reach concentrations of 500 $\mu g\ l^{-1}$ in localized patches beneath the ice, on the other.

The terrestrial ecosystem of the Vestfold Hills is sufficiently young and isolated to be species poor even when compared to many others of the world's desert systems. Algae, lichens and mosses are the known base of a short invertebrate food chain, comprising nematodes, mites and tardigrades. Nevertheless, the Vestfold Hills contain a diversity of ecological gradients related to the availability of freshwater in summer, salt buildup in the soil, proximity to bird colonies or seal wallows, exposure to wind, accumulation of windblown sand and the distribution of translucent rocks. These, in turn, reflect the interaction between factors such as the time since ice retreat, sea level and isostatic changes, surface geology, topography and climate.

The coast and outlying islands of the Vestfold Hills are lapped by the waters of Prydz Bay for part of each year. Sea ice breaks out in stages, usually reaching back to the outer islands in spring where it halts for a time before clearing altogether from the coast. This accessibility is a key factor in the existence of Adélie penguin rookeries along the Vestfold Hills' coast and on offshore islands (Fig. 5b), and also in the occurrence of the southernmost known Elephant seal wallows (Fig. 6a). In contrast, the larger Bunger Hills lie behind the Shackleton Ice Shelf and have no Adélie penguin rookeries. The Larsemann Hills area also has very few such sites and this may be a result of the stable sea-ice that makes nearby Amanda Bay a suitable site for winter-breeding Emperor penguins. The strip of coast from the Publication Ice Shelf to the Vestfold Hills includes the Larsemann Hills, Amanda Bay and the Rauer Islands and contains breeding sites of all ten bird species regarded as breeding on Antarctica proper, whereas the coast for some 400 km west and 600 km east of this stretch is largely devoid of bird breeding sites.

The fjords of the Vestfold Hills are ice covered for even longer than the coastal waters off Davis and may remain almost completely covered throughout the summer, as was the case for Ellis Fjord in 1985–1986. In other years, however, Ellis fjord can be completely free of ice for part of the summer (e.g. in 1974–1975). The longer lived ice cover in Tryne and Long Fjords offers protected breeding and moulting sites for Weddell seals in spring and early summer.

The nearshore marine area is also affected by the variability of ice cover, in that the break up and dispersal of ice effects primary physical factors such as light climate and mixing in the water column, as well as simply removing the ecosystem that develops with ice as its substrate. The nearshore marine ecosystem is biologically representative of other antarctic coastal sites, at least in terms of the taxa present. However, there is good evidence that the fjords provide environmental gradients with respect to ice cover and general mixing patterns and that their biology reflects this transition from nearshore marine to lake-like marine conditions. Contrasts between different fjords (e.g. Long and Ellis Fjords) are likely to be interesting in this context.

Three phases of biological research can be recognized for the Vestfold Hills. Prior to 1971, research and publication was the largely the province of interested expeditioners, particularly the station medical staff. From 1971–1984, the period reported here, professional biologists from within the Antarctic Division have wintered at Davis and others including some external researchers have spent summers there. From 1985, research in the Vestfold Hills has entered a third phase, with the reactivated involvement of Australian Universities funded by grants administered through the Antarctic Science Advisory Committee (ASAC). In this period the summer involvement of overseas researchers has also increased.

During the undeniably protracted genesis of this volume, biological research in the Vestfold Hills has continued in some of the directions indicated by the work presented here and there has been a pleasing acceleration of publication. Three primary directions have been followed by Antarctic Division scientists, and in cooperation with external scientists. First, seasonal aspects of saline lake biology have been investigated in lakes of different salinity, both for their intrinsic value in understanding the systems that exist today and for their contribution to interpreting the history of the Vestfold Hills' lakes. The history of saline lakes has been investigated from coring, particularly of the undisturbed sediments found in meromictic basins. Interpretation of this history is based on microfossil enumeration, mineralogy, carbon isotope measurements and studies of existing plankton communities in lakes of varying salinities. In cooperation with C.S.I.R.O. Fisheries and Oceanography chemists, geochemical 'biomarker' compounds have been sought from sediment and are of particular interest in these lakes, where important elements of the existing plankton are not preserved as microfossils. Microbiological investigations of saline lakes has progressed rapidly following the establishment of the Australian Collection of Antarctic Microorganisms (University of Tasmania) and chemical characterization of cultured bacteria has added to the geochemical work. Another project with implications for interpretation of lake development was the coordinated study of Ellis Fjord, in 1985, which provided data on seasonal aspects of the fjord's mixing pattern, phytoplankton abundance and zoobenthos. This study also demonstrated that formation of an anoxic layer can occur perhaps thousands of years before a basin is separated from the sea. A long term project arose from the observation of lake level rises between successive aerial photographic runs for the Vestfold Hills. The level of Deep Lake is known to have risen by about 1 m from 1975–1985 which indicates that salinity may undergo dynamic change over comparatively short periods and has implications for our understanding of the physical and biological process of this and other lakes in the region.

Second, the potential use of penguins as indicators of the condition of the wider ecosystem of Prydz Bay has motivated further studies of diet and energetics of Adélie and emperor penguins. These studies have been refined and extended to include other species of seabirds nesting in the region. Similar studies of elephant seals, involving anaesthetization and weighing of animals up to 2 tonnes and the use of a non-invasive technique (ultrasound) to quantify blubber reserves, may have like application to the adjacent subantarctic region in which the elephant seal population is known to be declining.

Third, the sulfur chemistry of stratified lakes and nearshore marine waters and its relationhip to microbiological processes has been another focus of research since the finding of very high concentrations of the sulfur compounds hydrogen sulfide and dimethyl sulfide in the bottom waters of Ellis Fjord and Organic Lake respectively. The seasonal measurement of reduced sulfur compounds in the sea, particularly, will contribute to the quantification of biologically mediated steps in the antarctic sulfur cycle.

Features of the Vestfold Hills that have been little investigated include the freshwater lakes, both in terms of their biota and their physical behaviour, a river system which carries meltwater from the Sørsdal glacier via Crooked and Druzhby Lakes to the head of Ellis Fjord (Fig. 2a), the fungal and bacterial microbiology of terrestrial sites, the effects of the still limited human presence and the ecological significance of year to year variation in formation and breakup of sea and lake ice. I believe that this volume will help provide direction to this research as it enters a more analytical phase and that it will also serve to point out areas in which much fundamental descriptive work remains to be done.

The future of the Vestfold Hills is the subject of much speculation at present. The potential for biological research is enormous as is the potential for research into upper atmosphere physics and disciplines such as geology. However, there are clear indications from public debate that tourism is likely to be the first land-based commercial exploitation of continental Antarctica. This will involve the biological resources represented by the unique fauna and flora and, consequently, oases such as the Vestfold Hills will be among the preferred sites of access. I feel that this presents the Antarctic Treaty nations and the scientists who have enjoyed the unique patronage of the treaty with a challenge – that is, to ensure that any such development is preceded by research specifically directed at assessing the impact of such developments and is guided by the strictest principles of environmental protection. I think it difficult to overstate the urgency of meeting this challenge.

In formulating this account I have drawn on my own experience, the much broader experience of H. R. Burton and conversations with other colleagues who have visited and worked at Davis, the Bunger Hills and at Law Base in the Larsemann Hills: D. Eslake, N. J. Gales, J. B. Gallagher, J. Gibson, D. Gillieson, K. Green, D. Hay, J. van den Hoff, R. Kirkwood, J. Kirkwood, T. D. Walker and M. Whitehead. I also acknowledge an unpublished report by R. Barker on his limnological reconnaissance of the Bunger Hills, the earlier account of Vestfold Hill's biology by G. W. Johnstone, D. J. Lugg and D. A. Brown (1973) which is referenced in most papers in this volume, and the papers delivered at the Vestfold Hills Symposium, both those reported here and the abstracts of others that could not be included.

Two aerial photographs, Fig. 4 following the Preface and Fig. 1 in Peterson *et al.* (page 222) are Crown copyright and are included here with permission of the Manager Surveying and Land Information Group, Department of Administrative Services, Australia.

Some figures in Masuda *et al.* (pages 103–114) and Seppelt *et al.* (pages 185–196) are modified from previous publications and are included with the permission of the Japanese National Institute of Polar Research and Academic Press Australia, respectively.

In many of the papers that follow there are acknowledgements of the role of expeditioners who have maintained Davis Station and also participated in the field work that underpins all the science reported here. I wish to add my sincerest thanks to the expeditioners, too numerous to name, who are fundamental to the conduct of life and work at Davis and Australia's other antarctic stations. To the referees who reviewed the papers presented here I extend my sincere thanks, their constructive criticism improved many of the texts. I owe a great debt to the Antarctic Division's typists (S. Birch, J. Cole, H. Cooper, M. Curran, V. Dickson, J. Foley, C. Nicholls) who produced the final copies of almost all papers that were sent to the publishers. I thank C. Jones for creating the index. I also acknowledge the assistance of the Division's graphic artists (J. Cox and B. Hill) and photographers (R. Reeves and G. Jacobson) who gave final form to many of the scientific figures. I thank my fellow editors for their support and the Antarctic Divison for encouraging my endeavours to complete this volume and giving me the opportunity to participate in two summer expeditions to Davis. Finally, I would like to record my thanks to Professor Dumont (the series editor) for his patience and to the publishers for their prompt and accurate attention to my many questions and requests.

John Ferris

The following four papers, originally delivered at the symposium, have appeared in Hydrobiologia in advance of the main text:

Bayly, I. A. E., 1986. Ecology of the zooplankton of a meromictic antarctic lagoon with special reference to *Drepanopus bispinosus* (Copepoda: Calanoida). Hydrobiologia 140: 199–231.

Everitt, D. A. & D. P. Thomas, 1986. Observations of the seasonal changes in diatoms at inshore localities near Davis Station, East Antarctica. Hydrobiologia 139: 3–12.

Perrin, R. A., P. Lu & H. J. Marchant, 1987. Seasonal variation in marine phytoplankton and ice algae at a shallow antarctic coastal site. Hydrobiologia 146: 33–46.

Thomas, D. P. & J. L. Jiang, 1986. Epiphytic diatoms of the inshore area near Davis Station. Hydrobiologia 140: 193–198.

Fig. 1a. A waterfall cascades down the northern face of Sørsdal Glacier on the south-eastern margin of the Vestfold Hills. 7 January 1980. G. W. Johnstone.

Fig. 1b. A 'dead ice moraine' and associated melt lake between the eastern end of Crooked Lake and the Sørsdal Glacier. 7 January 1980. G. W. Johnstone.

Fig. 2a. Ellis Rapids in full flood. Originating from the Sørsdal Glacier, meltwater flows through Chelnok, Crooked and Druhzby Lakes before debouching into the head of Ellis Fjord.
5 January 1980. G. W. Johnstone.

Fig. 2b. A view south along the eastern shore of Burton Lake. Part of the 'Marine Plain' is visible in the background.
December 1981. I. A. E. Bayly.

Fig. 3a. Lichen and moss at the western end of Lichen Valley. Bird bones have come from a skua nest territory above.
January 1979. R. D. Seppelt.

Fig. 3b. Hypersaline Club Lake, with white mirabilite deposits on its north-eastern shore. The surrounding marine beach terrace can be seen in the distance.
28 January 1972. G. W. Johnstone.

Fig. 4. Vertical aerial view of Lebed and Watts Lakes. A pair of prominent dykes (darker rock), run approximately north/south between these two lakes. Ellis Rapids are at the top, just left of centre.

Fig. 5a. A diving party at the ice edge near the mouth of Crooked Fjord. The Sørsdal Glacier is behind. 8 December 1981. G. W. Johnstone.

Fig. 5b. Part of the Adélie penguin colony on Magnetic Island. 7 December 1981. G. W. Johnstone.

Fig. 6a. Moulting male elephant seals wallow on the beach at Davis Station. Small clumps of snow adhere to the face and neck of the nearest seal.
April 1974. R. Williams.

Fig. 6b. The eroding carcass of an elephant seal beside Laturnula Lake.
4 February 1972. G. W. Johnstone.

Contents

Gavin Wildridge Johnstone (1941–1987)

Gavin was born on 15 May 1941 in Peeblesshire, Scotland. He completed his undergraduate degree (with Honours in Zoology) at Bristol University, followd by a Diploma in Conservation (with distinction) at University College, London and a Ph.D. entitled 'Ecology, dispersion and arena behaviour of black grouse *Lyrurus tetrix* (L.) in Glen Dye N-E Scotland' at Aberdeen University. Gavin held a Royal Society European Program postdoctoral fellowship at Groningen University (Netherlands) from 1968 to 1969. He subsequently joined the Australian National Antarctic Research Expedition (ANARE) to Macquarie Island; overwintering in 1970. Retained as deputy to the newly appointed senior biologist, Gavin remained in Australia where he married Kerry Nolan in 1972. Also in that year, he published a valuable summary of ANARE biological research for the period 1947–1971 (Johnstone, 1972). The Antarctic Division, having decided in 1971 to devote more resources to biological research on the continent, was faced with the necessity of choosing which of the three stations should be the focus for this effort. Gavin was given the task of assessing the Vestfold Hills area where Davis Station had operated almost continuously since 1957. His compilation of the observations made by previous expeditions and his impressions gained during a summer visit provided the first general account of the biology of the Vestfold Hills (Johnstone *et al.*, 1973), to which the present volume is sequel. These works formed a much needed basis for the subsequent direction of biological research by ANAREs and were instrumental in the choice of the Vestfold Hills as a focus for the Division's biological program in continental Antarctica.

In 1978 Gavin became full-time acting senior biologist and in 1980, when the Antarctic Division's growing biological section was informally split into land-based and marine subgroups, he became head of land-based biological research. In this capacity Gavin supervised many expeditioners who will remember his unstinting generosity with personal time, a trait widely appreciated and acknowledged by his scientific colleagues, his assiduous attention to detail and his dry sense of humour. Some of the fruits of Gavin's support and guidance appear in this volume. However, for all the rewards of successfully directing others, I believe Gavin felt a very real frustration when in the face of a substantial administrative load he found himself unable to devote as much time as he desired to his own research. Part of his responsibility in their period was to represent Australia at international fora and from 1983 he was a member of the SCAR BIOMASS working party on bird ecology and the Australian coordinator for the International Survey of Antactic Seabirds. On 15 April 1987 Gavin ended his own life - the culmination of some years of recurrent depression. An obituary appears in Polar Record 147.

At the time of Gavin's death, Australia's antarctic ornithological research was partly directed at long-term population studies. These involved banding various petrels and penguins at subantarctic and continental stations to provide information on longevity, age at sexual maturity, migration and dispersion, as well as monitoring breeding populations at continental stations to provide basic data on breeding success over time and as a possible indicator of changes in the wider marine ecosystem of Prydz Bay. Regular observations of numbers and behaviour of birds encountered by Australia's resupply ships was another long-term project. Short-term dietary studies of penguins, petrels and skuas were conducted at both continental and subantarctic stations by overwintering expeditioners.

Gavin's major contributions to the international scientific literature concern various aspects of biology and behaviour of the giant petrels (*Macronectes giganteus* and *M. halli*) which he studied intensively whilst overwintering at Macquarie Island and in subsequent summer trips to the island as well as to the Chatham Islands (New Zealand) and mainland Antarctica. He became a recognized authority on the southern and northern variants of these predator/scavenger petrels, originally regarded as one species (*Macronectes giganteus*). Gavin published on field recognition of the two species, which he described as 'extremely similar in appearance...' (Johnstone, 1979) and was able to show that *M. halli* tended to follow ships while *M. giganteus* did not (Johnstone, 1974). His data also supported the view that the two species were geographically distinct in summer, when *M. giganteus* was most often found south of the Antarctic Convergence while *M. halli* ranged to the north (Johnstone, 1974). Gavin also made an important early contribution to the understanding of the diet of the two species: examining defensively regurgitated material from chicks, he found that penguin and seal carrion were frequent components of both species diets (Johnstone, 1977). In conjunction with his observations of behaviour at seal carcasses (Johnstone, 1979), he also suggested that *M. halli* displayed a preference for seal carrion though it seemed to come off worst in disputes over carcasses with the more numerous and aggressive *M. giganteus*. Much of this work was directed at the question - is feeding behaviour related to the fact of *M. halli* beginning to breed about six weeks before *M. giganteus*? The question remains a matter of speculation, although recent work (Hunter, 1987) indicates that male *M. halli*, particularly, feed on seal carrion and that this species begins breeding at a time when seal carcasses are most readily available. Gavin's other published works on antarctic seabirds include a summation of bird observations made from ships resupplying Australia's antarctic and subantarctic stations. These observations, the work of many expeditioners over the years, provided '... a unique picture of the distribution of seabirds in the southeastern Indian Ocean.' (Johnstone & Kerry, 1976).

As part of a broader interest in ornithology, Gavin was Review Editor for the Royal Australasian Ornithologists Union from 1973–1979 and from 1977 was a member of the Editorial panel for Corella, the journal of the Australian Bird Study Association.

Specifically in relation to seabirds, Gavin journeyed to the great Australian Bight to participate in a general survey of birds along the Nullabor Cliffs and visited Albatross Island to observe the white-capped albatross (*Diomedea cauta*). His subsequent paper on this albatross (Johnstone *et al.*, 1975) reveals Gavin's ability at patient and detailed observation of bird behaviour, which was also an important part of his work on giant petrels. Indeed, Gavin was steeped in a naturalist tradition well suited to this observational and descriptive work which, though aptly described as 'unfashionable' by White (1984), remains essential to the understanding of such a remote and little investigated region as that of the Antarctic and subantarctic. Gavin also published more general articles on birds of this region, and was actively concerned with the preservation of their habitat and protecting them from the increasing human presence. He was a counsellor with the Australian Conservation Foundation from 1977–1979, participating in the first Australian National Wilderness Conference (Tierney & Johnstone, 1978). He was also Australia's representative on the International Council for Bird Preservation, Working Group on Seabirds, from 1978.

Gavin was known for involvement beyond his scientific contribution to the antarctic community. In University years, he took an active interest in mountaineering and participated in the 1965 Scottish Hindu Kush Expedition. In later years, he sang with the Tasmanian University Choral Society. He was well known for his photographs of birds, examples of which have been exhibited and appear in several books.

The symposium and this volume owe a great deal to Gavin's determination that the broad range of scientists engaged in biologically related research in the Vestfold Hills should be gathered together to share their findings and to provide a publication reflecting the unique variety of the region's biology. I hope that this work may also contribute to another of Gavin's active concerns: the conservation of the many organisms that rely on the Vestfold Hills and which make this rocky outcrop truly an 'Oasis' at the edge of the forbiddingly cold and dry desert that is the antarctic ice cap. I am pleased to dedicate this volume to a man who gave his utmost to Australia's antarctic research and without whom this volume would not have been possible.

John Ferris - with the
gratefully acknowledged
assistance of Gavin's
colleagues at the
Antarctic Division

References

Hunter, S., 1987. Species and sexual isolating mechanisms in sibling species of giant petrels Macronectes. Polar Biol. 7: 295–301.

Johnstone, G. W., 1972. A review of biological research by Australian National Antarctic Research Expeditions, 1947–71. Polar Record 16: 519–532.

Johnstone, G. W., 1974. Field characters and behaviour at sea of giant petrels in relation to their oceanic distribution. Emu 74: 209–218.

Johnstone, G. W., 1977. Comparative feeding ecology of the giant petrels Macronectes giganteus (Gmelin) and M. halli (Matthews). In: G. A. Llano (ed.), Adaptations within Antarctic Ecosystems. Smithsonian Institution, Wash., D.C.: 647–668.

Johnstone, G. W., 1979. Agonistic behaviour of the giant-petrels Macronectes giganteus and M. halli feeding at seal carcasses. Emu 79: 129–132.

Johnstone, G. W. & K. R. Kerry, 1976. Ornithological observations in the Australian sector of the Southern Ocean. In Proceedings of the 16th International Ornithological Congress, Canberra, August 1974. Academy of Science, Canberra: 725–738.

Johnstone, G. W., D. J. Lugg & D. A. Brown, 1973. The biology of the Vestfold Hills, Antarctica. ANARE Sci. Rep. Publ. 123. Antarct. Div., Dept. Sci., Melbourne, 62 pp.

Johnstone, G. W., D. Milledge & D. F. Dorward, 1975. The white-capped albatross of Albatross Island: numbers and breeding behaviour. Emu 75: 1–11.

Tierney, T. J. & G. W. Johnstone, 1978. Conserving Australia's wilderness: Antarctica as wilderness. In Proceedings of the First National Wilderness Conference, Canberra, October 1977. Australian Conservation Foundation, Melbourne: 118–126.

White, M. G., 1984. Marine benthos. In: R. M. Laws (ed.), Antarctic Ecology, 2. Academic Press, Lond.: 421–461.

List of participants*

D. Adamson (31), Macquarie University, Sydney, New South Wales, Australia
J. Baldwin, Monash University, Melbourne, Victoria, Australia
I. A. E. Bayly (24), Monash University, Melbourne, Victoria, Australia
M. N. Bester (26), University of Pretoria, Pretoria, South Africa
J. Bleasel (19), Australian Antarctic Division, Kingston, Tasmania, Australia
L. Bowling, University of Tasmania, Hobart, Tasmania, Australia
P. A. Broady (5), University of Melbourne, Melbourne, Victoria, Australia
M. D. Burch (2), Australian Antarctic Division, Kingston, Tasmania, Australia
C. M. Burke (1), Australian Antarctic Division, Kingston, Tasmania, Australia
H. R. Burton (38), Australian Antarctic Division, Kingston, Tasmania, Australia
E. C. V. Butler, University of Melbourne, Melbourne, Victoria, Australia
G. F. Claridge (8), Antarctic Policy and Transport Studies, Department of Science & Technology, Canberra, Australian Capital Territory, Australia
E. Colqhoun (7), University of Tasmania, Hobart, Tasmania, Australia
R. Croome, University of Tasmania, Hobart, Tasmania, Australia
S. Darby (13), Monash University, Melbourne, Victoria, Australia
A. Davey (25), Australian National University, Canberra, Australian Capital Territory, Australia
D. A. Everitt (29), C.S.I.R.O. Division of Oceanography, Hobart, Tasmania, Australia
J. M. Ferris, University of Tasmania, Hobart, Tasmania, Australia
B. Finlayson, University of Melbourne, Melbourne, Victoria, Australia
J. B. Gallagher (14), Australian Antarctic Division, Kingston, Tasmania, Australia
E. Guiler (34), University of Tasmania, Hobart, Tasmania, Australia
S. Harrington (33), University of New England, Armidale, New South Wales, Australia
C. W. Heath (10), Australian Antarctic Division, Kingston, Tasmania, Australia
H. Heatwole (40), University of New England, Armidale, New South Wales, Australia
M. Hindell (27), Australian Antarctic Division, Kingston, Tasmania, Australia
P. A. Horne (20), Latrobe University, Melbourne, Victoria, Australia
R. S. C. Horne, Australian Antarctic Division, Kingston, Tasmania, Australia

*Numbers in brackets refer to identities in group photograph. Addresses refer to August 1984; present addresses of contributing authors are given in the text.

G. W. Johnstone (30), Australian Antarctic Division, Kingston, Tasmania, Australia
K. Kerry (18), Australian Antarctic Division, Kingston, Tasmania, Australia
J. Kirkwood (21), Australian Antarctic Division, Kingston, Tasmania, Australia
Lu Peiding (23), First Institute of Oceanography, Qingdao, People's Republic of China
D. Lugg (16), Australian Antarctic Division, Kingston, Tasmania, Australia
B. G. Luke (36), Australian Government Analytical Laboratory, Victoria, Australia
H. Marchant (28), Australian Antarctic Division, Kingston, Tasmania, Australia
N. Masuda (32), Hokkaido University, Hakodate-city, Japan
T. A. McMeekin, University of Tasmania, Hobart, Tasmania, Australia
Meng Fan, First Institute of Oceanography, Qingdao, People's Republic of China
J. D. Miller (39), University of New England, Armidale, New South Wales, Australia
T. L. Montague (43), Australian Antarctic Division, Kingston, Tasmania, Australia
I. R. Morgan, "Attwood" Veterinary Research Laboratory, Westmeadows, Victoria, Australia
P. Nielsen, Hobart, Tasmania, Australia
D. Parer (15), Australian Broadcasting Commission, Melbourne, Victoria, Australia
R. Perrin, University of Tasmania, Hobart, Tasmania, Australia
J. A. Peterson (9), Monash University, Melbourne, Victoria, Australia
R. A. Puddicombe (12), Australian Antarctic Division, Kingston, Tasmania, Australia
P. G. Quilty (17), Australian Antarctic Division, Kingston, Tasmania, Australia
J. C. Richards (44), Australian Government Analytical Laboratory, Victoria, Australia
D. E. Rounsevell (6), Australian Antarctic Division, Kingston, Tasmania, Australia
R. D. Seppelt, Australian Antarctic Division, Kingston, Tasmania, Australia
A. P. Setty (37), National Institute of Oceanography, Dona-Paula, Goa, India
P. D. Shaughnessy, C.S.I.R.O., Institute of Biological Resources, Canberra, Australian Capital Territory, Australia
R. D. Simpson (4), University of New England, Armidale, New South Wales, Australia
M. Sudzuki (35), Nihon Daigaku University, Omiya-Shi, Saitama-Ken, Japan
D. P. Thomas, University of Tasmania, Hobart, Tasmania, Australia
M. J. Tucker (22), Australian Antarctic Division, Kingston, Tasmania, Australia

P. A. Tyler, University of Tasmania, Hobart, Tasmania, Australia
J. van den Hoff (3), Australian Antarctic Division, Kingston, Tasmania, Australia
J. K. Volkman (41), C.S.I.R.O., Division of Oceanography, Hobart, Tasmania, Australia
R. Williams (11), Australian Antarctic Division, Kingston, Tasmania, Australia
E. Woehler, Australian Antarctic Division, Kingston, Tasmania, Australia
S. Wright (42), Australian Antarctic Division, Kingston, Tasmania, Australia

Hydrobiologia 165: 1–11 (1988)
J. M. Ferris, H. R. Burton, G. W. Johnstone & I. A. E. Bayly (eds) Biology of the Vestfold Hills, Antarctica
Kluwer Academic Publishers

The ecology of photosynthetic bacteria in Burton Lake, Vestfold Hills, Antarctica

C. M. Burke[1] & H. R. Burton[2]
[1]*Department of Zoology, University of Western Australia, Nedlands, 6009, Australia;* [2]*Antarctic Division, Department of Science, Kingston, Tasmania, 7050, Australia*

Key words: photosynthetic bacteria, meromictic, Chlorobium, Antarctica, Burton Lake

Abstract

Photosynthetic bacteria in Burton Lake, a seasonally tidal, meromictic lake of maximum depth 18 m, located in the Vestfold Hills, Antarctica, were studied throughout 1983. The dominant species were *Chlorobium vibrioforme* and *Chlorobium limicola* (up to 5.4×10^6 cells ml^{-1}) and minor species were *Thiocapsa roseopersicina* ($<1.25 \times 10^5$ cells ml^{-1}) and *Rhodopseudomonas palustris* (<100 cells ml^{-1}). The *Chlorobium* spp. and *T. roseopersicina* were found throughout the anoxic water, which ranged in temperature from $-0.5\,°C$ to $-2.2\,°C$, but did not form discrete layers at the O_2-H_2S interface. The growth zone, however, of the *Chlorobium* spp. was delineated by the presence of light and H_2S and was restricted to less than 3 m below the O_2-H_2S interface. *R. palustris* was found in oxic and anoxic water. Available light, which varied from $0-4.9\mu E\ m^{-2}\ s^{-1}$ at the O_2-H_2S interface, was considered to be the major environmental factor controlling growth of the bacterial phototrophs. Growth was initiated in spring in low light levels ($<1\ \mu E\ m^{-2}\ s^{-1}$) following 3 months of darkness during winter. It is concluded that the dominance of the *Chlorobium* spp. was a result of their more efficient maintenance metabolism in winter and of their greater efficiency in utilizing low intensity light.

Introduction

The distribution of photosynthetic bacteria in meromictic lakes of temperate zones has been extensively studied (e.g. Gorlenko *et al.*, 1974; Sorokin & Donato, 1975; Parker *et al.*, 1983; Croome & Tyler, 1984) and the factors that control their growth *in situ* and *in vitro* analysed (e.g. Takahashi & Ichimura, 1970; Biebl & Pfennig, 1978; Abella *et al.*, 1980; Guerrero *et al.*, 1980; Parkin & Brock, 1980a, b; Montesinos *et al.*, 1983; Pedros-Alio *et al.*, 1983). A general conclusion from these works is that light intensity and quality are important factors in determining both the species capable of growth in a particular environment and those which will dominate. These factors would therefore be expected to be important in controlling photosynthetic bacterial growth in antarctic lakes, which have markedly seasonal light regimes. Herbert & Tanner (1977) described the isolation of photosynthetic bacteria from marine sediments around Signy Island (61°S) and found that the wavelength of the irradiating light selected the species that grew *in vitro*. However, in general, microbial studies in Antarctica (e.g. Kriss *et al.*, 1976; Hand, 1980; Hand & Burton, 1981; Tominaga & Fukui, 1981) have not concentrated on photosynthetic bacteria and so information is scant.

Burton Lake, which is in the Vestfold Hills (68°S 78°E), is meromictic and saline with a seasonal marine connection to Crooked Fjord. It has a maximum length of 2 150 m, mean width of 629 m, maximum depth of 18.3 m and mean depth of 7.16 m (Bayly, 1986). It is ice-covered for 10–11 months of the year and is isolated from Crooked Fjord for about 6–7 months during winter, spring and early summer. Snow cover on the ice is patchy. It has a permanent pycnocline at 11–13 m and a temporary pycnocline in the surface waters. There is a small, western lobe with a maximum depth of 2.75 m which is isolated by ice from the main body of the lake for about seven months of the year. The western lobe becomes mixed and well aerated during the ice-free period in summer, but develops hypersaline, cold and eventually totally anoxic conditions during the autumn to spring months. The lake's bathymetry is shown in Bayly (1986). During 1983 the lake was sampled to determine the species of photosynthetic bacteria present, their abundance and the factors influencing their development and seasonal changes.

Materials and methods

The lake was sampled on eleven occasions between January 1983 and January 1984. When present, ice and snow thicknesses were measured. Light penetration of the water column was measured using a LI-COR Quantum Photometer Li-185A and an underwater quantum sensor. Two litre Kemmerer bottle samples were collected at 1 m intervals through the water column. As the maximum measured tidal amplitude was 0.17 m, depth measurements may be in error by this amount while the lake was still linked to Crooked Fjord. Temperatures were measured by a mercury-in-glass thermometer installed in the transparent Kemmerer bottle. Prior to sampling of the anoxic water, the boundary of the oxic and anoxic waters was sampled using a close interval sampler to determine microstratification by taking 12×60 ml samples over 1.2 m. This sampler was adapted from previous samplers (Croome & Tyler, 1984) to fit through a 10 cm diameter SIPRE corer (Journal of Glaciology, 1958) ice hole. The syringes were mounted vertically, and helically arranged in a cylindrical plastic case.

As the zone to be sampled was no deeper than 13 m, the syringes were pressurized to 28 psi with compressed air, the sampler was submerged just below the surface to check for leaks and then gently lowered to the required depth. It was then sealed from the compressor and left *in situ* for 10 to 15 min to allow local disturbances to subside. After this, the suction intake of the compressor was used to evacuate the syringes and take simultaneous samples. During winter it was necessary to sample in a heated shelter to prevent samples freezing.

Dissolved oxygen was measured by Winkler titration according to Strickland & Parsons (1972). Sulphide was analysed at the Australian Government Analytical Laboratories, after fixation with cadmium chloride. The titration method of Fonselius (1983) was used, rather than the colorimetric method (Fonselius, 1983), because it was found that cadmium chloride interfered with colour development. Water samples were returned to Australia where density was determined to five decimal places, at 20 ± 0.02 °C using an Anton Paar DMA 55 precision calculating density meter. *In situ* density was calculated after Burton & Ferris (1983). Redox potential was measured initially on a Metrohm pH/Eh meter E588, but later a Radiometer Model 26 pH meter with a platinum electrode and a calomel reference electrode was used. All measurements were calibrated against Zobell solution (Zobell, 1946). pH was measured on either a Metrohm E588 or a Radiometer 26 pH meter, using a combined glass electrode.

Bacteriochlorophyll was determined by filtering samples through 0.2 μm mixed cellulose ester membrane filters (Schleicher & Schull); extracting these in 4 ml of methanol in the dark at 4 °C for 1 h, and finally centrifuging and scanning the supernatants in a Varian 635 spectrophotometer from 850 nm to 400 nm. The concentration of bacteriochlorophyll was calculated after Stanier & Smith (1960). *In vivo* spectra for species identification were determined by clearing the filters with cedarwood oil and scanning from 850 nm to 350 nm. Microscopy was performed on a Leitz Laborlux microscope with phase contrast optics; bacterial fluorescence was detected using dichroic mirrors (A and H2 by Leitz) with UV illumination. Total counts of bacteria were made by filtering 5–15 ml of water through a 0.2 μm membrane filter; staining the filters with 3% erythrosin in 5% aqueous phenol for at least 24 h; sequentially washing with sterile, distilled water and determining the mean count of five fields of view.

Sulphide-utilizing photosynthetic bacteria were enriched in broth media described by van Niel (1971) with an additional 2% sodium chloride. The media were further modified by adding 0.75 mg ml^{-1} sodium sulphide and adjusting to pH 6.8 for Chlorobiaceae and for Chromatiaceae by adding 0.25 mg ml^{-1} sodium sulphide and adjusting to pH 7.2. In neither medium was sodium thiosulphate used for primary enrichments. For some water samples a direct isolation technique of solidifying the media in tubes with 1.5% agar was used. The cultures were incubated at 15 °C and in the following light regimes:

1. continuous incandescent light of 200–600 lux for green Chlorobiaceae.
2. intermittent, 6 hourly, incandescent light at about 10 lux for bacteriochlorophyll *e* containing species of Chlorobiaceae.
3. intermittent, 6 hourly, incandescent light of less than 5 lux, filtered through a Kodak Wratten-gelatin filter 47 (for wavelengths greater than 750 nm) for slow-growing, large Chromatiaceae.
4. continuous, incandescent light of less than 5 lux filtered through a Kodak Wratten-gelatin filter 94A (for wavelengths greater than 780 nm) for fast-growing, small species of Chromatiaceae.

Rhodospirillaceae were enriched, or directly isolated on agar plates incubated in transparent Gaspak anaerobic jars (BBL), using the medium of Biebl & Pfennig (1981). They were incubated at 15 °C in continuous, incandescent light at 200–600 lux. Sulphate-reducing bacteria were enriched in Postgate's medium B (Postgate, 1979). Identification of phototrophic bacteria was according to Pfennig & Trüper (1974) and Trüper & Pfennig (1981). Identification of sulphate-reducing bacteria was according to Postgate (1979).

Results

Throughout 1983 the phototrophic bacterial flora was dominated by *Chlorobium* spp., in particular *Chlorobium vibrioforme* Pelsh and *Chlorobium limicola* Nadson. *Thiocapsa roseopersicina* Winogradsky and *Rhodopseudomonas palustris* Molisch were also present; but comprised only a minor part of the flora. The *Chlorobium* spp. and *T. roseopersicina* were primarily restricted to the anoxic zone, while *R. palustris* was found in oxic and anoxic waters.

The seasonal variations in temperature, *in situ* density, light penetration and bacteriochlorophyll *c*

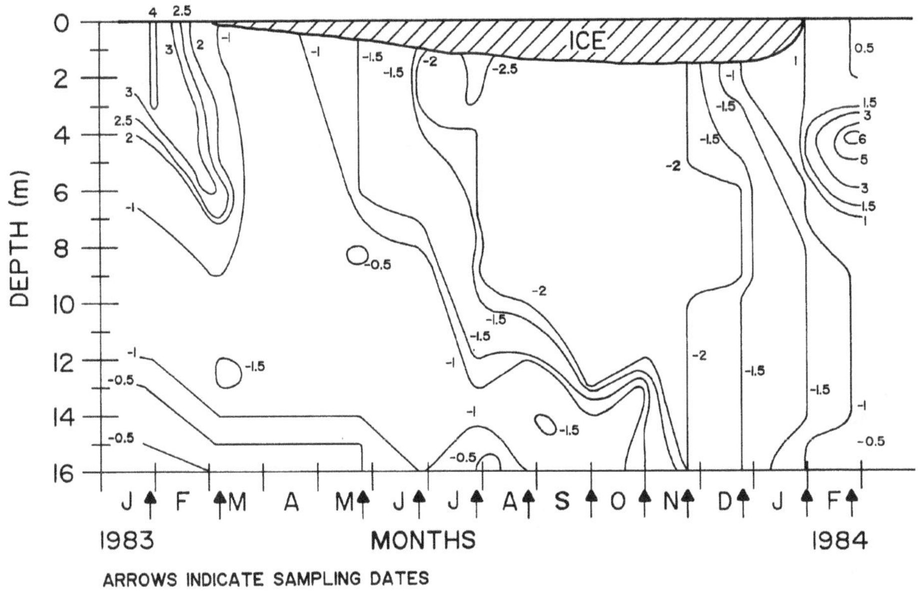

Fig. 1. Isotherms (°C) in Burton Lake during 1983.

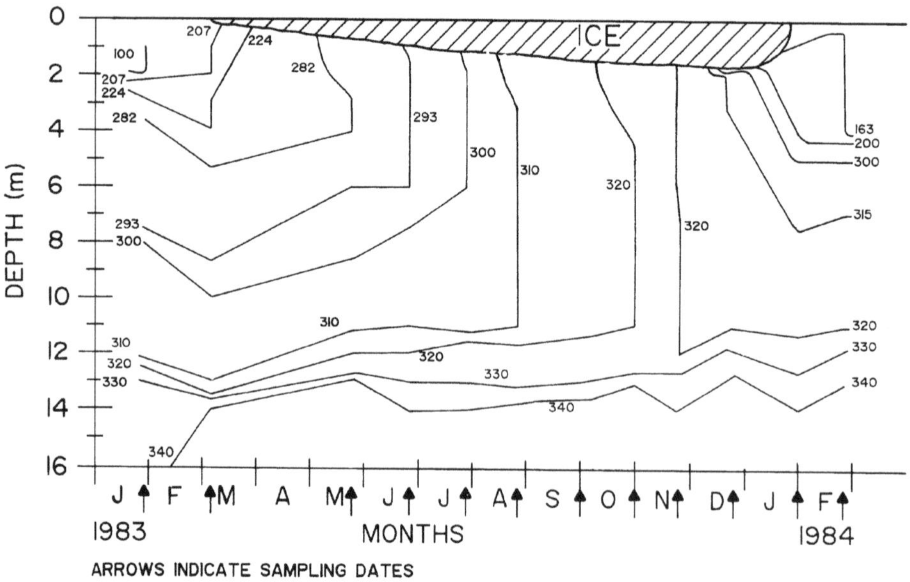

Fig. 2. Isopleths of *in situ* density in Burton Lake during 1983. Values are (density (g ml^{-1}) − 1) × 10^4.

concentration, for 1983, are shown in Figs. 1–4. The temperature of the anoxic zone (below 8 m) varied from −0.5 °C to −2.2 °C and thus the dominant component of the phototrophic bacterial flora (*Chlorobium* spp.) developed and maintained itself at temperatures consistently below zero (Fig. 1). The isopleths of *in situ* density (Fig. 2) show that a permanent pycnocline existed between 11 and 13 m and that a temporary density stratification between 2 and 8 m was eroded from January to November. This erosion was caused by convective mixing generated by salt exclusion during ice formation, and by August 26 the lake was mixed above 11 m. Salt exclusion, as exemplified by the increased densities recorded just below the ice on 30 October (Fig. 2), continued until November and was sufficient to mix the lake to 12 m. (Ice drilling prior to sampling may have accounted for a minor part of the increased densities by introducing brine from the ice into the water.)

This partial mixing of the lake directly affected the phototrophic bacterial flora by altering the distribution of oxygen (O_2) in the lake and by mixing some cells into the mixolimnion. The highest O_2 concentrations in the mixolimnion (about 15 mg l^{-1}) were recorded during the summer ice-free period and the lowest (<1 mg l^{-1}) were measured in late spring. The concentration of O_2 at all depths in the mixolimnion decreased from above 11 mg l^{-1} on 27 July (Fig. 5) to about 3.5 mg l^{-1} on 26 August. During the same period the interface of the oxic and anoxic zone (O_2-H_2S interface) deepened from 9 m to 11 m. These were the largest changes observed between consecutive sampling dates; but mixing continued to November (cf. Fig. 6). Chlorobiaceae lost to the mixolimnion remained temporarily viable in 3.5 mg l^{-1} of dissolved oxygen for in August they were enriched from depths between 8 and 16 m, but by 29 September were only enriched from 12 to 16 m.

The penetration of light in Burton Lake (Fig. 3) was affected predominantly by two factors. First, at least 80% of the incident light was absorbed by the snow and ice covering the lake and this was maximal in June–July when incident light was extinguished within 3 m of the surface. While snow and ice absorb red preferentially to blue and green wavelengths (Maykut & Grenfell, 1975) it is likely that when light penetrates to the O_2-H_2S interface all these wavelength groups would be present in a manner similar to nearby Ace Lake which has an O_2-H_2S in-

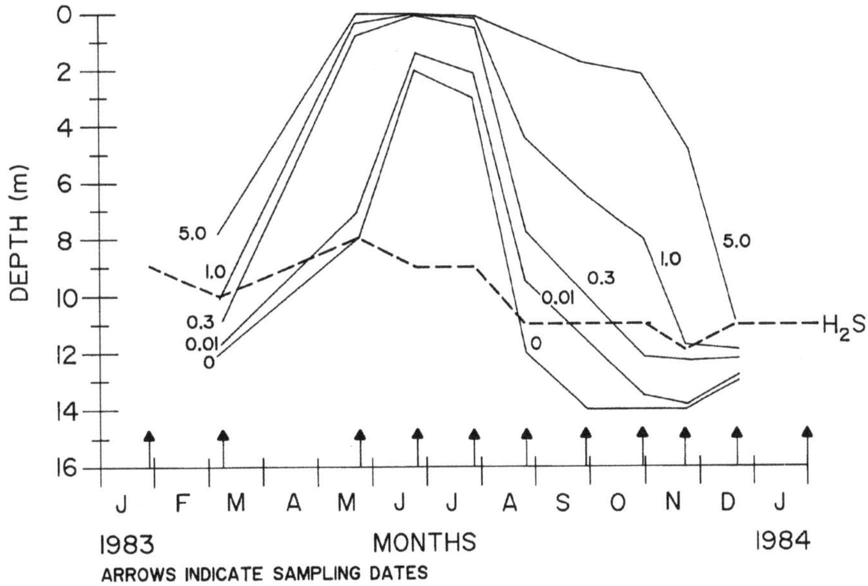

Fig. 3. Light penetration (microeinsteins m^{-2} s^{-1}) of Burton Lake during 1983.

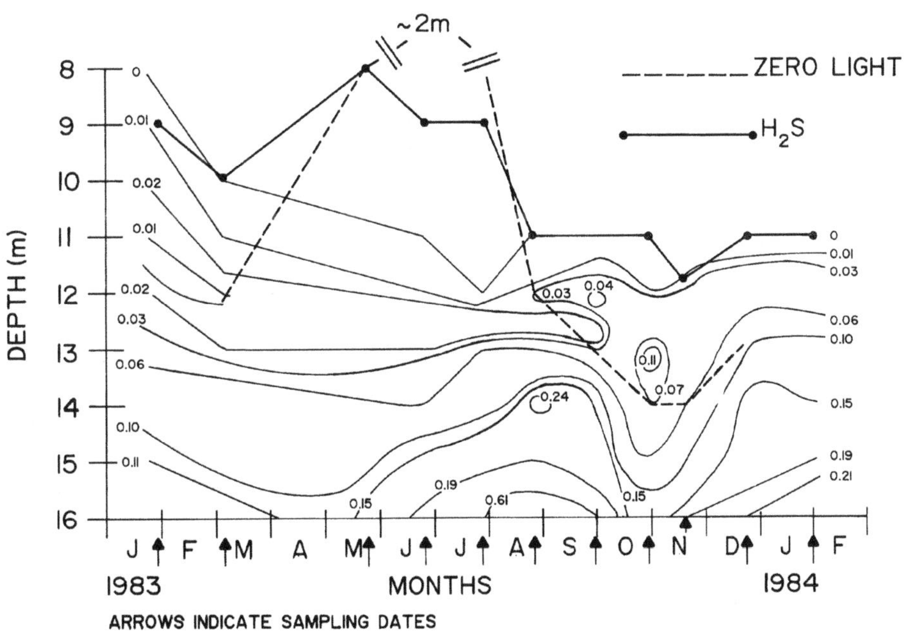

Fig. 4. Isopleths of bacteriochlorophyll *c* concentration (mg l^{-1}) in Burton Lake during 1983. Highest depth of H_2S and maximum depth of light penetration are indicated.

terface at 10–11 m and a similar salinity (Burke & Burton, this volume). Second, the remaining light was rapidly absorbed within 2 to 3 m of the O_2-H_2S interface by unicellular algae resident just above it (Bayly, 1986) and by bacterial phototrophs just below it, the latter becoming self-shading in late spring and summer (Figs. 3 & 4). The maximum light intensity measured at the O_2-H_2S interface during the ice-covered period was 4.9 microeinsteins m^{-2} s^{-1} (μE m^{-2} s^{-1}) and it is estimated that the maximum quantity to reach the interface during an ice-free period in late summer would be of the order of 10 μE m^{-2} s^{-1} as was measured on a bright day in the summer of 1981/82 (Bayly, 1986). In 1983, Burton Lake had three months of zero light at the O_2-H_2S interface (May–July); six months of light at 0–1 μE m^{-2} s^{-1} and three months of light at more than 1 μE m^{-2} s^{-1} (December–February).

The only bacteriochlorophyll found in the lake was bacteriochlorophyll *c* and isopleths of its concentration (Fig. 4) indicate that growth of Chlorobiaceae was initiated at 12 m in August, i.e. immediately light consistently penetrated the O_2-H_2S interface (Fig. 3). The highest concentration of bacteriochlorophyll was always found at the bottom of the lake, indicating considerable sedimentation. The sedimented cells were viable as they could be readily enriched from 16 m and there was generally more bacteriochlorophyll present than bacteriophaeophytin. However, once they sedimented below 14 m the cells were in darkness and so would not grow *in situ*. Winter and summer vertical profiles of selected physical, chemical and biological parameters are shown for the whole water column in Figs. 5 & 6, and for close interval sampling of the summer chemocline in Fig. 7. The winter profile (Fig. 5) on 27 July shows that O_2 was present to about 9 m. A density gradient between 5 m and 8 m prevented mixing to a greater depth. The permanent pycnocline at 12 m was well into the anoxic zone and at this depth the H_2S concentration was 20 mg l^{-1}. While there was an increase in bacteria across the

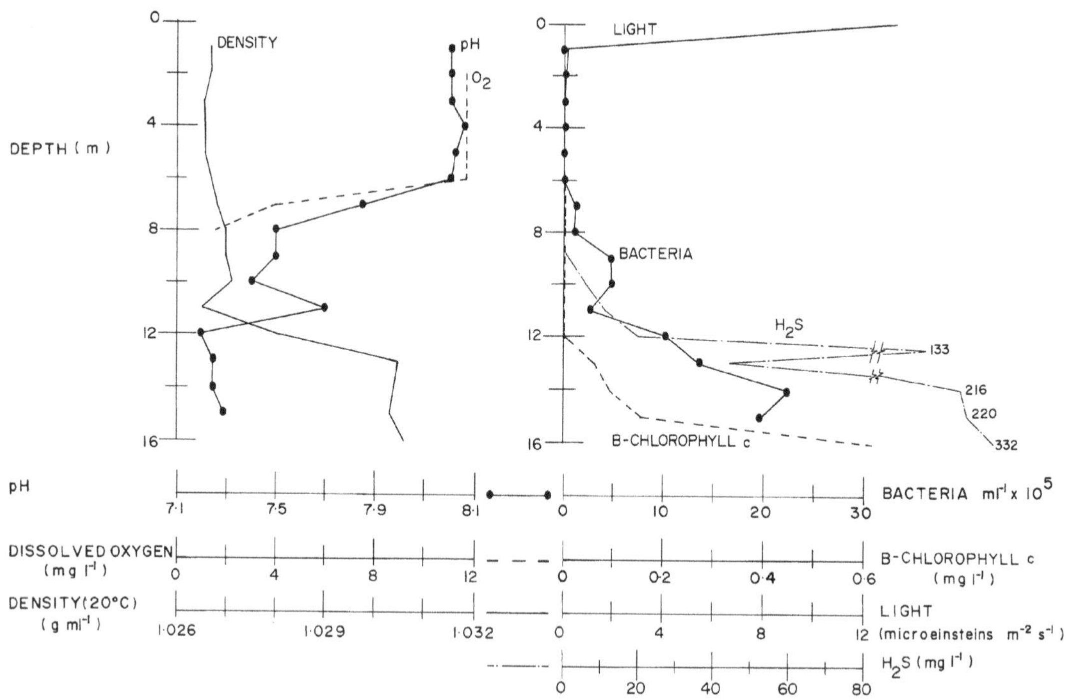

Fig. 5. Winter profiles of density, pH, light intensity, bacterial numbers, and concentrations of dissolved oxygen, hydrogen sulphide and bacteriochlorophyll *c* for Burton Lake on 27 July 1983.

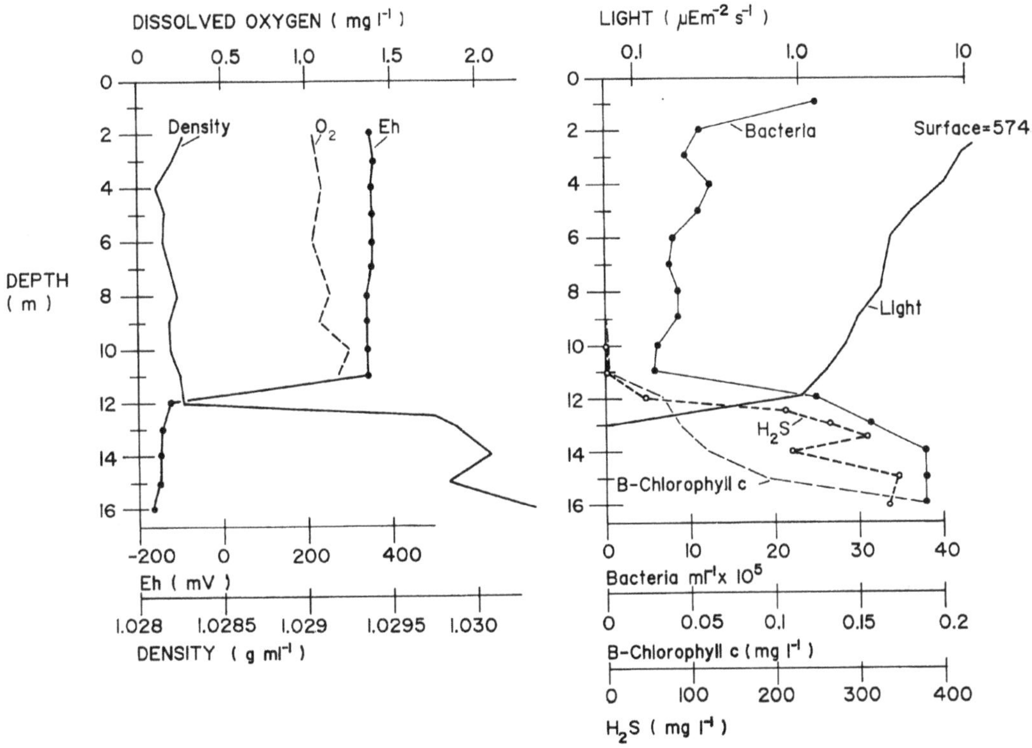

Fig. 6. Summer profiles of density, Eh, light intensity, bacterial numbers, and concentrations of dissolved oxygen, hydrogen sulphide and bacteriochlorophyll *c* for Burton Lake on 22 November 1983.

O_2-H_2S interface it is unlikely that phototrophic bacteria were a significant part of the flora here as there was no measurable bacteriochlorophyll and light intensity diminished to zero within 3 m of the water surface. The profile of bacteriochlorophyll indicates that the majority of the phototrophs had sedimented below 12 m. However enrichment cultures showed that viable cells of *Chlorobium* spp. were present from 6 m down to the sediments.

Partial mixing of the lake deepened the oxic zone to 11 m between 27 July and 26 August and further deepened it to 12 m by 22 November and on this date *Chlorobium* spp. were only enriched from 11 m and below. Thus the mixing process reduced the volume of the lake which provided a suitable environment for the Chlorobiaceae. The summer profile on 22 November (Fig. 6) shows that about 1 $\mu E\ m^{-2}\ s^{-1}$ of light was penetrating the O_2-H_2S interface and the phototrophic bacteria were growing in up to 200 mg l^{-1} of H_2S. Finer detail of this summer O_2-H_2S interface is given in Fig. 7, which shows the combined data of four close interval sampling runs from 22 November. The position of the O_2-H_2S interface is indicated by the redox cline. The change in redox potential with depth was abrupt: +320 mV to −125 mV within 10 cm. There was an increase in bacteriochlorophyll *c* across the redox cline and also a peak of bacteria. However, these two did not coincide, indicating the presence of a significant number of non-photosynthetic bacteria. This is further supported by microscopy which showed a predominance of non-fluorescent bacterial cells at 12.01 m with an increasing proportion of fluorescent cells from 12.11 m to a maximum at 12.76 m of 5.4×10^6 cells ml^{-1}. The majority of these were *Chlorobium* spp. as *Thiocapsa roseopersicina* counts were always $< 1.25 \times 10^5\ ml^{-1}$. Also, Chlorobiaceae were readily enriched from 12.11 m downwards with only one

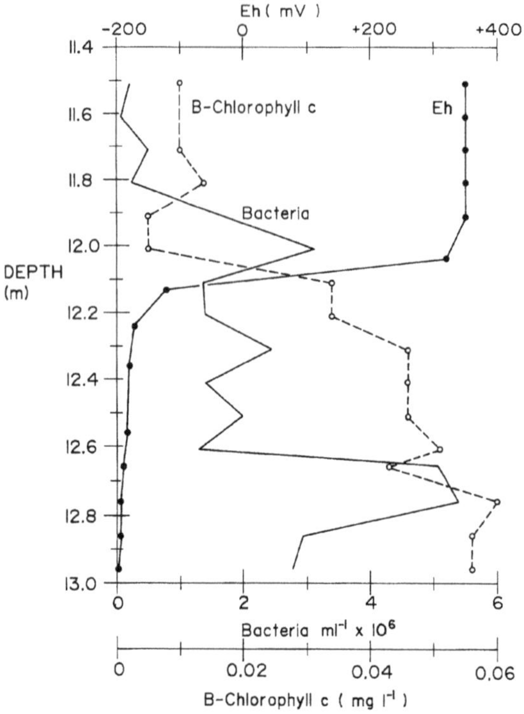

Fig. 7. Summer profiles of Eh, and concentrations of bacterial numbers and bacteriochlorophyll *c* across the O_2-H_2S interface for Burton Lake (22 November 1983), determined by close interval sampling.

enrichment above (at 11.81 m) perhaps originating from cells lost by the growing population through entrainment to the upper water mass during mixing. *Thiocapsa roseopersicina* and *Rhodopseudomonas palustris* were sporadically enriched from 11.51 m downwards. *R. palustris* was in fact isolated from water between 7 and 12 m on different dates, but was always present in low numbers (<100 cells ml^{-1}). There was no apparent discrete layering of *T. roseopersicina* above *Chlorobium* spp. in the anoxic zone. Numbers of both tended to increase with depth.

The western lobe of the lake was markedly different from the main water mass. During January–February of 1983 it was ice-free and completely oxic to the bottom; but following ice formation it became hypersaline from salt exclusion and became totally anoxic by September 29. Between August and November the temperature was in the range -4.4 °C to -5.2 °C, warming to -3 °C by 23 December. While H_2S was present in this period, photosynthetic bacteria were rarely enriched from the lobe and no bacteriochlorophyll was detected in the water column. *R. palustris* and *T. roseopersicina* were enriched occasionally in 1983, but *Chlorobium* spp. were enriched only from a single sample obtained in February 1982.

Discussion

The photosynthetic bacterial population of Burton Lake was dominated by *Chlorobium* spp. with only low numbers of *T. roseopersicina* and *R. palustris.* During winter (May–July) the weak incident light was completely absorbed above the anoxic water (usually within 2–3 m of the surface), mainly by the snow and ice. The darkness of the anoxic water presented the photosynthetic bacteria with a survival problem. Survival would be aided by the maintenance of high numbers for the commencement of growth in the following spring. This ability would give them some competitive advantage.

Van Gemerden (1980) experimentally determined the viability of *Chromatium vinosum* under zero and low light intensities. He reported that the viability rapidly declined in total darkness, although low light intensities (15–40 lux) were sufficient to maintain the cultures. He calculated the maintenance rate constant (a measure of the energy required to maintain cell integrity and structure) of several photosynthetic species and found that *Chlorobium* spp. were the most efficient at maintaining themselves in total darkness. Therefore, the dominance of *Chlorobium* spp. in Burton Lake could be the result of an earlier initiation of growth of these species compared to the others, following three months of winter darkness.

Herbert & Tanner (1977) and Parkin & Brock (1980b) have demonstrated the importance of light quality in the establishment of dominant species in an environment. They reported that wavelengths around 460 nm and 720–740 nm favoured Chlorobiaceae whereas wavelengths around 550–590 nm favoured Chromatiaceae. In Burton Lake, when light penetrates to the O_2-H_2S interface, wavelengths of 350 nm to 720 nm are thought to be present with the most intense light at 400–600 nm (Burke & Burton, this volume). Thus it is likely that there is light suitable for the growth of both Chlorobiaceae and Chromatiaceae and it seems unlikely that in Burton Lake selection of species occurred on the basis of light wavelengths.

It is evident from Figs. 3 & 4 that the concentration of bacteriochlorophyll *c* began increasing in the lake in August when the light intensity at the O_2-H_2S interface was less than 1 $\mu E\ m^{-2}\ s^{-1}$. As bacteriochlorophyll *c* was the only bacteriochlorophyll detected in the lake throughout the year, neither *T. roseopersicina* nor *R. palustris* (characterized by bacteriochlorophyll *a*) could have reached population numbers comparable to the *Chlorobium* spp. Furthermore, while both *T. roseopersicina* and *Chlorobium* spp. grew in culture in low light (10 lux), the former was only enriched from lake water when wavelengths of light greater than 780 nm were used to selectively inhibit the *Chlorobium* spp. Therefore, in Burton Lake, *Chlorobium* spp. appear to be more efficient at utilizing light of low intensities than *T. roseopersicina.* Biebl & Pfennig (1978) measured the growth at low light intensities (5–10 lux) of *Chromatium vinosum, Rhodopseudomonas capsulata, Prosthecochloris* sp. and *Chlorobium phaeovibrioides* and concluded that Chlorobiaceae grew faster than Chromatiaceae and Rhodospirillaceae. They ascribed this to greater efficiency of light harvesting by the bacteriochlorophyll-enriched chlorobium vesicles present in the Chlorobiaceae. In contrast, the photosynthetic apparatus of the Chromatiaceae and Rhodospirillaceae is disseminated throughout the cell cytoplasm and is not enriched with bacteriochlorophyll to the same extent as in the chlorobium vesicles. Thus both *T. roseopersicina* and *R. palustris* are unlikely to outgrow *Chlorobium* spp. in the low light intensities generally present in the anoxic zone of Burton Lake.

In Burton Lake, therefore, the critical light factors influencing the growth of the photosynthetic bacteria were related to the overall intensity of the light rather than to intensity at specific wavelengths. The photosynthetic flora in winter was in darkness for three months, during which time it had to maintain viability to initiate growth in spring. Growth of species of the Chlorobiaceae commenced as soon as light consistently penetrated the O_2-H_2S interface (in August, Fig. 4) and therefore occurred at low light intensities. It is concluded that the more efficient maintenance metabolism of the Chlorobiaceae gave them an advantage over the Chromatiaceae for utilizing the return of light in spring. Furthermore, they were then more efficient at using the available light. These two factors probably account for the observed predominance of *Chlorobium* spp. in Burton Lake.

While viable photosynthetic bacteria were present throughout the anoxic water of Burton Lake, light was extinguished within 3 m of the O_2-H_2S interface. Growth was therefore limited to this 3 m of the water column. Within this zone, there was no discernible stratification of the photosynthetic bacteria as has been reported in other lakes (e.g. Caldwell & Tiedje, 1975; Sorokin & Donato, 1975; Lawrence *et al.*, 1978; Gorlenko & Chebotarev, 1981; Croome & Tyler, 1984 and Lindholm *et al.*, 1984). This stratification has been reported as being very distinct and generally involving species that are either motile or gas vacuolate and are thus able to readily adjust their position within a habitat. Lindholm *et al.* (1984) reported diurnal vertical migrations of about 20 cm by *Chromatium* spp. and Clark & Walsby (1978) found that gas vacuoles can induce positive buoyancy. In contrast, in Burton Lake neither the *Chlorobium* spp. nor *T. roseopersicina* are motile or gas vacuolate and they were found throughout the anoxic zone (Figs. 5, 6 & 7) presumably because they could not readily adjust their position in the lake to find optimal growth conditions. This was clearly shown on 23 December, 1983 when the O_2-H_2S interface was at 11.36 m, 70 cm above its position on 22 November (see Fig. 7), yet the bulk of the photosynthetic bacteria were still below 12.3 m. The depth of the O_2-H_2S interface varied from a minimum of 8 m in May to a maximum of 12 m in November of 1983; it deepened in response to partial mixing of the lake, which may have hindered the development of stratified populations. The third bacterial phototroph of Burton Lake, *R. palustris*, was found in low numbers in both oxic and anoxic water. The absence of a mass accumulation of *R. palustris* at the interface is in agreement with Biebl & Pfennig

(1981) who state that this has never been observed for any of the Rhodospirillaceae.

In summary, the photosynthetic, bacterial flora of Burton Lake was dominated by *Chlorobium* spp. during 1983 with *T. roseopersicina* and *R. palustris* being only minor components. It is concluded that this dominance of *Chlorobium* spp. was due to their more efficient maintenance metabolism in winter and their greater efficiency of light harvesting in the very low light intensities prevalent at the O_2-H_2S interface in spring. They established at this time such a large population in comparison to *T. roseopersicina* that it is likely that their dominance simply continued on throughout summer when light was not limiting and they were not competitively advantaged.

Acknowledgements

This work was performed as part of the 1983 Australian National Antarctic Research Expedition to Davis Station. We thank members of that expedition for their assistance and in particular we thank R. Sibthorpe for his contribution to the fieldwork. Identification of a selection of the cultures was confirmed by Prof. N. Pfennig.

References

Abella, C., E. Montesinos & R. Guerrero, 1980. Field studies on the competition between purple and green sulphur bacteria for available light (Lake Siso, Spain). In M. Dokulil, H. Metz & D. Jewson (eds), Developments in Hydrobiology, 3. Dr W. Junk Publishers, The Hague: 173–181.

Bayly, I. A. E., 1986. Ecology of the zooplankton of a meromictic antarctic lagoon with special reference to Drepanopus bispinosus (Copepoda: Calanoida). Hydrobiologia 140: 199–231.

Biebl, H. & N. Pfennig, 1978. Growth yields of green sulphur bacteria in mixed cultures with sulphur and sulphate reducing bacteria. Arch. Microbiol. 117: 9–16.

Biebl, H. & N. Pfennig, 1981. Isolation of members of the Rhodospirillaceae. In M. P. Starr, H. Stolp, H. G. Truper, A. Ballows & H. G. Schlegel (eds), The Prokaryotes 1. Springer-Verlag, Berlin: 267–273.

Burke, C. M. & H. R. Burton, this volume. Photosynthetic bacteria in meromictic lakes and stratified fjords of the Vestfold Hills, Antarctica. Hydrobiologia.

Burton, H. R. & J. M. Ferris, 1983. Calculations of in situ density values for salt lakes. In A. R. Chivas & P. De Deckker (coordinators), Salt Lakes, Evaporites and Aeolian Deposits (SLEADS) workshop 83, Salt Lakes in Arid Australia. Australian National University, Canberra: 9–10.

Clark, A. E. & A. E. Walsby, 1978. The development and vertical distribution of populations of gas-vacuolate bacteria in a eutrophic, monomictic lake. Arch. Microbiol. 118: 229–233.

Caldwell, D. E. & J. M. Tiedje, 1975. The structure of anaerobic bacterial communities in the hypolimnia of several Michigan lakes. Can. J. Microbiol. 21: 377–385.

Croome, R. L. & P. A. Tyler, 1984. Microbial microstratification and crepuscular photosynthesis in meromictic Tasmanian lakes. Verh. int. Ver. Limnol. 22: 1695–1701.

Fonselius, S. H., 1983. Determination of hydrogen sulphide. In K. Grasshoff, M. Erhardt & K. Kremling (eds), Methods of Seawater Analysis, second edition. Verlag Chemie, GmbH Weinheim: 73–80.

Gorlenko, V. M. & E. M. Chebotarev, 1981. Microbiological processes in the meromictic Lake Sakovo. Microbiology 50: 98–102.

Gorlenko, V. M., E. N. Chebotarev & V. I. Kachalkin, 1974. Microbial oxidation of hydrogen sulphide in Lake Veisovo (Slavyansk Lake). Microbiology 43: 450–453.

Guerrero, R., E. Montesinos, I. Esteve & C. Abella, 1980. Physiological adaptation and growth of purple and green sulphur bacteria in a meromictic lake (Vila) as compared to a holomictic lake (Siso). In M. Dokulil, H. Metz & D. Jewson (eds), Developments in Hydrobiology, 3. Dr W. Junk, Publishers, The Hague: 161–171.

Hand, R. M., 1980. Bacterial populations of two saline antarctic lakes. In P. A. Trudinger & M. R. Walter (eds), Biogeochemistry of Ancient and Modern Environments. Aust. Acad. Sci., Canberra: 123–129.

Hand, R. M. & H. R. Burton, 1981. Microbial ecology of an antarctic saline meromictic lake. Hydrobiologia 82: 363–374.

Herbert, R. A. & A. C. Tanner, 1977. The isolation and characteristics of photosynthetic bacteria (Chromatiaceae and Chlorobiaceae) from antarctic marine sediments. J. appl. Bact. 43: 437–445.

Journal of Glaciology, 1958. Instruments and methods: Ice drills and corers. J. Glaciol. 3: 30.

Kriss, A. E., I. N. Mitskevich, E. P. Rozanova & L. K. Osnitskaya, 1976. Microbial investigations of Lake Vanda (Antarctica). Microbiology 45: 917–922.

Lawrence, J. R., R. C. Haynes & U. T. Hammer, 1978. Contribution of photosynthetic green sulphur bacteria to total primary production in a meromictic saline lake. Verh. int. Ver. Limnol. 20: 201–207.

Lindholm, T., K. Weppling & H. S. Jensen, 1984. Stratification and primary production in a small brackish lake studied by close-interval siphon sampling. Verh. int. Ver. Limnol. 22: 2190–2194.

Maykut, G. A. & T. C. Grenfell, 1975. The spectral distribution of light beneath first-year sea-ice in the Arctic Ocean. Limnol. Oceanogr. 20: 554–563.

Montesinos, E., R. Guerrero, C. Abella & I. Esteve, 1983. Ecology and physiology of the competition for light between Chlorobium limicola and Chlorobium phaeobacteroides in natural habitats. Appl. envir. Microbiol. 46: 1007–1016.

Parker, R. D., J. R. Lawrence & U. T. Hammer, 1983. A comparison of phototrophic bacteria in two adjacent saline meromictic lakes. Hydrobiologia 105: 53–61.

Parkin, T. B. & T. D. Brock, 1980a. Photosynthetic bacterial production in lakes: The effects of light intensity. Limnol. Oceanogr. 25: 711–718.

Parkin, T. B. & T. D. Brock, 1980b. The effects of light quality on the growth of phototrophic bacteria in lakes. Arch. Microbiol. 125: 19–27.

Pfennig, N. & H. G. Truper, 1974. The phototrophic bacteria. In R. E. Buchanan & N. E. Gibbons (eds), Bergey's Manual of Determinative Bacteriology, 8th edition. Williams & Wilkins, Baltimore: 24–64.

Pedros-Alio, C., E. Montesinos & R. Guerrero, 1983. Factors determining annual changes in bacterial photosynthetic pigments in holomictic Lake Ciso, Spain. Appl. envir. Microbiol. 46: 999–1006.

Postgate, J. R., 1979. The Sulphate Reducing Bacteria. Cambridge University Press, UK: 26–27.

Sorokin, J. I. & N. Donato, 1975. On the carbon and sulphur metabolism in the meromictic Lake Faro (Sicily). Hydrobiologia 47: 241–252.

Stanier, R. Y. & J. H. C. Smith, 1960. The chlorophylls of green bacteria. Biochem. Biophys. Acta 41: 478–484.

Strickland, J. D. H. & T. R. Parsons, 1972. A practical handbook of seawater analysis. Fish. Res. Bd Can., Bulletin 167: 21–26.

Takahashi, M. & S. Ichimura, 1970. Photosynthetic properties and growth of photosynthetic sulphur bacteria in lakes. Limnol. Oceanogr. 15: 929–944.

Tominaga, H. & F. Fukui, 1981. Saline lakes at Syowa Oasis, Antarctica. Hydrobiologia 82: 375–389.

Truper, H. G. & N. Pfennig, 1981. Characterization and identification of the anoxygenic phototrophic bacteria. In M. P. Starr, H. Stolp, H. G. Truper, A. Ballows & H. G. Schlegel (eds), The Prokaryotes, 1. Springer-Verlag, Berlin: 299–312.

van Gemerden, H., 1980. Survival of Chromatium vinosum at low light intensities. Arch. Microbiol. 125: 115–121.

van Niel, C. B., 1971. Techniques for the enrichment, isolation and maintenance of photosynthetic bacteria. In A. San Pietro (ed.), Methods in Enzymology, 23 (A). Academic Press, New York: 3–28.

Zobell, C. E., 1946. Studies on redox potential of marine sediments. Bull. am. Ass. pet. Geol. 30(4): 477–513.

Hydrobiologia 165: 13–23 (1988)
J. M. Ferris, H. R. Burton, G. W. Johnstone & I. A. E. Bayly (eds) Biology of the Vestfold Hills, Antarctica
Kluwer Academic Publishers

Photosynthetic bacteria in meromictic lakes and stratified fjords of the Vestfold Hills, Antarctica

C. M. Burke[1] & H. R. Burton[2]
[1]*Department of Zoology, University of Western Australia, Nedlands, Western Australia, 6009, Australia;* [2]*Antarctic Division, Department of Science, Kingston, Tasmania, 7050 Australia*

Key words: photosynthetic bacteria, meromictic, Antarctica

Abstract

Thirteen meromictic lakes and two permanently stratified fjords in the Vestfold Hills, Antarctica, were surveyed in 1983 for photosynthetic bacteria. Burton Lake and Ellis Fjord were sampled throughout the year to determine seasonal variations. Physical and chemical parameters were recorded and related to the species present. The dominant species in waters with salinities of ≤ 100.7 g kg^{-1} were *Chlorobium vibrioforme* and *Chlorobium limicola* with populations at the O_2-H_2S interface in the range 0.3 to 6.7×10^6 ml^{-1}. Neither of these species was found at higher salinities. *Thiocapsa roseopersicina* and a *Chromatium* sp. were found in low numbers ($<10^5$ ml^{-1}) in most of the same waters as the *Chlorobium* spp. These bacterial phototrophs developed in a narrow band below the O_2-H_2S interface where both light and H_2S were available. Very low numbers ($<10^2$ ml^{-1}) of *Rhodopseudomonas palustris* were found in both oxic and anoxic waters having salinity ≤ 148 g kg^{-1}. The dominance of the *Chlorobium* spp. is ascribed to their more efficient maintenance metabolism during the darkness, their faster growth at low light intensities (<1 μE m^{-2} s^{-1}) and the lack of selective filtering of incident light. The *Chlorobium* spp. grew well at -2 °C, but not -5 °C in hypersaline waters. The concentration of H_2S had no apparent effect on the development of the bacterial flora. Viable cells were found to depths of 100 m in Ellis Fjord indicating that viability in total darkness could have been maintained for periods of the order of 1700 days.

Introduction

The Vestfold Hills, (68°30′S, 78°E), contain 13 meromictic lakes which originated as relict sea water trapped in valleys by ice-sheet retreat (Adamson & Pickard, 1983). The lakes and fjords are ectogenically stratified with fresh meltwater in summer (Burton, 1981); and stability of this stratification has been aided by the short duration of ice-free periods, usually only four to six weeks each year, if at all. Over the past 8000 years they have had varying water budgets so that current salinities range from 12 g kg^{-1} to 197 g kg^{-1} (Burton, 1981). This study was initiated to identify the photosynthetic bacterial species in these lakes and fjords and to determine, where possible, the environmental parameters which limit their growth and distribution.

Study area

Figure 1 shows the Vestfold Hills and the location of the meromictic lakes (Abraxas, Ace, Anderson, Burton, Clear, Ekho, Farrell, Fletcher, McCallum, Organic, Oval, Pendant and Shield) and the two stratified fjords, Ellis (basins 1 and 2) and Taynaya Bay (basins 1, 2, 3). Burton Lake, including its western lobe, is seasonally tidal and is described in more de-

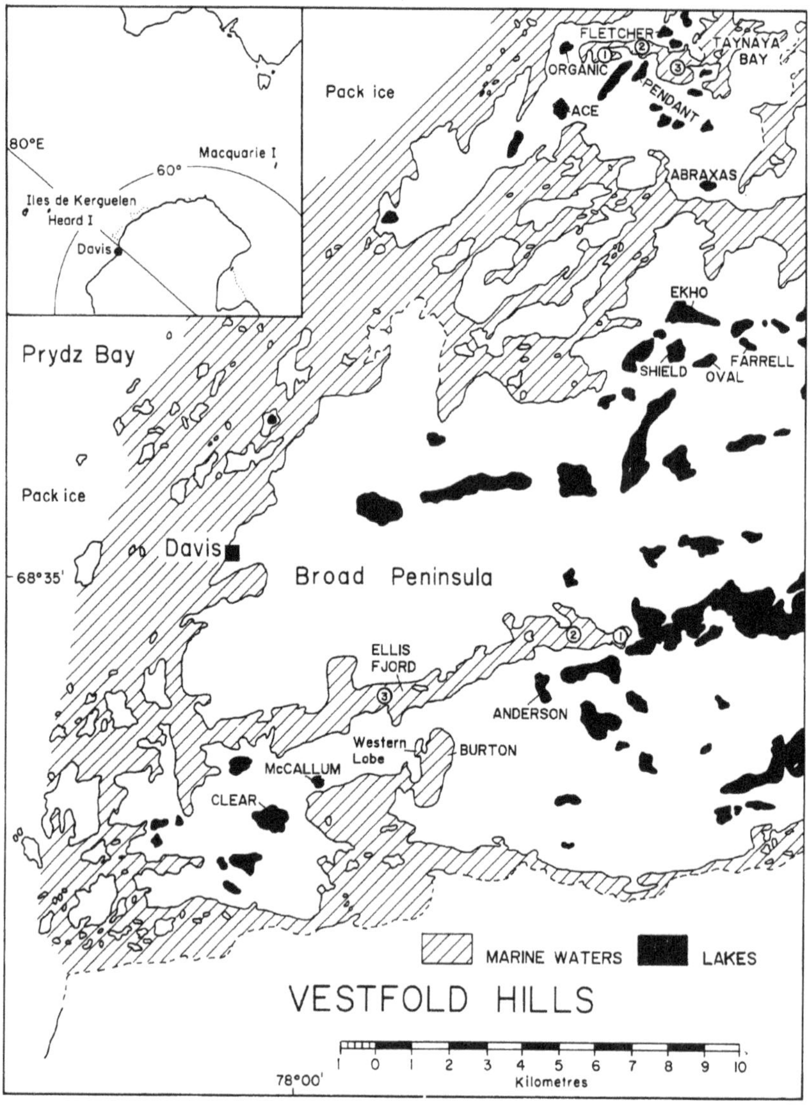

Fig. 1. Map of the Vestfold Hills showing the meromictic lakes and the permanently stratified fjords.

tail in Burke & Burton (this volume). The remaining lakes are all land-locked and some features of each are listed in Table 1. Ellis Fjord is approximately 10 km long and is over 100 m deep at several points in this length. It has a very restricted opening to the sea, less than 4 m deep, and a number of basins (1, 2 & 3 in Fig. 1) of which basins 1 and 2 are stratified. Basin 1, about 12 m deep, was stratified at 6–7 m during winter, 1983. It is separated from basin 2 by a sill less than 1 m deep. Basins 2 and 3 are separated by a sill 23–30 m deep which is adjacent to the innermost island of the fjord. Basin 2 had a chemocline at a depth of 36–43 m at different dates in 1983 and below this the water was anoxic. There was a very substantial flow of fresh water via Ellis Rapids into the head of the fjord during the summer melt. As a result, Ellis Fjord had three distinct layers of water in the stratified basins. In summer there was a surface layer 2 – 3 m deep of fresh water lying above approximately 36 m of tidal, oxic sea water (≃31‰) and then at least 60 m of slightly denser, anoxic bottom water. In winter the fresh water was replaced by up to 1.7 m of ice.

Taynaya Bay is more complex in morphology and has six basins, three of which were stratified (see Fig. 1). It is mostly shallower than Ellis Fjord, being less than 30 m deep; however basin 2 is greater than 65 m. Stratification occurred at 10, 15 and 20 m in basins 1, 2 and 3 respectively. As in Ellis Fjord, the bottom water was anoxic while the upper water was tidal and oxygenated. Taynaya Bay maintains a large amount of ice cover throughout most summers and this reduces wind mixing. This may also partially explain the shallower chemocline depths compared with Ellis Fjord.

Materials and methods

A Kemmerer bottle was used to sample the water column of Ellis Fjord from May to December 1983 and in January 1984 at one site in Basin 1, two sites in Basin 2, two sites in Basin 3 and one marine site just outside the fjord entrance. In basin 2, samples were taken at intervals of 10 m from the surface to the interface of the oxic and anoxic waters (O_2-H_2S interface). Five samples were taken in this region at intervals of 1 m and then two or three deep water samples were taken. In the other basins, surface and bottom samples were collected. Sampling of the surface water and O_2-H_2S interface of the lakes and Taynaya Bay was done irregularly in 1983. Field and chemical analyses, enrichment culturings and isolation of bacteria and bacterial counts were carried out as described in Burke & Burton (this volume). Salinity was calculated from density at 20 °C by an equation, Salinity = 1315.7895 × density (20 °C) − 1313.4211 derived from Appendix 1 of Whitfield & Jagner (1981). We recognize the potential weakness of using a linear regression well beyond the bounds of the data from which it was derived (e.g. to infer the salinity of hypersaline Organic Lake) and have given the equation so that measured density (at 20 °C) can be recovered, if desired. Some data collected in 1978 are also used.

Results

Table 1 lists some physical and chemical parameters of the lakes and fjords sampled. The sites are listed in increasing order of salinity of the water at the O_2-H_2S interface. A range in values indicates the extent of annual variation. There was a wide variation in salinity of the interface water samples from a minimum of 12.9 g kg^{-1} in Clear Lake to a maximum of 199.4 g kg^{-1} in Organic Lake. In some of the sites there was only a slight salinity difference between the upper and lower waters but, nonetheless, stratification was still permanent.

The depths of the O_2-H_2S interfaces in the permanently stratified water bodies varied from 3.5 – 43 m; but only four were deeper than 20 m. The principal effect of the deeper interfaces was a further reduction in the light intensity available to photosynthetic bacteria. For example, on 23 December 1983 the light incident at the surface of Burton Lake was 171 microeinsteins m^{-2} s^{-1} (μE m^{-2} s^{-1}) and the light intensity at the O_2-H_2S interface (11 m) was 4.9 μE m^{-2} s^{-1}. In contrast, on 5 January 1984 the light incident on Ellis Fjord basin 2 was 1316 μE m^{-2} s^{-1} but the intensity at the chemocline (41 m) was 0.14 μE m^{-2} s^{-1}.

The temperatures at the O_2-H_2S interfaces were generally above O °C throughout the year, the highest temperature recorded being 13.2 °C in Ekho Lake. However, Burton Lake (see also Burke & Burton, this volume: Fig. 1) and Ellis Fjord both had temperatures below zero (−0.5 to −2.2 °C) throughout the year. They were also tidal for at least five to six months of the year. A temperature of −0.7 °C was recorded in the anoxic water of Taynaya Bay (which is also intermittently tidal) in 1978, but no seasonal data are available. The coldest sites were the western lobe of Burton Lake (minimum temperature −5.2 °C) and Organic Lake (minimum temperature −6 °C); however both warmed to above 0 °C in summer. These sites were much shallower than the other lakes and fjords investigated. Organic Lake was the most saline.

Hydrogen sulphide (H_2S) was present in all the meromictic lakes except Organic lake. The western lobe of Burton Lake was not meromictic, but it did have H_2S throughout the water column by late

Table 1. Salinity, depth, temperature and H_2S and bacteriochlorophyll *c* concentration at the O_2-H_2S interfaces in stratified lakes and fjords of the Vestfold Hills.

Site	Salinity (g kg^{-1})	Depth (m)	Temp (°C)	H_2S (mg l^{-1})	Bacteriochlorophyll *c* (mg l^{-1})
Clear Lake	12.9	29	+4.8	+	0.08
McCallum Lake	17.9	19.5	+5.6	+	0.31
Abraxas Lake	18.1	18	+2.7	2.5	0
Pendant Lake	19.9	10	+5.4	2	0.086
Ace Lake	29.3	11	+7.7	2	0.11
Ellis Fjord 2	34.9	36 to 43	−1 to −2	20	0.01 to 0.06
Taynaya Bay 3	36.0	21	−0.7	2	0
Ellis Fjord 1	40.2	7	−1 to −2.2	2	0.1 to 0.23
Taynaya Bay 1	40.6	10.5	−0.7	797	4.6
Burton Lake	30.7 to 40.5	8 to 12	−0.5 to −2	1 to 268	0.1 to 0.11
western lobe[a]	40.0 to 89.5	0 to 2.75	−1.7 to −5.2	0 to 37	0
Fletcher Lake	100.7	6.5	+2.9	481	0.49
Anderson Lake	131.3	3.5	+2.3	+	0
Ekho Lake	139.0	23	+13.2	2	0
Shield Lake	148.0	18	+6.3	1	0
Farrell Lake	168.8	8	+5.5	2	0
Oval Lake	193.3	8	+5.5	2	0
Organic Lake	199.4	4	−6	0	0

[a] The western lobe is a part of Burton Lake which is isolated by ice formation from the main body of the lake in winter.

winter. In Lakes Farrell and Oval, H_2S was restricted to the very bottom of the water column and on some occasions was not detected (e.g. Farrell Lake on 28 February 1983). H_2S concentration just below the O_2-H_2S interface was usually <20 mg l^{-1}; but rose rapidly so that the photosynthetic bacteria were growing, at times, in several hundred mg l^{-1} of H_2S. For example, the concentration of H_2S in Burton Lake on 23 November 1983 at 11.96 m was zero while at 12.31 m it was 121.8 mg l^{-1} (see Burke & Burton this volume: Fig. 6). Sulphate-reducing bacteria were enriched from the water column of all lakes containing H_2S.

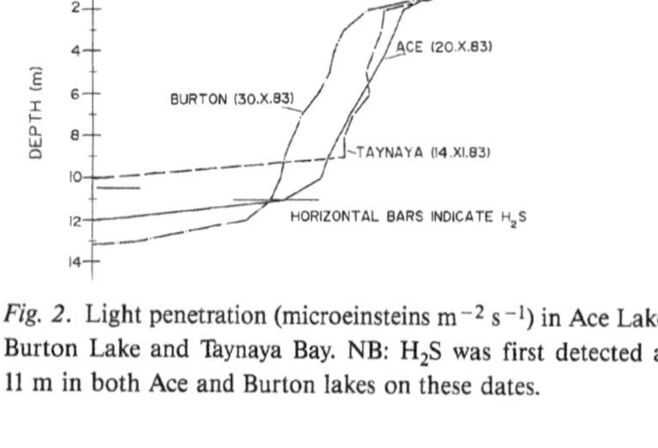

Fig. 2. Light penetration (microeinsteins m^{-2} s^{-1}) in Ace Lake, Burton Lake and Taynaya Bay. NB: H_2S was first detected at 11 m in both Ace and Burton lakes on these dates.

The only bacteriochlorophyll detected was bacteriochlorophyll *c* and while the waters were sometimes bright green at the O_2-H_2S interface, the concentrations were generally <0.2 mg l^{-1}. In basin 1 of Taynaya Bay, 4.6 mg l^{-1} of bacteriochlorophyll *c* was observed at 10.5 m; and though light was extinguished between 9 and 10 m (Fig. 2) it indicates that a considerable population of Chlorobiaceae was present in the photic zone of the anoxic water.

Table 2 shows from which lakes and fjords photosynthetic bacteria of the families Chlorobiaceae, Chromatiaceae and Rhodospirillaceae were enriched. Species of Chlorobiaceae (*Chlorobium vibrioforme* Pelsh and *Chlorobium limicola* Nadson) were the most commonly enriched bacterial phototrophs and were found in those sites with salinity ≤ 100.7 g kg^{-1} at the O_2-H_2S interface. *Thiocapsa roseopersicina* Winogradsky, *Chromatium* sp. Perty (Chromatiaceae) were enriched sporadically from sites with salinities ≤ 89.5 g kg^{-1}. They were only enriched in media exposed to light of wavelengths greater than 780 nm which selectively inhib-

Table 2. Photosynthetic bacteria found in stratified lakes and fjords of the Vestfold Hills.

Site	Chl.[a]	Chr[b]	R[c]
Clear Lake	+	+	−
McCallum Lake	+	+	−
Abraxas Lake	+	+	+
Pendant Lake	+	−	+
Ace Lake	+	+	−
Ellis Fjord 2	+	+	+
Taynaya Bay 3	+	−	−
Ellis Fjord 1	+	−	−
Taynaya Bay 1	+	−	−
Burton Lake	+	+	+
western lobe	+	+	+
Fletcher Lake	+	−	−
Anderson Lake	−	−	−
Ekho Lake	−	−	+
Shield Lake	−	−	+
Farrell Lake	−	−	−
Oval Lake	−	−	−
Organic Lake	−	−	−

[a] Chlorobiaceae; [b] Chromatiaceae; [c] Rhodospirillaceae.

ited the Chlorobiaceae. Both *T. roseopersicina* and *Chlorobium* spp. were enriched from samples taken at 100 m in Ellis Fjord where they were in permanent darkness. The only species of Rhodospirillaceae isolated, *Rhodopseudomonas palustris* Molisch, was present in both oxic and anoxic waters with salinities $\leq$148 g kg^{-1}. They numbered $<$100 ml^{-1}.

Total bacterial counts from Burton Lake and Ellis Fjord (basin 2) typically showed lowest counts in the mixolimnion with approximately an order of magnitude increase across the O_2-H_2S interface. In Burton Lake, the lowest count at the interface (0.5×10^6 ml^{-1} in July) was recorded in mid-winter and the highest in summer (5.4×10^6 ml^{-1} in December). In Ellis Fjord, the seasonal range was from 1.01×10^6 ml^{-1} in May to 6.7×10^6 ml^{-1} in December: but a much lower count of 0.3×10^6 ml^{-1} was observed in September. This was associated with a destratifying mixing process which increased the depth of the O_2-H_2S interface from 37 m in August to 43 m in September. The majority of the phototrophic cells were members of *Chlorobium* spp. as *T. roseopersicina* and *Chromatium* sp. totalled $<10^5$ cells ml^{-1} and *R. palustris* $<$100 cells ml^{-1}.

Figure 2 shows the penetration of light through

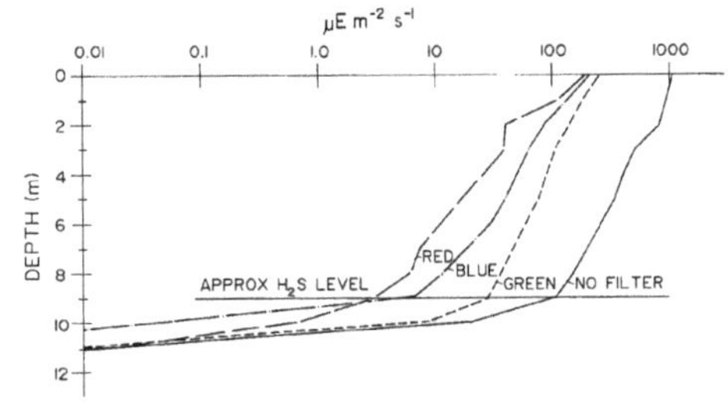

Fig. 3. Penetration of different wavelengths of light (microeinsteins m^{-2} s^{-1}) in Ace Lake on 8 February 1977.

the water column of three ice-covered sites in 1983. They show a similar pattern of high absorption through the snow and ice, and steady absorption in the water column culminating in extinction of the remaining light just below the O_2-H_2S interface. For Taynaya Bay, the bar indicating H_2S is shown 0.5 m below the depth of zero light. However, the concentration of H_2S at 10.5 m was 781 mg l^{-1} and of bacteriochlorophyll *c* was 4.6 mg l^{-1}. It is therefore reasonable to assume that H_2S was present in the lower region of the photic zone. The penetration of light into Ellis Fjord on 20 December 1983 is shown in Fig. 4. The intensity measured at the O_2-H_2S interface was always less in the fjord than in Burton Lake (see Burke & Burton, this volume) because of the fjord's greater depths. The measured intensity of light at different O_2-H_2S interfaces varied from 0–4.9 μE m^{-2} s^{-1} in 1983. However, greater intensities have been recorded in ice-free conditions as illustrated by Fig. 3, which shows the penetration of red, blue and green wavelengths in Ace Lake. The three wavelengths were present at the O_2-H_2S interface with green and blue predominating, but each were extinguished rapidly below this level. A profile taken on 15 November 1974, when about 2 m of ice covered the lake, showed a similar relative absorption of wavelengths with 0.2 μE m^{-2} s^{-1}, 7 μE m^{-2} s^{-1} and 22 μ m^{-2} s^{-1} of red, blue and green wavelengths respectively present at the interface.

The limnological characteristics of Burton Lake are described in Burke & Burton (this volume).

Figure 4 shows a typical profile of Ellis Fjord with a slight density stratification between 41 and 43 m. This is coincident with a decrease to zero of both dissolved oxygen and redox potential and the occur-

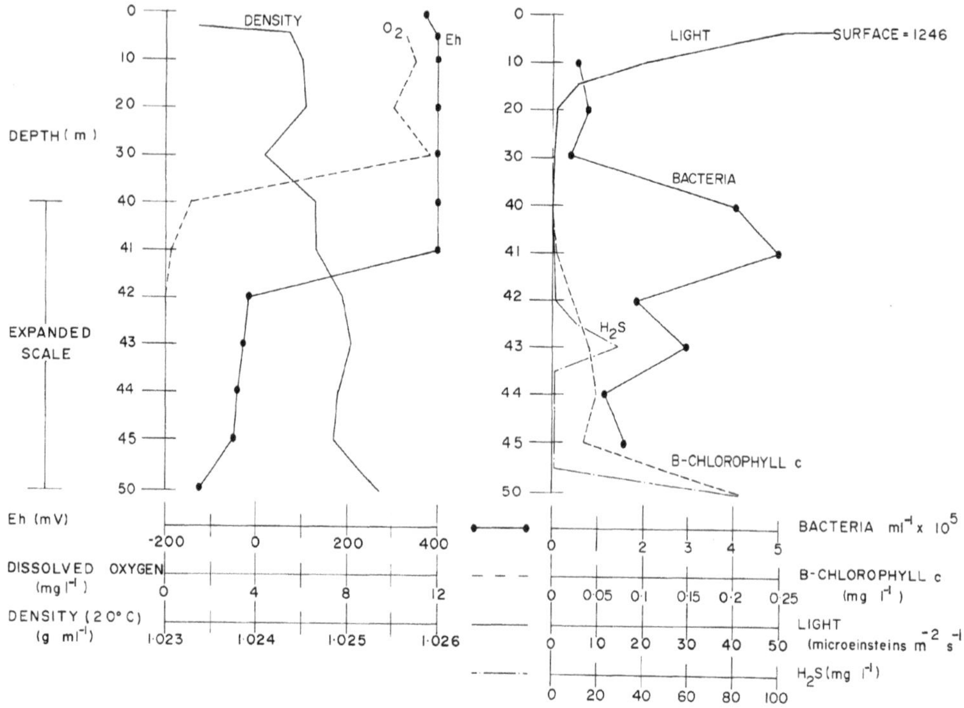

Fig. 4. Selected limnological and biological characteristics of the water column of Ellis Fjord (basin 2) on 20 December 1983.

rence of low concentrations of H_2S (2 mg l^{-1}). It is possible that both O_2 and H_2S are present for about 0.5–1 m because of the instability of the density stratification. There was a small, but measurable quantity of light at the oxycline (0.1 μE m^{-2} s^{-1}. The main peak of bacterial numbers was associated with the oxic zone just above the O_2-H_2S interface and there was a second peak coinciding with a peak in H_2S at 43 m from where *Desulfotomaculum* spp. and *Desulfovibrio* spp. were enriched. The concentration of bacteriochlorophyll *c* increased with depth from the interface. However, at most other times there was a small peak of up to 0.06 mg l^{-1} of bacteriochlorophyll *c* associated with an increased cell count just below the interface. It was difficult to determine when growth first occurred, though new bacteriochlorophyll *c* was noted in November–December when light intensity at the O_2-H_2S interface was less than 0.1 μE m^{-2} s^{-1}. At all times the concentration of pigment (bacteriochlorophyll *c* and/or bacteriophaeophytin *c*) was highest at depths of 60–70 m and below.

The temperature of all but the top 3 m of the water column was −1.5 °C, varying only in the range −0.8 to −2.0 °C throughout the year. A similar series of profiles to those in Fig. 4 were recorded in 1983 except that the depth of the O_2-H_2S interface varied from 36–43 m.

Discussion

A comparative study of the different habitats of certain photosynthetic bacteria has been made possible by the presence of lakes and fjords in the Vestfold Hills which differ in depth, morphology, dissolved salts and consequently, stratification. Previous studies have identified six factors that influence the development of photosynthetic bacterial populations: light intensity, light quality, temperature, the presence of other phototrophs (bacterial or algal), salinity and the concentration of sulphide (Takahashi & Ichimura, 1970; Cohen *et al.*, 1977; Parkin & Brock, 1980a, b; van Gemerden, 1980; Montesinos *et al.*, 1983; Pedros-Alio *et al.*, 1983).

Biota

A limited number of species of photosynthetic bacteria were found in the lake samples and further species could possibly be revealed by more extensive sampling. Hand (1980) reported *Rhodospirillum* sp. in Ace Lake. However, the methods used support the conclusion that while *Thiocapsa roseopersicina*, *Chromatium* sp. and *Rhodopseudomonas palustris* were present, the dominant species were *Chlorobium vibrioforme* and *Chlorobium limicola*. In accordance with the observations of Cohen *et al.* (1977), only one plate of photosynthetic bacteria existed in these meromictic lakes. *T. roseopersicina*, *R. palustris* and *Chromatium* sp. did not form a plate distinct from the populations of *Chlorobium* spp.

Light

While it is always the case that the light intensity at the O_2-H_2S interface of a lake will vary seasonally, such fluctuations are extreme in the lakes of the Vestfold Hills. For example, at the O_2-H_2S interface in Burton Lake there were three months of darkness, six months of intermediate intensity and three months of relatively high light intensity (estimated maximum of 10 $\mu E\ m^{-2}\ s^{-1}$; Burke & Burton, this volume). In Ace Lake, an intensity of 100 $\mu E\ m^{-2}\ s^{-1}$ has been recorded at the O_2-H_2S interface (Fig. 3); but periods of very high light intensity are likely to be brief in Antarctica and occur well after the establishment of the photosynthetic bacterial populations. Conceivably, high light intensities in summer could alter the species composition; but the data of Burke & Burton (this volume) do not appear to support this as *Chlorobium* spp. were dominant in Burton Lake throughout the ice free period in early 1983. The amount of snow and the transparency of the ice varied considerably between the sites sampled in this study, but their location at high latitude and the extended period of ice cover of all the sites produce a lengthy period of zero, or very low light intensity at their O_2-H_2S interfaces.

If all the lakes sampled were to have similar spectral distributions to that shown in Fig. 3 for Ace Lake, then both Chlorobiaceae and Chromatiaceae should grow well. However, Chlorobiaceae were dominant in all lakes in which they occurred. Burke & Burton (this volume) suggested that the more efficient maintenance metabolism of the Chlorobiaceae in darkness (van Gemerden, 1980), together with their faster generation times at low light intensities (Herbert & Tanner, 1977; Biebl & Pfennig, 1978), allowed them to dominate the monimolimnion of Burton Lake. If our assumption that the lakes have similar light regimes at their O_2-H_2S interfaces is valid, then the three months of winter darkness followed by three months of low light intensities become important factors in selecting for Chlorobiaceae.

These bacteria possess a light harvesting system which enables them to grow at lower light intensities than Chromatiaceae (Biebl & Pfennig, 1978). This system includes carotenoids which have absorption maxima in the range 450–515 nm; the wavelengths of light that penetrate water most deeply in the absence of biological filtering (Kirk, 1977; Truper & Pfennig, 1981). This ability is particularly important in Ellis Fjord where the depth of the O_2-H_2S interface varied from 36–43 m during 1983. Montesinos *et al.* (1983) reviewed the species of Chlorobiaceae found in different lakes and concluded that β-isorenieratene-containing species were dominant at depths greater than 9 m. This is not the case in the lakes and fjords of the Vestfold Hills which have deep O_2-H_2S interfaces (e.g. Ellis Fjord, Clear Lake, Abraxas Lake). In Ellis Fjord the dominant species were *Chlorobium vibrioforme* and *Chlorobium limicola* which contain chlorobactene, although some of the spectral scans of enrichment cultures had small peaks suggestive of the presence of β-isorenieratene as well. Biebl & Pfennig(1978) concluded that in the absence of selective filtering of incident light by photosynthetic organisms there would be no advantage for species containing β-isorenieratene. Therefore, it is not simply increased depth below the surface which favours species with β-isorenieratene, rather increased depth together with selective filtering of incident light. In Ellis Fjord the growth of chlorobactene-containing species indicates that there was little or no selective filtering of incident light by photosynthetic organisms.

Salinity

The Chromatiaceae and Chlorobiaceae were detected only in lakes with salinities ranging from 12.9 g kg^{-1} (Clear Lake) to 100.7 g kg^{-1} (Fletcher Lake). Therefore, the upper limit for these bacteria was between 100.7 and 131.3 g kg^{-1} (the salinity of the monimolimnion of Anderson Lake where neither were found) which is a lower maximum than reported elsewhere (e.g. Cohen *et al.*, 1977). The Rhodospirillaceae had a greater salt tolerance in these lakes, being isolated from Shield Lake which had a salinity of 148 g kg^{-1}. However, their low numbers suggest they were only a minor component of the microbial flora of these lakes.

Cohen *et al.* (1977) reported the occurrence of *Chromatium violescens* and *Prosthecochloris* sp. in Solar Lake in salinities of about 180 g kg^{-1}; but in other respects Solar Lake is very different from the lakes of the Vestfold Hills. It is situated at low latitude and can reach temperatures of 60 °C and is shallow enough for light to penetrate the entire water column during stratification. These factors may allow photosynthetic bacteria to develop at a salinity which inhibits them in areas such as the Vestfold Hills which are much colder and with comparatively very low light intensities.

Apart from direct salinity effects on viability, the position of the photosynthetic bacterial populations in the water column was affected by salinity gradients. Most lakes and fjords studied here had very stable density stratifications due to the morphology of the drainage basins, the lengthy periods of ice cover, and the large density gradients that commonly existed. This implies that the O_2-H_2S interface, and therefore the populations of photosynthetic bacteria, were stable. In Burton Lake and Ellis Fjord however, there was a variation of up to 5 m in the depth of the O_2-H_2S interface in 1983. In Burton Lake this was attributed to salt exclusion during ice formation (Burke & Burton, this volume); but though likely, this was less certain in Ellis Fjord. The end result in both cases was that the initial populations at the O_2-H_2S interfaces, when mixed with oxic waters, were lost, H_2S was chemically oxidized and new populations of photosynthetic bacteria developed at the deeper O_2-H_2S interface.

Temperature

An indirect effect of hypersalinity in cold environments is a reduction in the temperature at which the water freezes. Very low temperatures occur because heat gained during summer is rapidly lost when air temperatures are well below zero and there is no ice cover. Kerry *et al.* (1977), Tominaga & Fukui (1981) and Torii & Yamagata (1981) reported antarctic lakes or ponds that do not freeze over even though surface water temperatures may reach −20 °C. Organic Lake in the Vestfold Hills had a temperature of −6 °C in November 1978 and the western lobe of Burton Lake was −5.5 °C in October 1983. Therefore organisms growing in these environments must be adapted to both extreme cold and extreme salinity. Wright & Burton (1981) concluded that the biochemical adaptations required for either cold or salt tolerance in bacteria seemed to be incompatible and therefore cold, hypersaline lakes were likely to have an impoverished bacterial flora. This may explain why the western lobe of Burton Lake had a much reduced photosynthetic bacterial flora in comparison with Burton Lake. In the main lake bacteriochlorophyll *c* was detectable and Chlorobiaceae in particular were frequently found by enrichment, whereas in the colder, more saline western lobe no bacteriochlorophyll *c* was detectable and enrichments rarely led to growth of any photosynthetic bacterial species. However, most environments sampled had temperatures above −2 °C at the O_2-H_2S interface, which Burke & Burton (this volume) found did not inhibit the growth of *Chlorobium* sp. in Burton Lake. The higher temperatures could be attributed to insulation during the lengthy periods of ice-cover and, in Ekho Lake, to a heliothermal effect (Sonnenfeld & Hudec, 1980) as well.

Hydrogen sulphide concentration

Many authors have reported that the concentration of H_2S is unlikely to affect the development of either Chlorobiaceae or Chromatiaceae (e.g. Biebl & Pfennig, 1978; Parkin & Brock, 1980b; Pedros-Alio *et al.*, 1983).

Three factors are given to explain this. First, the Chromatiaceae can store elemental sulphur intracellularly; second, the Chlorobiaceae have a pronounced capacity for syntrophic growth with sulphate reducing bacteria; and third, H_2S is often abundant. However, Takahashi & Ichimura (1970) and Steenbergen & Korthals (1982) concluded that sulphide concentration could limit growth particularly if sulphate reduction was insignificant at the O_2-H_2S interface. The concentrations of H_2S at the interfaces of those sites of the Vestfold Hills which possessed photosynthetic bacteria were often high (up to 781 mg l^{-1}) and sulphate reducing bacteria were enriched from the interface waters of all but one of the sites. The exception was Organic Lake in which the anoxic bottom water was both cold and salty during winter and in which no photosynthetic bacteria were found. So, while H_2S together with the available light delineated a zone in which *Chlorobium* spp. and *T. roseopersicina* grew, no evidence was found that they were H_2S-limited within this zone. *R. palustris* was found in both oxic and anoxic water. It is known to use H_2S as a photosynthetic electron donor, but is also able to grow photoorganotrophically or heterotrophically, so even low concentrations of H_2S are not required for growth. Hence H_2S concentration was not significant in determining the photosynthetic bacterial flora of these sites.

Interactions

By definition, phototrophic species will be in competition for the available light. While their respective light-harvesting pigments reduce this competition by using different wavelengths (Stanier *et al.*, 1981), the species composition of photosynthetic bacterial flora is known to be influenced by both algal and bacterial filtration (absorption) of light (Pedros-Alio *et al.*, 1983; Montesinos *et al.*, 1983). However, in the lakes and fjords of the Vestfold Hills the species composition of photosynthetic bacteria did not appear to be affected by selective light filtration brought about by populations of either algae or bacteria closer to the surface. In Ace Lake, where there was a plate of flagellate algae just above the O_2-H_2S interface (Burch, this volume) there were still wavelengths of light available for the growth of all photosynthetic bacteria (see Fig. 3). The species of Chlorobiaceae that might be expected to grow beneath an algal plate (those containing β-isorenieratene as the major carotenoid (Montesinos *et al.*, 1983) were absent and chlorobactene-containing species were dominant instead. The absence of any vertical separation of photosynthetic bacteria into distinct plates implies that there was no selection for a particular species by the selective filtration of light by another species.

Because the dominant photosynthetic bacteria, *Chlorobium* spp., produced high *in situ* concentrations of pigments, light was rapidly extinguished within 2–3 m of the O_2-H_2S interfaces in these lakes and fjords, Bacteriochlorophyll *c* concentrations in Table 1 are average to high when compared with values from the literature (e.g. Lawrence *et al.*, 1978; Tominaga & Fukui, 1981; Parker *et al.*, 1983; Croome, 1984; Lindholm *et al.*, 1985). The concentration of 4.6 mg l^{-1} in Taynaya Bay basin 1 is exceeded only by a concentration of bacteriochlorophyll *d* of 6.6 mg l^{-1} reported by Tominaga & Fukui (1981) in Lake Nurume in the Syowa Oasis, Antarctica. Other workers (e.g. Takahashi & Ichimura, 1968; Caldwell & Tiedje, 1975; Parkin & Brock, 1980a; Steenbergen & Korthals, 1982) have reported concentrations of different bacteriochlorophylls in the range 0.015–0.8 mg l^{-1}. The high concentrations of bacteriochlorophyll were very likely to be a response to low light intensities (van Gemerden, 1980) and in turn cause self-shading with the result that only a narrow growth zone was found just below the O_2-H_2S interface.

The highest concentrations of bacteriochlorophyll and/or bacteriophaeophytin occurred at great depth below the photic zone in Ellis Fjord, indicating very slow degradative processes. This is supported by the finding of viable cells of *Chlorobium* spp. and *T. roseopersicina* at 100 m in Ellis Fjord. They had maintained viability in total darkness for the time it took the sediment approximately 60 m from the growth zone at the O_2-H_2S interface. If there was no turbulent mixing in the monimolimnion, then this time could be of the order of 1700 days, using the fastest sinking rate of 3.54 cm day^{-1} determined by Clark & Walsby (1978) for cells of *Lamprocystis roseopersicina* with collapsed gas vesicles. Pfennig & Truper (1981) reported similar occurrences in the depths of the Black Sea, but weren't able to say how photosynthetic bacteria survived in total darkness, nor whether they were autochthonous or allochthonous. In Ellis Fjord, at least, they are autochthonous, but the mechanism of survival in total darknes is still unknown.

Conclusions

Of the sites examined in the Vestfold Hills, *Chlorobium* spp. were found in large populations ($>10^6$ ml^{-1}) in waters with salinities ≤ 100.7 g kg^{-1}. Low numbers ($<10^5$ ml^{-1}) of *T. roseopersicina* and *Chromatium* sp. were also found in most of these waters with the *Chlorobium* spp. These bacterial phototrophs developed in a narrow band below the O_2-H_2S interface where light and H_2S were available. H_2S concentration did not limit growth. Low numbers of *R. palustris* (<100 ml^{-1}) were also found in both oxic and anoxic waters of salinity ≤ 148 g kg^{-1}. The dominance of species of Chlorobiaceae over species of Chromatiaceae was considered to be a result of the greater efficiency of their maintenance metabolism in total darkness during winter, of their faster growth in the low light intensities prevailing in spring, and of the lack of selective filtering of the incident light. Growth was observed in temperatures as low as $-2\,°C$, but $-5\,°C$ appeared unsuitable. Very low light intensities (<1 μE m^{-2} s^{-1}) sustained the growth of photosynthetic bacteria. Sedimentation of inactive, but viable cells continually occurred and these cells maintained viability for extended periods of total darkness possibly of the order of 1700 days.

Acknowledgements

This work was performed as part of the 1983 Australian National Antarctic Research Expedition to Davis Station. The authors gratefully acknowledge the assistance of the staff of the Antarctic Division and of the expeditioners at Davis Station. In particular we thank R. Sibthorpe for his assistance with the fieldwork. Identification of a selection of the cultures was confirmed by Prof. N. Pfennig.

References

Adamson, D. & J. Pickard, 1983. Late quartenary ice movement across the Vestfold Hills, East Antarctica. In R. L. Oliver, P. R. James & J. B. Jago (eds), Antarctic Earth Science. Aust. Acad. Sci., Canberra: 465–469.

Biebl, H. & N. Pfennig, 1978. Growth yields of green sulphur bacteria in mixed cultures with sulphur and sulphate reducing bacteria. Arch. Microbiol. 117: 9–16.

Burch, M. D., this volume. Annual cycle of phytoplankton in Ace Lake, an ice covered, saline meromictic lake. Hydrobiologia.

Burke, C. M. & H. R. Burton, this volume. The ecology of photosynthetic bacteria in Burton Lake, Vestfold Hills, Antarctica. Hydrobiologia.

Burton, H. R., 1981. Marine lakes of Antarctica. In Antarctic Symposium, The University of Tasmania, Hobart. Aust. & N.Z. Ass. Adv. Sci., Tasmanian Div., Hobart, Australia: 51–58.

Caldwell, D. E. & J. M. Tiedje, 1975. The structure of anaerobic bacterial communities in the hypolimnia of several Michigan lakes. Can J. Microbiol. 21: 377–385.

Clark, A. E. & A. E. Walsby, 1978. The development and vertical distribution of populations of gas-vacuolate bacteria in a eutrophic, monomictic lake. Arch. Microbiol. 118: 229–233.

Cohen, Y., W. E. Krumbein & M. Shilo, 1977. Solar Lake (Sinai) 2. Distribution of photosynthetic microorganisms and primary production. Limnol. Oceanogr. 22: 609–620.

Croome, R. L., 1984. A physicochemical and microbial study of two meromictic lakes in South-West Tasmania, with observations on Australian chrysophytes. Ph.D. Thesis, Univ. of Tasmania, Hobart, Australia: 120 pp.

Hand, R. M., 1980. Bacterial populations of two saline antarctic lakes. In P. A. Trudinger & M. R. Walter (eds), Biogeochemistry of Ancient and Modern Environments. Aust. Acad. Sci., Canberra: 123–129.

Herbert, R. A. & A. C. Tanner, 1977. The isolation and some characteristics of photosynthetic bacteria (Chromatiaceae and Chlorobiaceae) from antarctic marine sediments. J. Appl. Bact. 43: 437–445.

Kerry, K. R., D. R. Grace, R. Williams & H. R. Burton, 1977. Studies on some saline lakes in the Vestfold Hills, Antarctica. In G. A. Llano (ed.), Adaptations within Antarctic Ecosystems. Smithsonian Institution, Wash., D.C.: 839–858.

Kirk, J. T. O., 1977. Attenuation of light in natural waters. Aust. J. Mar. Freshwat. Res. 28: 497–508.

Lawrence, J. R., R. C. Haynes & U. T. Hammer, 1978. Contribution of green photosynthetic bacteria to total primary production in a meromictic saline lake. Verh. int. Ver. Limnol. 20: 201–207.

Lindholm, T., K. Weppling & H. S. Jensen, 1985. Stratification and primary production in a small brackish lake studied by close-interval siphon sampling. Verh. int. Ver. Limnol. 22: 2190–2194.

Montesinos, E., R. Guerrero, C. Abella & I. Esteve, 1983. Ecology and physiology of the competition for light between Chlorobium limicola and Chlorobium phaeobacteroides in natural habitats. Appl. envir. Microbiol. 46: 1007–1016.

Parker, R. D., J. R. Lawrence & U. T. Hammer, 1983. A comparison of photosynthetic bacteria in two adjacent saline meromictic lakes. Hydrobiologia 105: 53–61.

Parkin, T. B. & T. D. Brock, 1980a. The effects of light quality on the growth of phototrophic bacteria in lakes. Arch. Microbiol. 125: 19–27.

Parkin, T. B. & T. D. Brock, 1980b. Photosynthetic bacterial production in lakes: The effects of light intensity. Limnol. Oceanogr. 25: 711–718.

Pedros-Alio, C., E. Montesinos & R. Guerrero, 1983. Factors determining annual changes in bacterial photosynthetic pigments in holomictic Lake Ciso, Spain. Appl. envir. Microbiol. 46: 999–1006.

Pfennig, N. & H. Truper, 1981. Isolation of members of the families Chromatiaceae and Chlorobiaceae. In M. P. Starr, H. Stolp, H. G. Truper, A. Balows & H. G. Schlegel (eds), The Prokaryotes, 1. Springer-Verlag, Berlin: 279–289.

Sonnenfeld, P. & P. P. Hudec, 1980. Heliothermal lakes. In A. Nissenbaum (ed.), Hypersaline Brines and Evaporitic Environments. Developments in Sedimentology, 28. Elsevier, Amsterdam: 91–100.

Stanier, R. Y., N. Pfennig & H. G. Truper, 1981. Introduction to the phototrophic prokaryotes. In M. P. Starr, H. Stolp, H. G. Truper, A. Balows & H. G. Schlegel (eds), The Prokaryotes, 1. Springer-Verlag, Berlin: 197–211.

Steenbergen, C. L. M. & H. J. Korthals, 1982. Distribution of photosynthetic microorganisms in the anaerobic and microaerophilic strata of Lake Vechten (The Netherlands). Pigment analysis and role in primary production. Limnol. Oceanogr. 27: 883–895.

Takahashi, M. & S. Ichimura, 1968. Vertical distribution and organic matter production of photosynthetic sulphur bacteria in Japanese lakes. Limnol. Oceanogr. 13: 644–655.

Takahashi, M. & S. Ichimura, 1970. Photosynthetic properties and growth of photosynthetic bacteria in lakes. Limnol. Oceanogr. 15: 929–944.

Tominaga, H. & F. Fukui, 1981. Saline Lakes at Syowa Oasis, Antarctica. Hydrobiologia 82: 375–389.

Torii, T. & N. Yamagata, 1981. Limnological studies of saline lakes in the Dry Valleys. In L. D. McGinnis (ed.), The Dry Valley Drilling Project. Antarct. Res. Ser., 33, Am. Geophys. Union, Wash., D.C.: 141–159.

Truper, H. G. & N. Pfennig, 1981. Characterization and identification of the anoxygenic phototrophic bacteria. In M. P. Starr, H. Stolp, H. G. Truper, A. Balows & H. G. Schlegel (eds.), The Prokaryotes, 1. Springer-Verlag, Berlin: 299–312.

van Gemerden, H., 1980. Survival of Chromatium vinosum at low light intensities. Arch. Microbiol. 125: 115–121.

Whitfield, M. & D. Jagner, 1981. Marine Electrochemistry: a practical introduction. J. Wiley & Sons, N.Y. 529 pp.

Wright, S. W. & H. R. Burton, 1981. The biology of antarctic saline lakes. Hydrobiologia 82: 319–338.

Hydrobiologia 165: 25–33 (1988)
J. M. Ferris, H. R. Burton, G. W. Johnstone & I. A. E. Bayly (eds) Biology of the Vestfold Hills, Antarctica
Kluwer Academic Publishers

Sulfate reduction rates and some aspects of the limnology of four lakes and a fjord in the Vestfold Hills, Antarctica

P. D. Franzmann[1], G. W. Skyring[2], H. R. Burton[3] & P. P. Deprez[3]
[1]*Department of Agricultural Science, University of Tasmania, GPO Box 252C, Hobart, Tasmania, 7001, Australia;* [2]*Baas Becking Geobiological Laboratory, PO Box 378, Canberra, ACT, 2601, Australia;* [3]*Antarctic Division, Department of Science, Kingston, Tasmania, 7050, Australia*

Key words: Sulfate reduction, meromictic, saline lakes, Antarctica

Abstract

Sulfate reduction rates were measured in waters and sediments from four antarctic lakes and an antarctic fjord basin by a radiometric technique. There was generally a linear correlation between the period of incubation and sulfate reduced; the average of the correlation coefficients was 0.76 ± 0.1. The rates at 6 °C were very low ($0.0-1.1\ \mu mol\ kg^{-1}\ d^{-1}$) when compared to most other marine and non-marine environments for which sulfate reduction rates have been reported. Lactate and acetate did not stimulate sulfate reduction. Temperatures of the sediments selected from the different sites varied from -0.4 to 4.5 °C and the chloride and sulfate concentrations of the sediments varied from 0.19 to 0.83 mol kg^{-1} and 0.04 to 41.01 mmol kg^{-1} respectively. Sulfate reduction rates did not correlate with the chlorosity of sediment porewaters.

Introduction

Dissolved sulfide occurs in the bottom waters of several freshwater lakes, saline meromictic lakes and fjord basins in the Vestfold Hills, Antarctica (Burke, 1983). Sulfide plays an important role in the ecology of these lakes and fjord basins. In Burton Lake, Ace Lake and Ellis Fjord, its presence limits the distribution of copepods to the epilimnion. The high reducing capability of sulfide maintains anoxic conditions in the hypolimnion. Populations of photosynthetic sulfur bacteria (Burke & Burton, this volume) and chemolithotrophic thiosulfate-oxidizing bacteria (Hand, 1980), which occur in these lakes, utilize sulfide or its oxidation product, thiosulfate, as electron donors.

Biogenic sulfide arises from either heterotrophic catabolism of sulfur-containing organic compounds or the bacterial reduction of oxidized sulfur compounds such as sulfate, thiosulfate or elemental sulfur. At the temperatures and pressures associated with coastal environments, hydrogen sulfide is formed only by biological mechanisms; as the sulfate-reducing bacteria are probably responsible for most of the sulfide formation at temperatures below 100 °C (Trudinger, 1979).

The sulfide concentration in the hypolimnion of Ace Lake was reported at 7.0 mM (Hand & Burton, 1981), and sulfate concentration decreased from 2.3 mM to below their level of detection (0.03 mM) over a 12 m interval below the chemocline which suggests that the lake contained an active population of sulfate-reducing bacteria. As sulfide production is an important step in the biogeochemical cycling of sulfur and has obvious effects on the ecology of these lakes, sulfate reduction rates were determined for sediments and waters of several lakes in the Vestfold Hills. A major objective was to determine the effects of ionic compositions of the lakes on sulfate reduction rates.

Materials and methods

Sample sites

Sites sampled (with depth from lake surface to sediment in parentheses) in February 1984 were Burton Lake (17.1 m), Shield Lake (37.6 m), Ellis Fjord (58.6 m), Watts Lake (27.5 m) and Ace Lake (24.0 m) (Fig. 1). Burton Lake was also sampled in May 1984. The sampler for the collection of sediment was a cylinder with a fully extended internal piston, the inwards withdrawal of which sucked in lake mud (King & Everitt, 1980). Water samples were collected with a Kemmerer bottle.

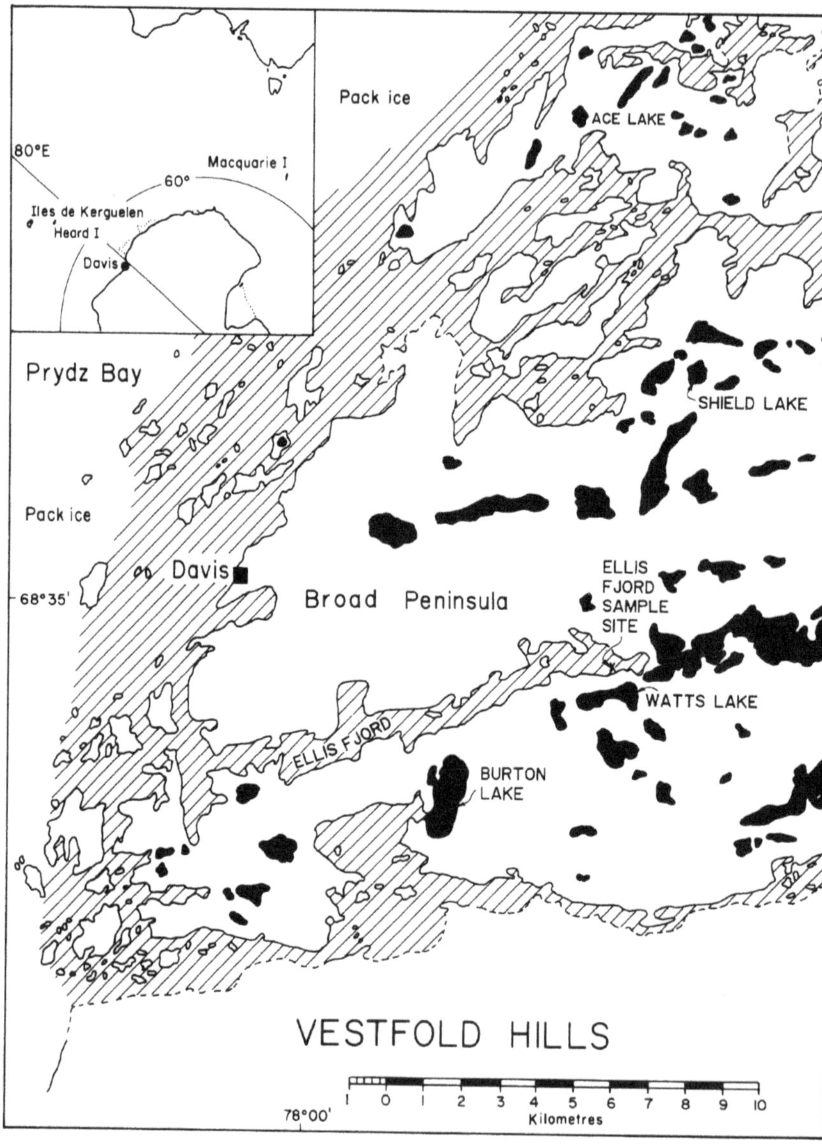

Fig. 1. Map of the Vestfold Hills.

Chemical determinations

Water samples were diluted with deionized distilled water and 0.1 ml of each of the diluted solutions was injected into a model 10 Dionex ion chromatograph. To determine the sulfate and chloride concentrations, the columns used were an S2 anion separator (P/N 31019) with an S2 anion concentrator (P/N 31022) and an anion suppressor column (P/N 30828). The eluent, 3 mM $NaHCO_3$ and 2.4 mM Na_2CO_3 in distilled water, had a flow rate of 2.3 ml min^{-1}. Separated ions were detected by specific conductivity. The concentration of each ion was determined from a standard curve of conductivity versus concentration for each ion. Mud (50 g) was mixed with 30.0 ml of 5.0% zinc acetate and 5.0 ml of chloroform and refrigerated overnight to prevent reoxidation of sulfide and further reduction of sulfate by the sulfate-reducing bacteria. An aqueous mud extract was obtained after centrifugation, and the sulfate and chloride concentrations were determined. The percentage porewater in each sediment was determined from the difference in weight after drying an untreated sediment sample at 105 °C. Sulfate and chloride concentrations, (kg^{-1} sediment wet weight), were calculated after adjustment of each ion's concentration in the porewaters by their respective percentage porewater values. Sulfide concentrations in the waters were determined colorimetrically using a Varian spectrophotometer model 635 after reaction of the sample with 1 mM 5,5-dithiobis-(2-nitrobenzoic acid) (Ellman, 1959).

The concentrations of acetic acid and lactic acid in sediment porewaters were determined for each site on a monthly basis between June and December 1984. Porewaters were collected from fresh sediment samples by centrifugation. The porewaters were dispensed into glass ampoules which were sealed and sterilized at 121 °C for 15 min. The ampoules were sent to the Australian Government Analytical Laboratories, Kingston, Tasmania, Australia for analysis. Analysis was performed by two systems of High Performance Liquid Chromatography after filtration through a 0.45 μm membrane filter. In system 1, a 25 μl sample was passed through a Bio Rad Organic Acid Analysis Column (300 mm × 7.8 mm Aminex HPX-87H) in an eluent of degassed 0.005 N H_2SO_4 with a flow rate of 0.6 ml min^{-1}. Peaks were detected by UV absorption at 210 nm. In system 2, the column used was a Reversed Phase C18 radial Pak 5 μ (Waters-Millipore); the eluent was degassed 0.5–1.0% $(NH_4)_2PO_4$ (pH 3–4), and the flow rate was 1–1.2 ml min^{-1}. The sample size and detection method were the same as used in system 1.

Sulfate reduction experiments

Sediment and water samples were immediately returned to the laboratory after collection. Approximately 10 g quantities were distributed into 13 mm × 130 mm silicone coated glass tubes fitted with silicone rubber stoppers. In the February and May experiments, 1.0 ml of a 0.44 mM solution of $Na_2{}^{35}SO_4$ (Amersham) and a 0.5 mM solution from a different batch of $Na_2{}^{35}SO_4$ was injected into the samples respectively. In each case the zero time radioactivity in the samples was approximately 10^6 counts min^{-1} (cpm) g^{-1} sample. In the May experiments, sodium lactate and sodium acetate were added to some samples to a final concentration of 6 mmol kg^{-1} sediment. These organic acids were dissolved in the solutions of radioactive sulfate. The additions caused a rise in Eh. For example, with the addition of the radioactive sulfate, the Eh of a sample of 12 m Burton Lake water rose from −170 mV to −111 mV in the May experiments. In the February experiments, after addition of the radioactive sulfate, the samples were incubated for 6 weeks at 6 °C. Triplicate samples were withdrawn every 7 days and 1.0 ml of a saturated aqueous solution of $CdCl_2$ was added to each sample to precipitate the sulfide. The tubes were then frozen until analysed. During the May experiments, incubation was for 4 weeks and duplicate samples were withdrawn every 7 days.

Extraction of sulfide and determination of sulfate reduction rate

The N_2 which was used to sparge H_2S from the samples was bubbled through a trap which contained a mixture of water, granular zinc, and methylviologen to remove traces of oxygen. The trap was replaced when the mixture was no longer blue. A four port reaction flask with an internal magnetic stirring rod was flushed with N_2 for 5 min before the sample was added through one of the flask ports. The sparging gas was delivered from the bottom of the reaction flask and subsequently passed through a reflux condenser, an empty 35 mm × 250 mm glass trap, and finally bubbled through two 35 mm × 250 mm glass traps (connected in series) which contained 40.0 ml of 2.0 M NaOH. After addition of the sample, 60 ml of 5 M HCl was added through an injection port of the reaction flask and the mixture was stirred and flushed with N_2 for 1 h, after which 5 g of granular tin and 30 ml of 20% $SnCl_2$ in concentrated HCl was added to the chamber. The mixture was continuously refluxed and flushed with N_2 for a further 1 h. After distillation, radioactivity was measured in (1) the slurry remaining in the reaction flask, (2) the sulfide collected in the NaOH traps and (3) the $Na_2{}^{35}SO_4$ added to the sample. Radioactivity was measured in an LKB Rackbeta Model 1215 scintillation counter in the February experiments and in a Packard Tricarb Model 2002 scintillation counter in the May experiments. Luma gel (Lumac Systems A G) was used as the scintillation fluid. Counts were corrected for background and for quench by the use of an internal standard. The cpm of each solution were determined in triplicate.

The rate (R) at which $Na_2{}^{35}SO_4$ was reduced to $H_2{}^{35}S$ kg^{-1} sediment (wet weight) d^{-1} was determined by the following equation:

$$R = \frac{1000C}{K.T.W.}$$

where C is the total cpm in the two NaOH traps, K is the total cpm μmol^{-1} of $Na_2{}^{35}SO_4$ at the completion of the experiment, T is the incubation time in days, and W is the sample weight in g. After determination of the sulfate concentration in the sediment, the total reduction rate (TR) in $\mu mol\ kg^{-1}$ (wet weight) d^{-1} was calculated according to:

$$TR = \frac{R.\ SO}{ST}$$

where SO is the concentration (mmol) of sulfate in the sample and ST is the concentration (mmol) $Na_2{}^{35}SO_4$ added to the sample.

Results

The sulfate to chloride ratio, temperature, and sulfide and sulfate concentrations throughout the water columns of the five sites sampled in February are given in Figs. 2 and 3. Sulfide was not detected in the water column of Watts Lake, which was the only non-meromictic lake sampled. A temperature profile was not obtained from Ellis Fjord.

Sulfide concentration increased with depth in the meromictic basin of Ellis Fjord and in all the meromictic lakes. In the hypolimnia of all these sites, except Shield Lake, there was a continual decrease in the sulfate concentration and in the sulfate to chloride ratio. In Shield Lake, both parameters increased with increasing depth in the upper levels of the hypolimnion but they decreased below 30 m. Chlorosity of water from the epilimnion and hypolimnion of each of the sites is given in Table 1.

An example of the results of a time course experiment is given in Fig. 4. Sulfate reduction did not show a linear relationship with time for Ace Lake samples. For other sites, correlation coefficients for linear regression computations were; Burton Lake sediment (February), 0.86; Watts Lake sediment, 0.69; Burton Lake sediment (May), 0.61; Ellis Fjord sediment, 0.73; Shield Lake sediment, 0.68; Burton Lake water (12 m, May), 0.74; Burton Lake sediment

Table 1. Chlorosity of bottom water and water from the epilimnion for four antarctic lakes and an antarctic fjord.

Site	Depth to Chemocline (m)	Epilimnion[a] Cl^- (mol l^{-1})	Hypolimnion[b] Cl^- (mol l^{-1})
Watts Lake	NA	0.029	0.034
Burton Lake	11	0.516	0.691
Ellis Fjord	42	0.536	0.578
Shield Lake	18	1.706	2.814
Ace Lake	10	0.327	0.691

[a] Chlorosity of water from a depth of 5 m; [b] Chlorosity of water from 1 m above the sediment; NA = Not applicable.

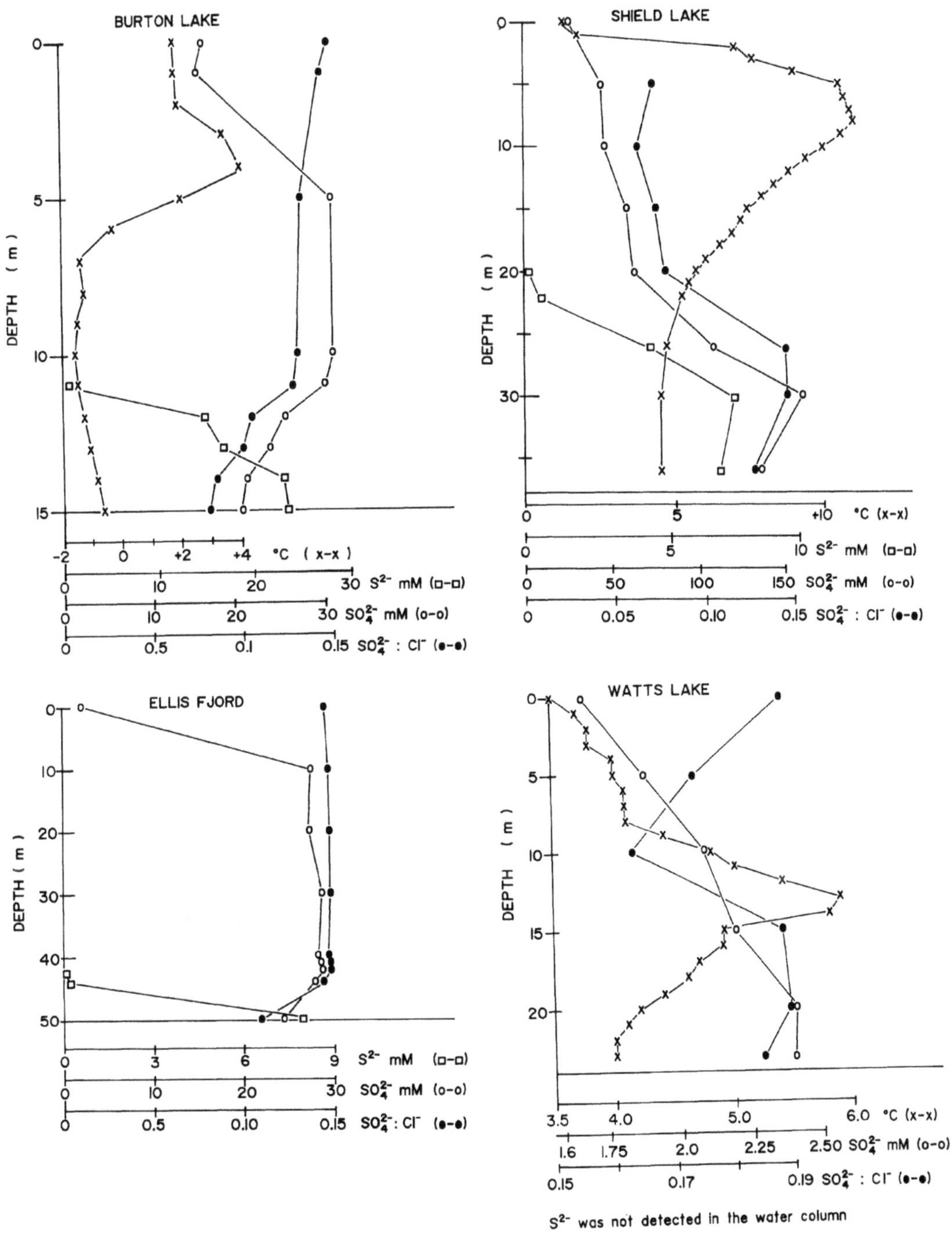

Fig. 2. Vertical profiles of temperature, sulfate concentration, sulfide concentration and the ratio of sulfate concentration to chloride concentration in Shield Lake, Burton Lake, Watts Lake and Ellis Fjord.

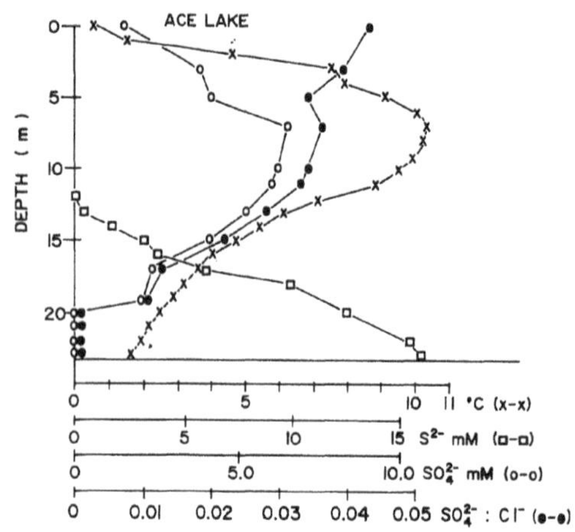

Fig. 3. Vertical profiles of temperature, sulfate concentration, sulfide concentration and the ratio of sulfate concentration to chloride concentration in Ace Lake.

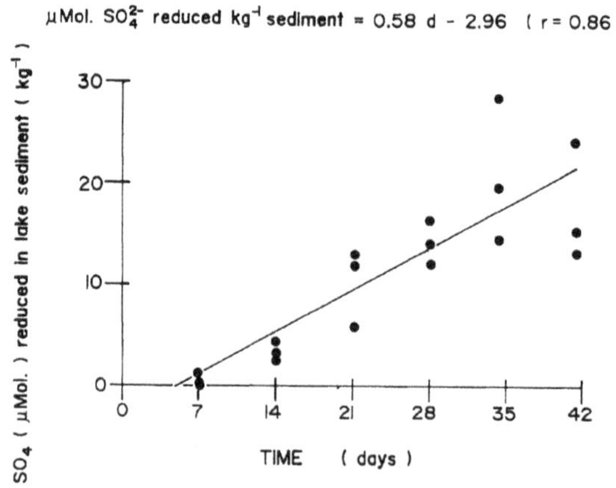

Fig. 4. Plot of μM SO_4^{2-} reduced (kg^{-1} sediment wet weight) versus time in days for sediment collected from Burton Lake in February 1984.

plus 6 mM acetate, 0.85; Burton Lake sediment plus 6 mM lactate, 0.91. Intercepts for the regressions varied from 2.14 to −1.8 days. Sulfate reduction rates for the February experiments are given in Table 2. Table 3 shows the sulfate reduction rate in water from 12 m in Burton Lake and the response of the sulfate reduction rate in Burton Lake sediments to the addition of 6 mM sodium lactate or acetate. The mean of the percentage ^{35}S recovery in the radiometric experiments, calculated from the ratio of total cpm recovered after distillation to the total cpm added $sample^{-1}$, was 104 ± 9%.

The concentration of acetic and lactic acids in porewater samples collected between June and December was variable and did not show any seasonal trend of concentration increase or decrease at each study site over the period sampled. The maximum and minimum concentrations of acetic and lactic acids measured in the porewaters are given in Table 4.

Table 3. Sulfate reduction rates in Burton Lake water and sediments supplemented with artifical substrates.

Location	sulfate reduction $\mu mol\ kg^{-1}\ d^{-1}$
Sediment (Feb. 1984)	0.58
Sediment (May 1984)	0.24
Sediment plus 6 mM acetate (May 1984)	0.37
Sediment plus 6 mM lactate (May 1984)	0.34
Water from 12 m (May 1984)	0.59

Table 2. Temperature, some ionic composition and sulfate reduction rates for sediments of four antarctic lakes and an antarctic fjord.

Site	T °C	Cl^{-1} mol kg^{-1}	SO_4^{2-} mmol kg^{-1}	S^{2-} mmol kg^{-1}	sulfate reduction[a] $\mu mol\ kg^{-1}\ d^{-1}$
Watts Lake	3.6	0.19	0.47	72.2	1.13
Burton Lake	−0.4	0.55	13.97	130.0	0.58
Ellis Fjord	ND	0.35	18.68	133.0	0.67
Shield Lake	4.5	0.83	41.01	95.1	0.69
Ace Lake	1.4	0.45	0.04	174.4	0.00

[a] These rates were calculated from the regression analyses relating sulfate reduction with time over an incubation period of 6 weeks at 6 °C. Each result is the average of 18 determinations; ND = Not determined.

Table 4. The range of concentrations of acetic and lactic acids detected in porewaters collected between June and December from the sediments of four antarctic lakes and an antarctic fjord.

Site	Acetic μmol l^{-1}	Lactic μmol l^{-1}
Watts Lake	17 – 350	11
Burton Lake	17 – 120	11 – 66
Ellis Fjord	33 – 50	11 – 33
Shield Lake	230 – 633	33 – 333
Ace Lake	17 – 416	11 – 222

Discussion

The lakes of the Vestfold Hills are thought to have originated from sea water trapped in basins after isostatic land uplift caused by the recession of the continental ice cap from the region (Johnstone *et al.*, 1973). Since that time, water has evaporated from Shield Lake and concentrated the ions with respect to the original sea water. Watts Lake, flushed by fresh water at some period in the past, has the lowest ionic concentration of the lakes studied.

Ellis Fjord and Burton Lake have connections to the sea. The chloride concentrations in these water bodies (Table 1) are similar to that of seawater of salinity 34.8‰, which has a chloride concentration of 0.562 mol l^{-1} (Walton Smith, 1974). Ace Lake has a greater variation in chloride concentration between epilimnetic and hypolimnetic water than in the other sites. The chlorosity of the hypolimnetic water of Ace Lake would indicate that the original seawater was concentrated by evaporation or freeze-concentration (Matsubaya *et al.*, 1979) and this was later overlain by fresher water which formed the epilimnion. Temperature measurements taken for Watts Lake over one year (Heath, this volume) indicate that the lake turns over once a year. The other lakes studied are permanently stratified with respect to salinity (Burton, 1981) which prevents lake turnover. Stagnation of the hypolimnia ensured that once oxygen was consumed from these waters, other electron acceptors such as sulfate were necessarily utilized in bacterial respiration. Chemical profiles (Figs. 2 and 3) show that sulfate has been depleted in the bottom waters of each of the meromictic basins. This has proceeded to such an extent in Ace Lake, that sulfate is totally depleted in the water column below a depth of 21 m and as a result methanogenesis occurs in the lake (Hand & Burton, 1981).

One reason for the selection of the five study sites was to discern the effect of ionic composition of the sediment on sulfate reduction rates. Skyring *et al.* (1983), in their study of sulfate reduction in intertidal sediments of Spencer Gulf, showed a negative correlation between sulfate reduction rate and salinity of the porewaters. In the present study of antarctic environments, a relationship between sulfate reduction and chlorosity was not observed. There are, however, extreme variations between the sediments from the various sites. For example, the sulfate concentrations in the sediments of Ace and Watts Lakes (0.04 and 0.47 mmol kg^{-1} respectively) would be expected to limit sulfate reduction (Postgate, 1951). The maximum and minimum concentrations of lactic acid and acetic acid in sediment porewaters from the sites sampled are given in Table 4. It is likely that microbially available acetate concentrations in sediments are larger than concentrations determined from porewaters (Sansone & Martins, 1982). The concentration of acetate detected in each antarctic sample was higher than concentrations determined by Ansbaek & Blackburn (1980) in porewaters from Limfjordin, Denmark (0.1 – 6.0 μmol l^{-1}) where the sulfate reduction rate was significantly higher (7 mmol m^{-2} day^{-1}) than the rates determined for the antarctic sites. This suggests that sulfate reduction rates in the antarctic sediments were not limited by the concentration of organic acids. Additions of artificial organic substrates such as lactate and acetate did not appreciably stimulate sulfate reduction rates (Table 2). Rates of sulfate reduction in nature are undoubtedly governed by a number of factors.

Watts Lake sediment, which had the highest sulfate reduction rate of the sites, also had the lowest average sulfide concentration. It is possible that sulfate reduction was suppressed by product inhibition in Burton Lake and Ellis Fjord where sulfate was not in rate-limiting concentrations but sulfide content averaged 130 mmol kg^{-1}. It is known that *Desulfovibrio desulfuricans* will tolerate hydrogen sulfide to concentrations of 2.5 g l^{-1} (Krouse & McCready, 1979) or 71 mmol l^{-1}. But in their taxonomic review of dissimilatory sulfide-reducing bacteria, Pfennig & Biebl (1981) state that *Desulfuromonas* spp., which can also utilize sulfate as an electron acceptor, should be sub-cultured when the sulfide concentration in cultures reaches 10–12 mmol l^{-1}. This would suggest the organism is susceptible to product inhibition or toxicity. From growth rate experiments, it is known that hydrogen sulfide does appear to depress the growth of sulfate-reducing bacteria (Trudinger *et al.*, 1972). In Lake Belovod, Sorokin (1970) recorded sulfate reduction rates in bottom waters of about 5 μmol l^{-1} day^{-1}. He found a second reduction zone, 1 μmol l^{-1} day^{-1} just below a layer where photosynthetic bacteria and chemosynthetic bacteria were in high numbers. This zone had a stable negative redox potential of −60 mV to −100 mV. Burton Lake (May experiments), exhibited a similar magnitude of sulfate reduction in 12 m water as in the sediments (Table 3). At the 12 m level in Burton Lake, photosynthetic bacteria of the genus *Chlorobium* were present in numbers up to 5×10^7 ml^{-1} (Burke, pers. commun.) and the oxidation reduction potential was always less than −100 mV. The free sulfide concentration in water from the 12 m level in Burton Lake was 14.8 mmol l^{-1} whereas bottom water contained 23.5 mmol l^{-1}. It is possible that the *Chlorobium* population depleted the sulfide by photosynthetic oxidation and provided organic substrates for the sulfate reducers.

The two most significant environmental parameters which have been found to influence sulfate reduction rates in non-sulfate limiting anaerobic sediments are temperature and organic substrate supply (Nedwell & Abram, 1979; Skyring *et al.*, 1983). A comparison of the sulfate reduction rates (Table 1) with rates determined for non-antarctic sites (Nedwell, 1982; Tsou *et al.*, 1973; Krouse & McCready, 1979; Skyring, in press) show that the antarctic rates are significantly lower than those obtained for most coastal environments, but comparable to some low temperature, deep sea rates and some rates in Russian freshwater lakes and reservoirs. Although sulfate reduction rates are low when compared to most coastal environments, sulfide concentrations of the sediments (Table 2) are high. It may be that the bulk of the sulfide was formed during a period of more intense sulfate reduction and is being maintained by the present slow reduction of sulfate and by sulfide produced by bacterial heterotrophic degradation of organic sulfur compounds.

All of the lakes are ice-covered for at least nine months of the year, and this prevents wind induced turbulence in the water column in winter whilst the stratified waters of the chemoclines (where the water abruptly increases in density) inhibits mixing in summer. The upward translocation and subsequent oxidation of the hydrogen sulfide produced in the hypolimnia of these separated ecosystems must be at least partially limited by vertical molecular diffusion, the same processes Yusa (1979) determined to be limiting heat and salt transportation in other antarctic meromictic lakes.

Theoretical calculations of diffusion in an antarctic meromictic lake show it to be a very slow process of mixing. Wilson (1964) calculated that if all the salts were concentrated in a layer of negligible thickness in the bottom of Lake Vanda, it would have taken 1200 years to reach the current lake chemical stratification. Diffusion coefficients for sodium chloride and hydrogen sulfide are of a similar order of magnitude, 1.5×10^{-5} and 1.7×10^{-5} cm^2 sec^{-1} respectively (Weast, 1973–1974), although the molecular diffusion coefficient of an ion varies with the temperature of water, the concentration of other ions as well as its own concentration. The interpretation of hydrogen sulfide diffusion is further complicated by sulfate reduction in the water column. Further evidence for the slow loss of sulfide from the hypolimnia of these lakes comes from Ace Lake. In the bottom of this lake there is a large pool of sulfide, however negligible sulfate is available for its further production if sulfide is lost. The constant mean ^{34}S values throughout Ace Lake (Burton & Barker, 1979) indicate it is a closed system with respect to sulfur and very little mixing occurs between depths.

Acknowledgements

The authors thank Tony Chaffey of the Australian Government Analytical Laboratories, Kingston, Tasmania, Australia for the determination of the concentration of organic acids in sediment porewaters. The Baas Becking Geobiological Laboratory is supported by the Bureau of Mineral Resources, the Commonwealth Scientific and Research Organization and the Australian Mineral Industries Research Association Ltd.

References

Ansbaek, J. & T. H. Blackburn, 1980. A method for the analysis of acetate turnover in a coastal marine sediment. Microb. Ecol. 5: 253–264.

Burke, C. M., 1983. Photosynthetic sulphur bacteria in the Vestfold Hills. ANARE Annual Report. Antarctic Division, Kingston, Aust.

Burke, C. M. & H. R. Burton, this volume. Photosynthetic bacteria in meromictic lakes and stratified fjords of the Vestfold Hills, Antarctica. Hydrobiologia.

Burton, H. R., 1981. Marine lakes of Antarctica. In Antarctic Symposium, October 2–3, 1981 at the University of Tasmania, Hobart, Australia. Australian and New Zealand Association for the Advancement of Science: 51–58.

Burton, H. R. & R. J. Barker, 1979. Sulfur chemistry and microbiological fractionation of sulfur isotopes in a saline antarctic lake. Geomicrobiol. J. 1: 329–340.

Ellman, G. L., 1959. Tissue sulphydryl groups. Arch. Biochem. Biophys. 82: 70–74.

Hand, R. M., 1980. Bacterial populations of two saline antarctic lakes. In P. A. Trudinger & M. R. Walter (eds), Biogeochemistry of Ancient and Modern Environments. Australian Academy of Science: 123–129.

Hand, R. M. & H. R. Burton, 1981. Microbial ecology of an antarctic saline meromictic lake. In W. D. Williams (ed.), Salt lakes, Proceedings of the International Symposium on Athalassic (Inland) Salt Lakes. Dr W. Junk Publishers, The Hague: 362–274.

Heath, C. W., this volume. Annual primary productivity of an antarctic continental lake: phytoplankton and benthic algal mat production strategies. Hydrobiologia.

Johnstone, G. W., D. J. Lugg & D. A. Brown, 1973. The biology of the Vestfold Hills, Antarctica. ANARE Scientific Reports Series B(1), Publication No. 123. Antarctic Division, Kingston, Aust.

King, E. W. & D. A. Everitt, 1980. A remote sampling device for under-ice water, bottom biota, and sediments. Limnol. Oceanogr. 25: 935–938.

Krouse, H. R. & R. G. L. McCready, 1979. Reductive reactions in the sulfur cycle. In P. A. Trudinger & D. J. Swain (eds), Biochemical Cycling of Mineral-Forming Elements. Elsevier Scientific Publishing Co. N.Y.: 315–368.

Matsubaya, O., H. Sakai, T. Torii, H. Burton & K. Kerry, 1979. Antarctic saline lakes – stable isotopic ratios, chemical compositions and evolution. Geochim. Cosmochim. Acta. 43: 7–25.

Nedwell, D. B., 1982. The cycling of sulfur in marine and freshwater sediments. In D. B. Nedwell & C. M. Brown (eds), Sediment Microbiology. Academic Press, N.Y.: 73–106.

Nedwell, D. B. & J. W. Abram, 1979. Relative influence of temperature and electron donors and electron acceptors on bacterial sulfate reduction in salt marshes. Microb. Ecol. 5: 67–72.

Pfennig, N. & H. Biebl, 1981. The dissimilatory sulfur-reducing bacteria. In M. P. Starr, H. Stolp, H. G. Truper, A. Balows & H. G. Schlegel (eds), The Prokaryotes. Springer-Verlag, Berlin: 941–947.

Postgate, J. R., 1951. The reduction of sulfur compounds by Desulfovibrio desulfuricans. J. Gen. Microbiol. 5: 725–738.

Sansone, F. J. & C. S. Martins, 1982. Volatile fatty acid cycling in organic-rich marine sediments. Geochim. Cosmochim. Acta. 46: 1576–1589.

Skyring, G. W., (in press). Sulfate reduction in coastal environments. Geomicrobiology.

Skyring, G. W., L. A. Chambers & J. Bauld, 1983. Sulfate reduction in sediments colonised by cyanobacteria, Spencer Gulf, South Australia. Aust. J. Mar. Freshw. Res. 34: 359–374.

Sorokin, Y. I., 1970. Interrelations between sulfur and carbon turnover in meromictic lakes. Arch. Hydrobiol. 66: 391–466.

Trudinger, P. A., 1979. The biological sulfur cycle. In P. A. Trudinger & D. J. Swain (eds), Biochemical Cycling of Mineral-Forming Elements. Elsevier Scientific Publishing Co. N.Y.: 293–314.

Trudinger, P. A., I. B. Lambert & G. W. Skyring, 1972. Biogenic sulfide ores: a feasibility study. Econ. Geo. 67: 1114–1127.

Tsou, J. L., D. Hammond & R. Horowitz, 1973. Interstitial water studies, Leg 15. Study of CO_2 release from stored deep sea sediments. In B. C. Heezen (ed.), Initial Reports of Deep Sea Drilling Project Vol. XX. US Govt Printing Office. Washington D.C.: 851–863.

Walton Smith, F. G. (ed.), 1974. Handbook of Marine Science Volume 1. CRC Press, Ohio: p 5.

Weast, R. C. (ed.), 1973–1974. Handbook of Chemistry and Physics. CRC Press. Ohio: F55–F56.

Wilson, A. T., 1964. Evidence from chemical diffusion of a climatic change in the McMurdo Dry Valleys 1200 years ago. Nature. 201: 176–177.

Yusa, Y., 1979. Analysis of thermosolutal phenomena observed in McMurdo saline lakes (Extended Abstract). In T. Nagata (ed.), Proceedings of the Seminar III on Dry Valley Drilling Project, 1978. Memoirs of National Institute of Polar Research Special Issue No. 13. Tokyo: 42–48.

Hydrobiologia 165: 35–40 (1988)
J. M. Ferris, H. R. Burton, G. W. Johnstone & I. A. E. Bayly (eds) Biology of the Vestfold Hills, Antarctica
Kluwer Academic Publishers

Preliminary observations on psychrotrophic and psychrophilic, heterotrophic bacteria from antarctic water samples

T. A. McMeekin
Department of Agricultural Science, University of Tasmania, G.P.O. Box 252C, Hobart, Tasmania 7001, Australia

Key words: antarctic bacteria, curved bacteria, temperature relations, identification

Abstract

Bacteria isolated from antarctic sea and lake water samples were examined for growth at 5 °C and 25 °C. Of 250 isolates, 249 were psychrotrophic and one was psychrophilic. The effect of temperature on the growth rate of the latter (strain 755) and a typical psychrotroph (strain 203A) was described by the square root of the growth rate constant. T_{MIN} values calculated from this model confirmed the strains as psychrophile and psychrotroph.

Identification of 93 isolates showed the majority to be *Pseudomonas* (82%), with *Flavobacterium* and *Moraxella* representing 13% of the isolates. The remainder were coryneforms and yeasts. Isolate 611 was studied in detail because of its unusual morphology. This strain has features in common with the genera *Spirosoma, Runella* and *Flectobacillus*, but cannot be assigned with certainty. It may represent a new taxon of curved, gram negative bacteria.

Introduction

Bacteria may be classified according to their temperature relationships as thermophiles, mesophiles, psychrotrophs and psychrophiles, the last two categories being of particular interest in the antarctic ecosystem. Psychrotrophs are bacteria that grow at chill temperatures but have optima approaching 30 °C, whereas psychrophiles also grow well at 0 °C but have an optimum at approximately 15 °C and a maximum temperature for growth of 20 °C (Morita, 1975). Psychrotrophs may represent the zymogenous microflora of cold environments and psychrophiles the autochthonous flora (Baross & Morita, 1978).

The effect of temperature on the rate of bacterial growth is normally described by the Arrhenius equation, but Arrhenius plots give curves rather than the predicted linear relationship, indicating a poor description of experimental data. In place of the Arrhenius equation, Ratkowsky *et al.* (1982, 1983) proposed a linear relationship between the square root of the growth rate constant (r) and absolute temperature (T):

$$\sqrt{r} = b(T - T_{MIN})\{1 - \exp[c(T - T_{MAX})]\}$$

where b and c are constants and T_{MIN} and T_{MAX} are notional temperatures which may be intrinsic properties of the organism and may therefore be a useful means to categorize bacteria according to their temperature relations. A graduation of T_{MIN} values for categories from psychrotroph to mesophile and thermophile has been demonstrated (Ratkowsky *et al.*, 1982) but data for psychrophiles is limited with a discontinuity in T_{MIN} values from 248 °K for the psychrophile described by Harder & Veldkamp (1971) to approximately 263 °K for psychrotrophic isolates (Ratkowsky *et al.*, 1982). It was therefore of interest to obtain bacteria from antarctic water samples which might be used to fill the gap between psychrophiles and psychrotrophs.

In addition, samples were expected to yield gram negative, yellow pigmented bacteria useful for taxonomic studies of *Flavobacterium* spp. Bergey *et al.*, and related genera. These organisms form a large proportion of the microflora of temperate waters, both fresh and saline (Shewan & McMeekin, 1983), but environmental isolates remain poorly characterized despite recent taxonomic advances which allow accurate speciation of isolates from clinical sources (Holmes *et al.*, 1984).

Materials and methods

1. Sample collection and treatment

Water samples were obtained from several sites in the vicinity of the Australian Antarctic Base at Davis (Table 1). These were maintained at refrigeration temperatures during storage and transport to avoid thermal injury to psychrophiles.

2. Isolation procedures

Two isolation procedures were used; a direct plating technique on plate count agar (Oxoid), marine agar, and Zobell's agar (Difco) with incubation at 5 °C, and an enrichment technique in marine broth (Difco), Zobell's broth and 10^{-1} and 10^{-2} dilutions of the latter. Isolates were obtained from the enrichment cultures by streaking on the appropriate solid medium at weekly intervals for up to four weeks. A total of 250 isolates were obtained and incubated at 5 °C and 25 °C. Those growing at both temperatures were designated psychrotrophs, while those growing at 5 °C but not 25 °C were psychrophiles (Morita, 1975).

Table 1. Description of sample sites.

Site	Depth	Salinity	Temperature
Sea water (Davis)	Surface	34‰	1.0 °C
Oval Lake	Surface	70‰	0.5 °C
Oval Lake	5 m	160‰	6.0 °C
Burton Lake	0.1 – 2 m	10 – 20‰	1.0 °C
Ellis Fjord	Surface	34‰	4.0 °C

3. Identification procedures

Selected isolates were identified on the basis of colony and cell morphology, gram reaction, motility, presence of catalase and oxidase and oxidative or fermentative dissimilation of glucose according to the scheme of Shewan *et al.* (1960) as modified by Thomas & McMeekin (1980).

Because of unusual morphology, strain 611 was examined in greater detail. Morphological features were revealed by electron microscopy. Suspensions prepared from growth on Zobell's agar were placed on parlodion-coated grids and stained with 1% phosphotungstic acid at pH 7.0 for 15 s. Specimens were examined in a JEOL-1200EX transmission electron microscope operated at 80 kV. Additional biochemical reactions were determined after 7 days at 20 °C using the AP120E system and Flow Laboratories N/F (non-fermenter) system (details of tests are given in Results section). Antibiotic sensitivities were tested on heart infusion agar at 22 °C (Table 2).

4. Determination of temperature relations of psychrophilic strain (755) and psychrotrophic strain (203A)

The effect of temperature on the rate of bacterial

Table 2. Antibiotic sensitivity profile of strain 611.

Antibiotic	Concentration (μg)	Inhibition zone (mm)
Ampicillin	25	64
Cephrodine	30	90
Chloramphenicol	30	54
Erythromycin	15	58
Gentamicin	10	14
Nitrofurantoin	200	65
Nalidixic acid	30	48
Penicillin	1.5 i.u.	35
Sulphafurazole	24	24
Tetracycline	30	75
Trimethoprim	2.5	90

i.u. = international units.

growth was examined in a temperature gradient incubator (Toyo Kagaku Sangyo Co. Ltd., Tokyo, Japan). Growth was examined by nephelometry over the temperature range 0–35 °C at approximately 1 °C intervals. The growth medium (marine broth) was inoculated with an actively growing culture of each strain and growth rate at each temperature was calculated as the reciprocal of the time required for the culture to reach 30% or 50% turbidity.

Results and discussion

1. Isolation of psychrotrophs and psychrophiles

Of the 250 isolates incubated at 5 °C and 25 °C only one failed to grow at the upper temperature, indicating a preponderance of psychrotrophic types according to the definition of Morita (1975). The one psychrophilic strain isolated (strain 755) was obtained from a water sample taken from Ellis Fjord (Table 1).

Morita (1975) gave two major reasons why microbiologists are not successful in isolating psychrophilic bacteria. These are improper selection of source material, and improper handling procedures in the laboratory. The source material must be carefully chosen and the past history is critical since it must never reach warm temperatures for an extended period of time. This may explain why psychrotrophs are often more common than psychrophiles in polar soil samples (Cameron *et al.*, 1972). However, if the environment is constantly cold, psychrophiles are normally found, albeit in low numbers (Wiebe & Hendricks, 1974; Morita, 1975). Considerable care must also be exercised during the isolation procedure to ensure that the source material is kept cold at all times. These two requirements were fulfilled in this study and it is therefore surprising that only one psychrophile was found.

A further factor to be considered is the nutritional status of the organisms. It is now well documented that bacteria in oligotrophic environments can adopt a starvation survival strategy (Novitsky & Morita, 1976, 1977, 1978) which was termed exogenous dormancy by Stevenson (1978). In this state environmental influences, such as nutrient deprivation, cause marked decreases in size and activity. Release of the stress factor may permit a return to 'normal' development; the cells may remain viable, but non-recoverable on laboratory media (Xu *et al.*, 1982) or they may be killed as a result of substrate accelerated death (Postgate & Hunter, 1962). The ability to adopt this dormant state and to respond to nutrient availability is of obvious ecological advantage to organisms normally found in nutrient depleted waters.

In further work it may be an advantage to add water samples directly to transport media containing low concentrations of media or vitamins. These would act as resuscitation/enrichment cultures during transport from the sampling site to the laboratory. Addition of vitamin solution to marine agar has been shown to markedly enhance the viability of stored marine bacteria (Lewis, 1984).

2. Identification

Ninety three isolates obtained by direct plating technique were identified to the genus level. Nearly all were gram negative rods which is consistent with the findings of others on polar water samples (Baross & Morita, 1978). Of these, 82% of the isolates were identified as *Pseudomonas* Migula, i.e. gram negative, motile, oxidase positive, oxidative rods; 13% were gram negative, non-motile, oxidase positive, oxidative rods, approximately half of which were pigmented (*Flavobacterium*) and half non-pigmented (*Moraxella* Lwoff). Small numbers of gram positive rods (coryneforms) and yeast were also found.

It is of interest that no fermentative bacteria were recovered in this study, particularly members of the genus *Vibrio* Pacini which represent the majority of the psychrophiles isolated from polar water samples (Baross & Morita, 1978). These authors put forward the hypothesis that vibrios may represent the indigenous marine 'coliforms' as they have been constantly isolated from the guts of marine fishes (Liston, 1957) and invertebrates (Liston & Colwell, 1963; Baross & Liston, 1970). The lack of vibrios recovered corresponds with the lack of psychrophiles and in further work it may be useful to use a medium selective for vibrios, as would ex-

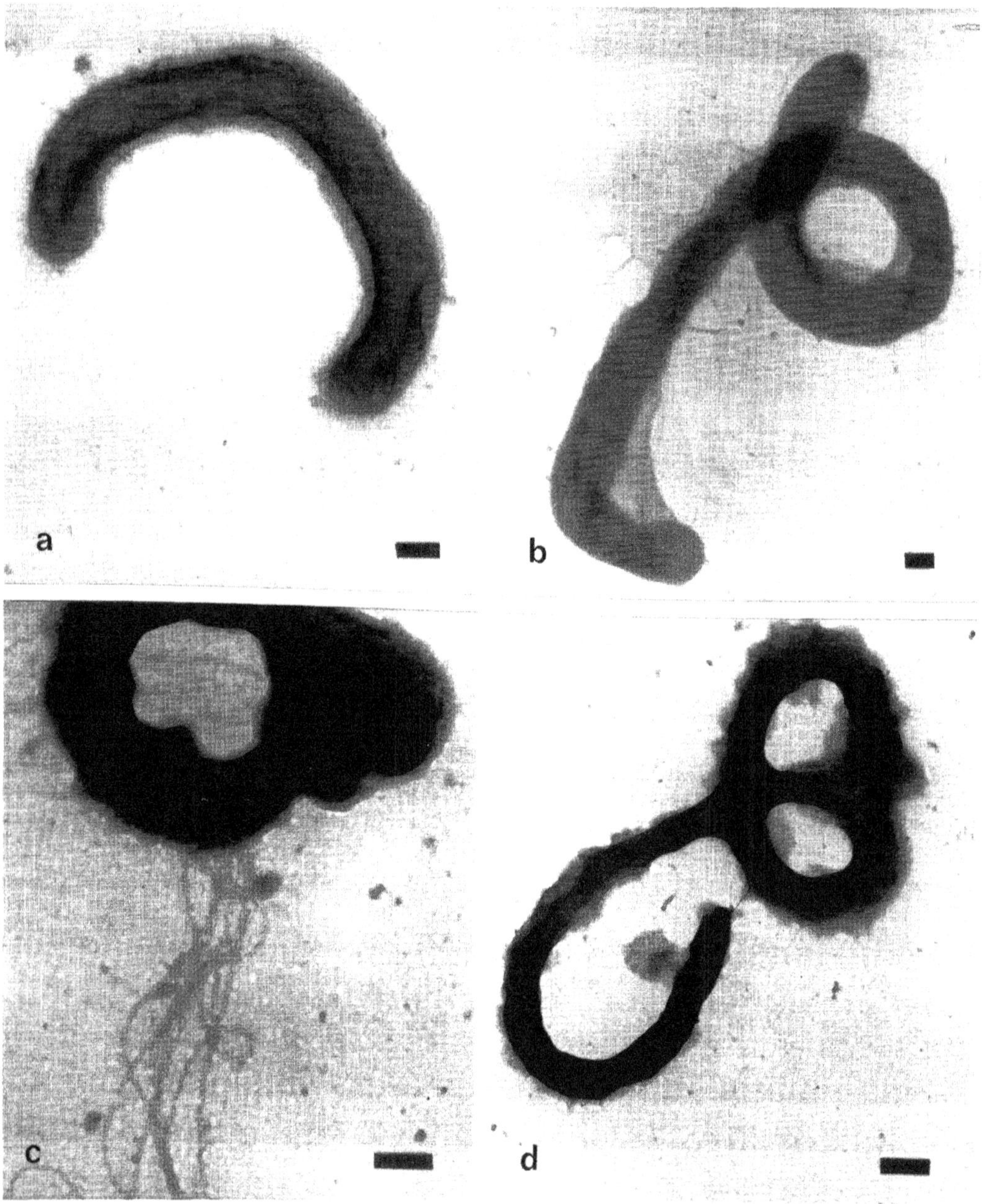

Fig. 1. Isolate 611, From seawater near Davis, Vestfold Hills. a. Typical C shape morphology of strain 611. A single filamentous appendage rises at one pole of the cell. Bar = 0.2 μm. b. Ring shaped structure with extension. The ring can be clearly seen to be formed by overlapping. Extracellular material is also visible. Bar = 0.2 μm. c. Ring shaped cell with bundle of filamentous appendages. Bar = 0.2 μm. d. 'Pretzel' shaped cell comprising two rings and an extension. Bar = 0.2 μm.

amination of the digesta of aquatic fauna and sediment samples.

From a taxonomic viewpoint, several strains are of particular interest and require more detailed study. These include the psychrophilic isolate 755 which fits the description of the revised genus *Flavobacterium* (Holmes *et al.*, 1984). Although psychrotrophic members of this genus have been described, there are no previous reports of psychrophiles.

Isolate 611, from a seawater sample obtained near Davis, is also a pigmented, gram negative bacterium which might be classified as a *Flavobacterium* sp. except for its unusual morphology. The cells, although not flexible, are extremely curved and may form rings and other aberrant shapes as shown in Fig. 1a–d. They are gram negative and non-motile, but 'flagella-like' appendages are seen by electron microscopy.

Colonies on Zobell's agar are approximately 2 mm in diameter and orange pigmented. Pigment does not change colour on addition of 20% KOH and may therefore be a carotenoid and not a flexirubin-type pigment. Colonies also grow on blood agar and on heart infusion agar but not on diagnostic sensitivity test agar or CLED agar (Oxoid). Poor growth is obtained in liquid media including Zobell's broth. Growth occurs at 5 °C and 25 °C but not at 37 °C.

The organism is not fermentative, does not grow anaerobically and is catalase and oxidase positive. Negative reactions were obtained in each of the following tests: beta-galactosidase, arginine dihydrolase, lysine decarboxylase, ornithine decarboxylase, citrate utilization, H_2S production, urease, tryptophan deaminase, indole, Voges Proskauer test, gelatin liquefaction, acetamide utilization, aesculin hydrolysis, DNAse, fluorescent pigment production, pyocyanin production, growth at 42 °C, acid from glucose, mannose, inositol, sorbitol, rhamnose, salicin, melezitose, amygdylin, arabinose, xylose, lactose and maltose.

Antibiotic sensitivity results are shown in Table 2. The isolate was fully sensitive to all antibiotics tested except gentamicin.

The taxonomy of this strain remains uncertain, but it fits the description of non-motile or rarely motile, gram negative, curved bacteria given in Bergey's Manual of Systematic Bacteriology (Larkin & Borrall, 1984). However, the characteristics of strain 611 so far determined are not consistent with the species of *Spirosoma* Migula *Runella* Larkin & Williams and *Flectobacillus* Larkin, Williams & Taylor and the isolate may represent a previously undescribed taxon of curved bacteria.

Table 3. Cardinal temperatures (°K) of psychrotrophic strain 203A and psychrophilic strain 755.

	Strain 203A	Strain 755
T_{MIN}	264	251
T_{OPT}	300	288
T_{MAX}	305	293

3. Temperature relations of psychrotroph (strain 203A) and psychrophile (strain 755)

The effect of temperature on the growth rate of both strains was found to be accurately described by the $\sqrt{r}$ model and the cardinal temperatures for each are shown in Table 3. These indicate that strain 203A is a typical psychrotroph, whilst strain 755 has a psychrophilic nature.

Acknowledgements

I am indebted to C. Burke and H. R. Burton for sample collection and transport, to P. Pennington and S. Cooke for expert technical assistance, to G. Nash for assistance with electron microscopy and K. Thomson for antibiotic sensitivity testing.

References

Baross, J. A. & J. Liston, 1970. Occurrence of Vibrio parahaemolyticus and related hemolytic vibrios in marine environments of Washington State. Appl. Microbiol. 20: 179–187.

Baross, J. A. & R. Y. Morita, 1978. Microbial life at low temperatures: Ecological aspects. In D. J. Kushner (ed.), Microbial Life in Extreme Environments. Academic Press, N.Y. & Lond. 10–57.

Cameron, R. E., F. A. Morelli & R. M. Johnson, 1972. Bacterial species in soil and air of the Antarctic Continent. Antarct. J. U.S. 7: 187–189.

Holmes, B., R. J. Owen & T. A. McMeekin, 1984. The genus Flavobacterium. In N. R. Kreig (ed.), Bergey's Manual of Systematic Bacteriology, I. Williams & Wilkins, Baltimore: 353–361.

Larkin, J. M. & R. Borrall, 1984. Spirosomaceae. In N. R. Kreig (ed.), Bergey's Manual of Systematic Bacteriology, I. Williams & Wilkins, Baltimore: 125–132.

Lewis, T. E., 1984. Microflora of coralline algae. B.Sc.(Hons.) Thesis. University of Tasmania: 36–38.

Liston, J., 1951. The occurrence and distribution of bacterial types on flatfish. J. gen. Microbiol. 16: 205–216.

Liston, J. & R. R. Colwell, 1963. Host and habitat relationships of marine commensal bacteria. In C. H. Oppenheimer (ed.) Symposium on Marine Microbiology. C. C. Thomas, Springifield, Illinois: 611–624.

Morita, R. Y., 1975. Psychrophilic bacteria. Bacteriol. Rev. 39: 144–167.

Novitsky, J. A. & R. Y. Morita, 1976. Morphological characterisation of small cells resulting from nutrient starvation of a psychrophilic marine Vibrio. Appl. Environ. Microbiol. 32: 617–622.

Novitsky, J. A. & R. Y. Morita, 1977. Survival of a psychrophilic marine Vibrio under long-term nutrient starvation. Appl. Environ. Microbiol. 33: 635–641.

Novitsky, J. A. & R. Y. Morita, 1978. Possible strategy for the survival of marine bacteria under starvation conditions. Mar. Biol. 48: 289–295.

Postgate, J. R. & J. R. Hunter, 1962. The survival of starved bacteria. J. gen. Microbiol. 29: 233–263.

Ratkowsky, D. A., J. Olley, T. A. McMeekin & A. Ball, 19[illegible] Relationship between temperature and growth rate of bacte[illegible] cultures. J. Bacteriol. 149: 1–5.

Ratkowsky, D. A., R. K. Lowry, T. A. McMeekin, A. N. Sto[illegible] & R. E. Chandler, 1983. Model for bacterial culture growth [illegible] throughout the entire biokinetic temperature range. J. Bact[illegible]ol. 154: 1222–1226.

Shewan, J. M., G. Hobbs & W. Hodgkiss, 1960. A determina[illegible] scheme for the identification of certain genera of bacteria w[illegible] special reference to the Pseudomonadaceae. J. appl. Bacter[illegible] 23: 379–390.

Shewan, J. M. & T. A. McMeekin, 1983. Taxonomy (and ecolo[illegible] of Flavobacterium and related genera. Ann. Rev. Microbiol. [illegible] 233–252.

Stevenson, L. H., 1978. A case for bacterial dormancy in aqu[illegible] systems. Microb. Ecol. 4: 127–133.

Thomas, C. J. & T. A. McMeekin, 1980. Contamination of br[illegible]er carcass skin during commercial processing procedures: [illegible] electron microscopic study. Appl. Environ. Microbiol. [illegible] 133–144.

Wiebe, W. J. & C. W. Hendricks, 1974. Distribution of hete[illegible]trophic bacteria in a transect of the Antarctic Ocean. In R. [illegible] Colwell & R. Y. Morita (eds) Effect of the Ocean Environm[illegible] on Microbial Activities. University Park Press, Baltim[illegible] 524–535.

Xu, H. S., N. Roberts, F. L. Singleton, R. W. Attwell, D. [illegible] Grimes & R. R. Colwell, 1982. Survival and viability of n[illegible] culturable Escherichia coli and Vibrio cholerae in the estuar[illegible] and marine environment. Microb. Ecol. 8: 313–323.

Hydrobiologia 165: 41–57 (1988)
J. M. Ferris, H. R. Burton, G. W. Johnstone & I. A. E. Bayly (eds) Biology of the Vestfold Hills, Antarctica
Kluwer Academic Publishers

Pigment and lipid compositions of algal and bacterial communities in Ace Lake, Vestfold Hills, Antarctica

J. K. Volkman[1], H. R. Burton[2], D. A. Everitt[1] & D. I. Allen[3]
[1]*CSIRO Marine Laboratories, Division of Oceanography, GPO Box 1538, Hobart, Tasmania, 7001, Australia;* [2]*Antarctic Division, Department of Science, Channel Highway, Kingston, Tasmania, 7050, Australia;* [3]*Chemistry Department, University of Tasmania, GPO Box 252C, Hobart, Tasmania, 7001, Australia*

Key words: organic geochemistry, saline lake, Antarctica, lipids, pigments, plankton

Abstract

The compositions of carotenoids, chlorophylls and lipids at four depths in Ace Lake have been determined as a means of studying the vertical zonation of species in the lake and for comparison with the lipids found in the bottom sediments. The four major species of phytoplankton found in the lake were identified by electron microscopy. The most abundant phytoplankter was *Pyramimonas gelidicola* McFadden (Chlorophyta, Prasinophyceae) which occurred in greatest numbers at 10 m, the base of the oxylimnion. The pigments and lipids at this depth were mainly derived from this alga. At 11 m (the top of the anoxylimnion) only traces of lipids and pigments attributable to *P. gelidicola* were found, indicating only limited settling of algal cells through to the anoxylimnion, at least in summer. The pigments at 11 m were dominated by bacteriochlorophylls *c* derived from green photosynthetic bacteria *Chlorobium* spp. These pigments were also abundant at 23 m suggesting the presence of intact bacterial cells which had settled out from higher in the water column. Major non-polar lipid classes in the sediments included sterols, alcohols, hydrocarbons and an unusual suite of very long-chain unsaturated ketones and esters which have not previously been reported from antarctic environments. Several novel compounds, not found previously in either sediments or organisms, are reported. These include tri- and tetra-unsaturated straight-chain C_{39} methyl ketones and C_{40} ethyl ketones and the methyl ester of a tetra-unsaturated straight-chain C_{36} fatty acid. The distributions of lipids in the sediment were markedly different from those in the water column indicating extensive bacterial degradation and recycling of labile lipids.

Introduction

The Vestfold Hills region of Antarctica contains hundreds of lakes and ponds with distinctive chemical compositions and biota. The remoteness of the area and the scarcity of terrestrial sources of organic matter make it ideal for studying the production and utilization of autochthonous organic matter in aquatic ecosystems. The lakes were formed as a result of isostatic uplift of the land following retreat of the ice cap about 5000 to 8000 years ago. Some of the lakes are freshwater but many contain relict seawater with some having salinities up to eight times that of seawater (Kerry *et al.*, 1977). The chemical compositions of these lakes have been considerably altered by evaporation, ablation, dilution with freshwater, precipitation and biological modification. The biological communities have also undergone continual change, to the extent that many lakes now contain few living species (Campbell, 1978).

Sediments deposited in a lake provide an historical record of the biological and chemical changes which have occurred in the lake since its formation. Phytoplankton species which form cysts or have hard skeletons or scales, such as diatoms, some dinoflagellates and coccolithophorids, leave recognizable remains in the sediment, but many antarctic lakes contain populations dominated by small, naked flagellates and bacteria which leave few identifiable morphological remains (Vincent & Vincent, 1982). The techniques of organic geochemistry provide one means of assessing changes in the abundance and species compositions of these organisms over time. Aquatic organisms biosynthesize a wide variety of organic compounds, some of which have distinctive structures that are found in a limited number of species. An example is the sterol dinosterol (4α,23,24-trimethylcholest-22E-en-3β-ol), which is biosynthesized by most dinoflagellates but not found in other phytoplankton species. From the occurrence of such 'biomarker' compounds in a sediment it is sometimes possible to infer the types of organisms present when the sediments were deposited. One complication is that many of these compounds are degraded or transformed to other lipids, particularly at the sediment-water interface, and this needs to be considered when using lipid data to interpret the depositional history of the lake from its sediment record.

In this paper we present some initial results concerning the lipid and pigment compositions of particulate matter in waters of Ace Lake. This is a small, saline, meromictic lake in the Vestfold Hills of eastern Antarctica. The phytoplankton and bacterial communities in the lake are strongly stratified and we show that this is reflected in the lipids and pigments found. The lipid composition of a surface sediment was also determined to establish which of those lipids produced in the water column were preserved in the sediment and thus could be used as biomarkers.

Description of the study site

The Vestfold Hills occupy an area of approximately 400 km^2 on the north-eastern side of Prydz Bay, Princess Elizabeth Land, eastern Antarctica. Ace Lake is located at 68° 28.4′ S, 78° 11.1′ E on Long Peninsula which is flanked by Prydz Bay to the north and Long Fjord to the south. A map of the area and detailed description are given by Hand & Burton (1981). Although the lake is very close to the sea (152 m at its closest point), seawater is prevented from entering the lake by low hills. Ace Lake has an area of 1.3×10^5 m^2 and drains a catchment area of about 5×10^5 m^2 consisting of low rocky hills devoid of plant life, except for scattered and scarce mosses and lichens. Its maximum depth in February 1984 was 24.5 m, which represents a deepening of the lake by 1.5 m since measurements were made in 1977 (Burton & Barker, 1979). The lake is ice covered at least from March to December, which severely limits wind driven mixing of the lake waters thus aiding long-term stratification. In some years, such as 1984, only a wide moat melts out.

The oxylimnion extends to a depth of about 8.5 m (7 m in 1977), but over the next 2 m the oxygen content and light penetration drop to zero producing a sharp oxycline. In the underlying anoxylimnion, the waters are permanently dark and anoxic, and increasing concentrations of CH_4 and H_2S are found near the lake bottom (Hand & Burton, 1981). In summer, temperatures range from 0° C at the surface to about 6° C at the oxycline. However, the temperature is about 1° C throughout the year at the sediment surface. Salinities range from about 80% that of seawater at 3 m to 30% greater than that of seawater at the bottom (Burton & Barker, 1979).

Materials and methods

Water and sediment samples

Samples of lake water were collected on 24 February 1984 from depths of 5 m, 10 m, 11 m and 23 m using a 2 l, 50 cm long, plastic Kemmerer water sampler, designed to fit through a 10 cm diameter ice hole. Each water sample thus contained material from a 50 cm depth interval spanning the nominal depth. The water samples were returned immediately to Davis Station (by helicopter) and filtered in the laboratory through pre-extracted glass fibre filters (Whatman GF/F, nominal pore size 0.7 μm). These filters retained a variable proportion of small bacterial cells (depending on the extent of clogging of the pores), but should have quantitatively retained phytoplankton cells including small flagellates. Several samples were taken at each depth to provide filtered material for pigment and lipid analysis and identification of phytoplankton species.

Sediment corres could not be obtained in February 1984 due to the unstable broken ice cover at that time. Analyses were therefore carried out on a core collected using a Züllig corer in June 1978. This was obtained from anoxic waters near the centre of the lake at a water depth of 23 m. Sediment was extruded from the core liner onto the ice surface and sectioned at 5 cm intervals. These samples were then stored frozen, in the dark, at −15° C until extracted in January 1984.

Lipid analyses

The total solvent extractable (TSE) lipids in particulate matter from the water column were obtained by extracting the GF/F filters with propan-2-ol (7 ml, 10 min) and $CHCl_3$:MeOH (2:1, 2 × 7 ml, 2 × 10 min) at room temperature with the aid of ultrasonification. The combined extracts were washed with distilled water and filtered through a Whatman GF/F filter to remove particulate matter. The aqueous layer was back-extracted with $CHCl_3$ to maximize lipid recoveries, and this extract was also filtered through the GF/C filter. The combined TSE lipids were reduced to dryness by rotary evaporation and redissolved in 500 μl of $CHCl_3$:MeOH (2:1), for further analysis. The TSE lipids in the sediment were obtained in a similar fashion, except that 20 ml of propan-2-ol followed by four extractions with $CHCl_3$: MeOH (4 × 20 ml) was used to extract 11 g of wet sediment. The initial extraction with propan-2-ol was included to minimize transesterification which can occur when $CHCl_3$:MeOH is used. The TSE lipids were then washed with dilute saline solution and filtered through a 0.2 μm PTFE filter.

The concentrations of each of the major lipid classes were determined using an Iatroscan TH-10 thin layer chromatography-flame ionization detector (TLC-FID) analyser (Iatron Laboratories, Japan), using methods similar to those described previously (Sipos & Ackman, 1978; Kaitaranta & Nicolaides, 1981; Parrish & Ackman, 1983; Volkman *et al.*, 1986). Peak areas were measured using a Shimadzu Chromatopac CR2AX integrator and converted to concentration values using response factors and calibration curves determined in our laboratory from appropriate lipid standards (Kaitaranta & Nicolaides, 1981). Portions of the TSE lipids (6% for 10, 11 and 23 m samples; 18% for 5 m sample and 0.05% for the sediment sample) were applied to the origin of S-II chromarods (5 μm silica in CH_2Cl_2 (4 μl). The rods were developed with hexane:diethyl ether:acetic acid (60:17:0.15 v/v/v), dried at 100 °C for 5 minutes, and then scanned through the flame ionization detector at 30 s $scan^{-1}$. Chromatograms were recorded with a YEW 3056 pen recorder at chart speed of 20 cm min^{-1} and a sensitivity of 0.5 mV cm^{-1}.

A portion of the TSE lipids was separated into various lipid classes by column chromatography on silicic acid (see Volkman *et al.*, 1983 for details). Free fatty acids were eluted in a fraction containing alcohols and sterols, and were separated by extraction into aqueous KOH. Free fatty acids were converted to methyl esters using 14% BF_3/MeOH for capillary gas chromatographic (GC) analysis. Detailed compositional data were not obtained for other lipid classes in the particulate matter samples owing to the small amounts of lipids isolated, however more detailed analyses were possible with the sediment lipids.

Lipid fractions were analysed by capillary gas chromatography using a Hewlett Packard 5890A gas chromatograph fitted with a purged splitless injector operated at 320 °C. The column used was a Hewlett Packard high performance grade bonded phase dimethylsilicone fused silica capillary column (25 m × 0.2 mm ID), with H_2 as the carrier gas at 35 cm s^{-1} linear velocity. In a typical analysis the sample in CH_2Cl_2 (0.5 μl) was injected at 30 °C. After 60 s the oven was heated to 150 °C at 30 °C min^{-1} and then to 330 °C at 4 °C min^{-1}. A flame ionization detector was used at 350 °C. Compounds were identified from coinjection with standards, and from their electron-impact mass spectra obtained from a VG 70/70F magnetic sector mass spectrometer (MS) coupled to a Pye 204 capillary gas chromatograph using similar chromatographic conditions to the above. The GC-MS data were acquired and processed using a VG 2000 data system based on a PDP-8 computer. The MS was scanned from 600 to 20 amu with a recycle time of 2.2 s.

Pigment analyses

Chlorophylls and carotenoids were analysed by high performance liquid chromatography (HPLC) based on the method of Wright & Shearer (1984). Pigments in solution are unstable in the presence of light and oxygen, so they were extracted rapidly with cold MeOH using ultrasonication followed by centrifugation. The crude extract was purified by passing the solution through an octadecyl silica cartridge (C_{18} SEP-PAK) followed by 2 × 1.5 ml of ethyl acetate. After filtration through a 0.4 μm Millipore FH filter, the extract was analysed using a Waters Associates liquid chromatography system consisting of M6000 and M45 pumps, model 660 solvent programmer, U6K injector and two RCM-100 radial compression modules in series. The columns used were two reverse phase Radial-PAK A octadecyl silica cartridges and the pigments were eluted using a linear gradient from 90% acetonitrile to 100% ethyl acetate over 20 min with a flow rate of 2 ml min^{-1}. Pigments were detected using the sum of absorbances at 405 and 436 nm from a Waters Associates Model 440 two channel absorbance detector. Absorption spectra were also obtaining using a Hewlett Packard 8450A spectrophotometer fitted with a Helma 15 μl flow cell, connected in series after the absorbance detector.

Electron microscopy

Phytoplankton species in lake water and sediments were identified by scanning electron microscopy (SEM) and transmission electron microscopy (TEM). For SEM, a few drops of nanoplankton suspension were placed on glass cover slips coated with polylysine as an adhesive (Marchant & Thomas, 1983). The sample was allowed to settle for 1 h, washed and then dehydrated with washes containing 10% to 100% acetone in 10% increments. After critical point drying with CO_2, the cells were sputter coated with gold and examined using a JEOL JSM-840 scanning electron microscope. For TEM, the suspension was placed on formvar coated copper grids treated with polylysine and then fixed with osmium tetroxide vapour for 30 s. The grids were rinsed in distilled water, air dried, stained with 1% uranyl acetate for 1 min and examined using a JEOL JEM-1200EX transmission electron microscope.

Results and discussion

Phytoplankton, zooplankton and bacteria

Most of the saline lakes in the Vestfold Hills contain only a few species of eukaryotic organisms (Wright & Burton, 1981), and very little is known about the bacterial populations (Hand & Burton, 1981). In February 1984 the phytoplankton population in Ace Lake was dominated by the small green flagellate *P. gelidicola*, identified from its cell morphology and characteristic body and flagellar scales (Fig. 1). This alga was most abundant at 10 m (at the bottom of the oxylimnion) while few cells were detected at 5 m. Only three other algal species were commonly found in the lake and these exhibited marked differences in their vertical distribution. A small biflagellate alga, tentatively identified as a *Cryptomonas* sp. (Fig. 1), predominant in the 5 m sample. Also present was a smaller biflagellate alga of unknown identity (Fig. 1), and a few cells of a large unarmoured dinoflagellate related to *Gymnodinium*. Very few diatom cells were found in the 5 m or 10 m water samples.

Burch (this volume) carried out a seasonal study of the phytoplankton of Ace Lake during 1979 and reported the same four species of algae. He showed that algal biomass began to increase in October just under the ice and continued to increase until January and February. *P. gelidicola* was numerically the most abundant and by February maximum cell numbers were concentrated at 10 m as was found in our study. The *Cryptomonas* sp. also started its growth cycle beneath the ice and migrated downwards as

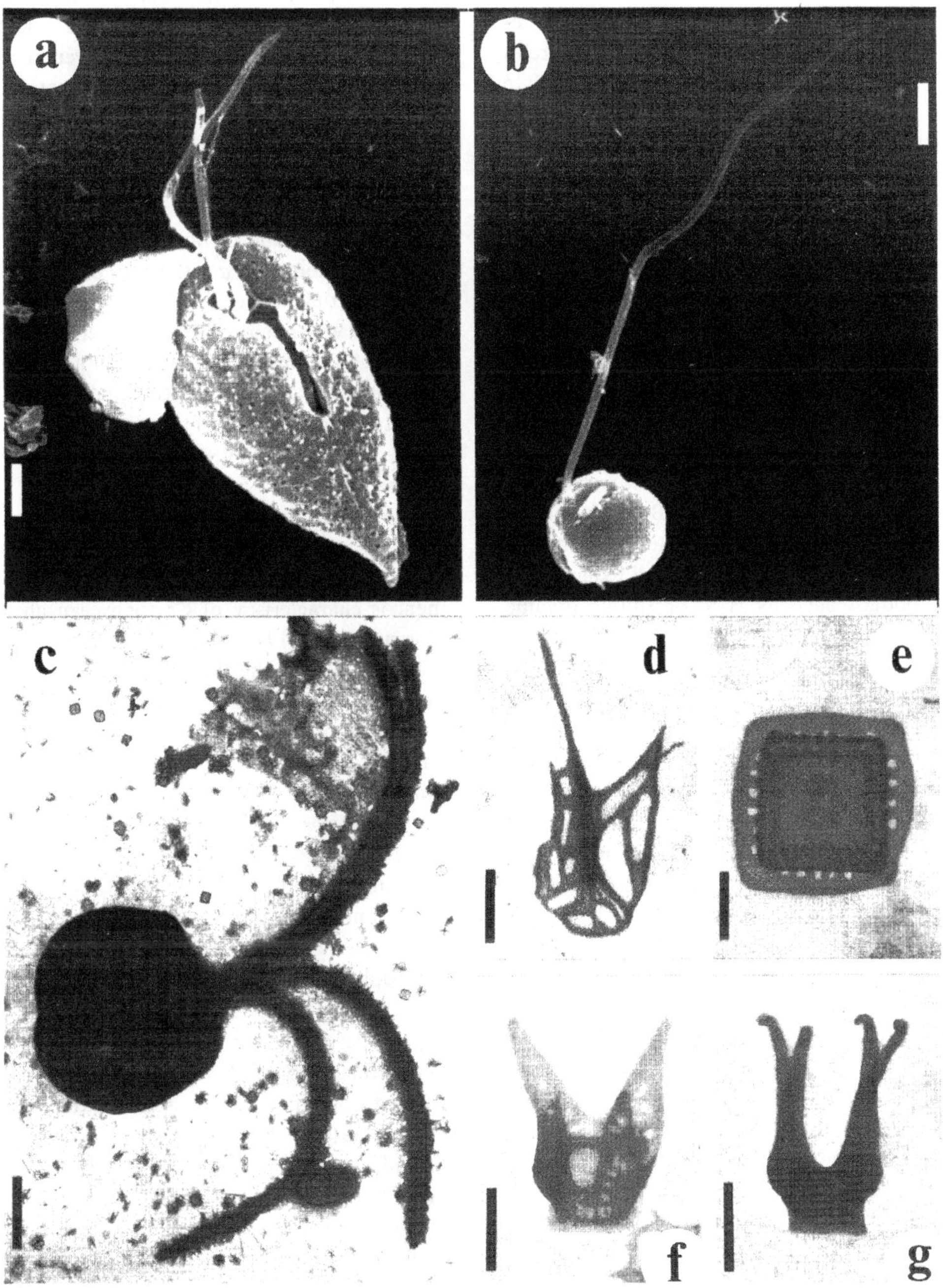

Fig. 1. Phytoflagellates of Ace Lake a. SEM. *Cryptomonas* sp. from 2 m depth on 23 September, 1979. Scale bar = 1 μm; b. SEM. Unidentified small biflagellate from same sample (second flagellum broken). Scale bar = 1 μm; c. TEM. *P. gelidicola* from 10 m depth on 24 February 1984. Scale bar = 2 μm; d.&e. TEM. Scales of *P. gelidicola* from same sample as 1c. Scale bar = 0.1 μm; f.&g. TEM. Scale of unknown organism from 0.65–0.70 m depth in the lake sediments. Scale bar = 0.2 μm.

light intensities increased, but unlike *P. gelidicola*, its preferred habitat appeared to be at 5 m where it remained throughout most of the season. Burch's study showed that the unidentified microflagellate (maximum cell numbers of 1.4×10^4 cells ml^{-1}) was restricted to depths above the thermocline (2-5 m) during 1979, which is consistent with its low abundance in our sample of water from 5 m. These shallower waters were not sampled in our study.

The microbial ecology of Ace Lake has been discussed in earlier publications (Burton & Barker, 1979; Hand, 1980; Hand & Burton, 1981), but many bacterial species have yet to be identified. No light is transmitted below the first 50 cm of the anoxylimnion due to high concentrations of photosynthetic bacteria at this depth. The major species are the green sulfur bacteria *Chlorobium vibrioforme* Pelsh and *Chlorobium limicola* Nadson (Burke & Burton, this volume). These anoxygenic bacteria use bacteriochlorophylls for photosynthesis, and reduced sulfur compounds such as H_2S, or simple organic compounds as electron donors. Species of photosynthetic purple bacteria (Rhodospirillaceae) and *Chromatium* sp. have also been reported from Ace Lake (Hand, 1980). Large populations of methanogenic bacteria are found in the sediments and profiles of methane concentration suggest that methane production may also occur in the water column (Burton, 1980). Sulfate reducing bacteria are also present in Ace Lake but numbers are thought to be low due to low sulfate ion concentrations in the sediment and bottom waters (Burton & Barker, 1979). Recent incubation studies have confirmed that sulfate reduction does occur in the sediments of Ace Lake albeit at very low rates (Franzmann *et al.*, this volume).

The zooplankton of Ace Lake consists of a single species of calanoid copepod, *Paralabidocera antarctica* (I.C. Thompson) (Bayly, 1978). A previous report of a second species of copepod, thought to be *Acartia* sp. (Bayly, 1978) has not been confirmed by more recent collections.

Composition of pigments in water column particulate matter

Chromatograms of chlorophyll and carotenoid pigments extracted from particulate matter from four depths in Ace Lake are shown in Fig. 2. Identifications of the major constituents are given in Table 1. Differences in the chromatograms indicate that the populations of phytoplankton and photosynthetic bacteria in the lake are restricted to narrow depth intervals.

The pigments at 5 m consist of a simple mixture of chlorophylls and carotenoids. The major chlorophyll is chlorophyll *a* with traces of chlorophyll *b* indicating that few green algae are present. Chlorophyll *c* was not detected, nor were carotenoids typical of diatoms such as fucoxanthin, diadinoxanthin and diatoxanthin indicating that diatoms are at most trace constituents of the phytoplankton population. Examination by SEM revealed only a few diatom cells.

The major carotenoid (peak 11) was identified from its retention index and absorption spectrum (Wright & Shearer, 1984) as alloxanthin. This pigment is characteristic of Crytophycean algae (e.g. Norgard *et al.*, 1974) which is consistent with our identification of the major phytoplankter at this depth as a species of *Cryptomonas* (Fig. 1). High concentrations of β,ϵ-carotene are also typical of the Cryptophyceae (Wright & Shearer, 1984). Carotenoids characteristic of photosynthetic dinoflagellates, such as peridinin, were not detected although a few dinoflagellate cells were found in this sample.

An unusual feature of the HPLC chromatogram of the 5 m sample is the presence of a pigment (peak 15) which has a very similar absorption spectra to chlorophyll *a* and elutes just before it. Wright & Shearer (1984) observed minor amounts of this chlorophyll derivative in various algal species and suggested that it could be an allomer of chlorophyll *a* formed when methanol is used to extract pigments from algae (Mantoura & Llewellyn, 1983). This compound was not present in the extract of cultured *P. gelidicola* (Fig. 2e), nor would we expect such extensive allomerization. Gieskes & Kraay (1983) reported high concentrations of an apparently identical chlorophyll derivative in oceanic samples which was mainly associated with particles less than 1 μm suggesting an origin from picoplankton or detritus. Further work to establish the identity and source of this chlorophyll derivative would obviously be valuable.

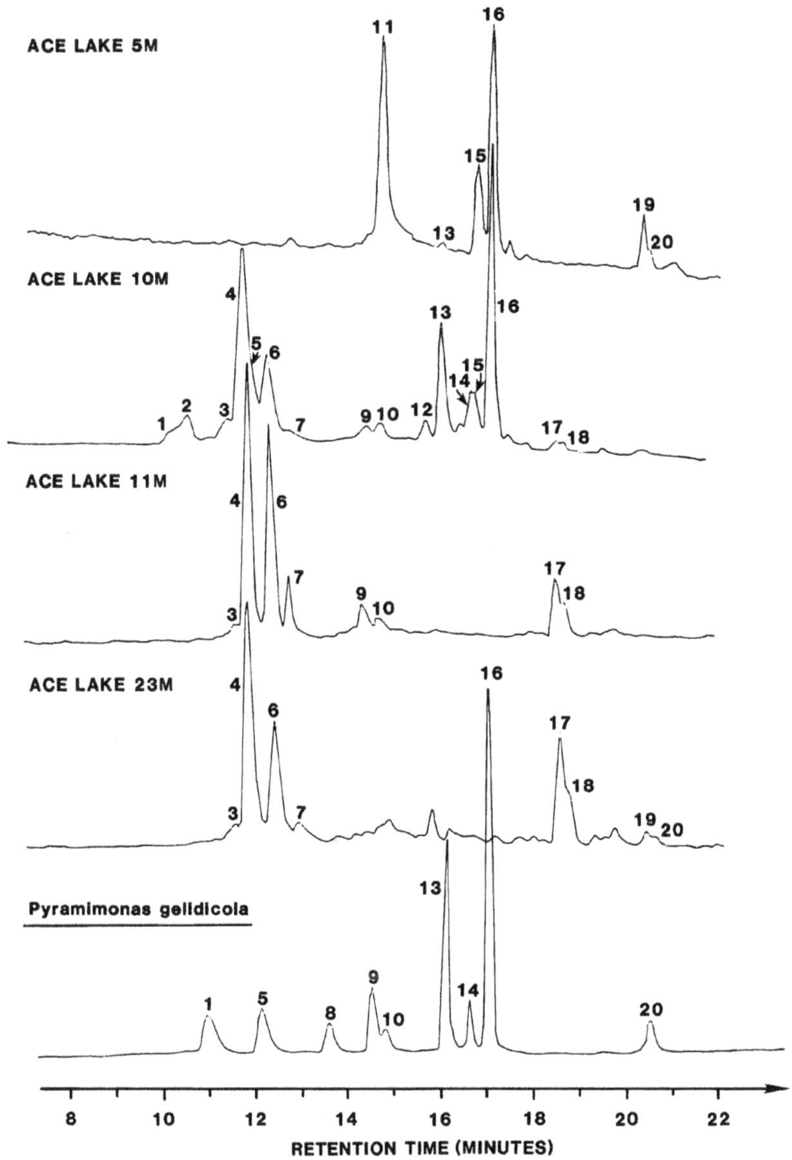

Fig. 2. High-performance liquid chromatograms of chlorophylls and carotenoids in particulate matter from Ace Lake, and in the green alga *P. gelidicola* cultured from Burton Lake, Peak identifications are given in Table 1.

At 10 m, the pigment profile is dominated by chlorophylls *a* and *b* and the carotenoids neoxanthin, violaxanthin, lutein and zeaxanthin (Fig. 2b). This pigment distribution closely matches that of *P. gelidicola* cultures (compare Figs. 2b and 2e), and of other *Pyramimonas* species (Ricketts, 1970; Gillan, 1981), confirming the importance of this alga as a source of pigments (and lipids) at this depth. Chlorophyll *b* is a useful marker pigment for green algae and its absence from the deeper samples and low abundance at 5 m, confirms that *P. gelidicola* is confined to a narrow depth range, being most abundant

Table 1. Identifications of pigments in Ace Lake.

Peak No.[a]	Pigment	Peak No.	Pigment
1	Neoxanthin	12	Chlorophyll *b* type
2	unknown chlorophyll type	13	Chlorophyll *b*
3	Bacteriochlorophyll *c* isomer?	14	unknown carotenoid
4	Bacteriochlorophyll *c* isomer	15	Chlorophyll *a* type
5	Violaxanthin	16	Chlorophyll *a*
6	Bacteriochlorophyll *c* isomer	17	Chlorobactene
7	Bacteriochlorophyll *c* isomer	18	unknown
8	Antheraxanthin	19	β, ϵ-carotene
9	Lutein	20	β,β-carotene
10	Zeaxanthin		
11	Alloxanthin		

[a] Peak numbers refer to Fig. 2.

just above the oxycline. The pigment distribution at 11 m is markedly different from that at 10 m in that it contains no chlorophylls or carotenoids typical of green algae, indicating the absence of living *P. gelidicola*. The absence of chlorophyll *a* suggests that there are no other photosynthetic algae living at this depth. Degraded chlorophyll pigments such as phaeophytin or phaeophorbide etc. are also very low in abundance suggesting that there is also little degraded algal material present.

The chromatogram of the 11 m sample is dominated by three pigments eluting at 11.82, 12.37 and 12.85 min. Each of the absorbance spectra are superimposable (Fig. 3) and show absorbance maxima at 412, 432, 628 and 664 nm. These pigments were identified as Chlorobium chlorophylls '660' (bacteriochlorophylls *c*), which are the main chlorophyll pigments in green sulfur bacteria such as *Chlorobium* (Holt, 1966; Holt *et al.*, 1966). Note that the absorbance spectra shown in Fig. 3 were recorded in the HPLC eluant, and thus there is a slight shift to longer wavelengths compared with spectra recorded in diethyl ether which is less polar (Goedheer, 1966). Some species of green photosynthetic bacteria contain bacteriochlorophylls *d*, but these were not detected in the lake.

Until recently there was some uncertainty as to the detailed structures of bacteriochlorophylls *c*. Holt (1966) and Holt *et al.* (1966) proposed six structures which differed in the type and arrangement of alkyl substituents on the macrocyclic porphyrin ring. However, Jackson (1976) concluded that there were only four different pigments (Fig. 4), of which two structures (III and IV) have been confirmed (Archibald *et al.*, 1966). These divergent views were reconciled when Smith *et al.* (1983) established that the 2-(1-hydroxyethyl) group in I and II exists as two chromatographically separable R and S diasteriomers. These authors showed that the elution order on reverse-phase columns was IV, III, II(R), II(S), I(R), and I(S). Possible identifications for peaks 3, 4, 6 and 7 in our chromatrogram (Fig. 2) are thus IV, III, II(R) and I(S or R) (Fig. 4) but further work is needed to confirm this. An interesting feature of the structures of these pigments is that the alcohol sidechain is the C_{15} isoprenoid farnesol and not phytol which is found in chlorophyll *a*. Farnesol is rarely found in oceanic sediments or water column particulate matter but in environments such as Ace Lake it could prove to be a useful marker for organic matter derived from these photosynthetic bacteria.

The predominance of bacteriochlorophylls *c* at 11 m is consistent with studies in 1978 which showed high concentrations of bacteria at 10 m just below the oxycline in Ace Lake (Burton & Barker, 1979; Burton, 1980). Deepening of the lake in recent years due to a change in its water budget has apparently shifted this maximum to about 11.5 m. The photosynthetic bacteria at this depth are dominated by *Chlorobium* spp. (Burke & Burton, this volume), which produce either bacteriochlorophylls *c* or *d* (Stanier & Smith, 1960). Bacteriochlorophylls *c* are

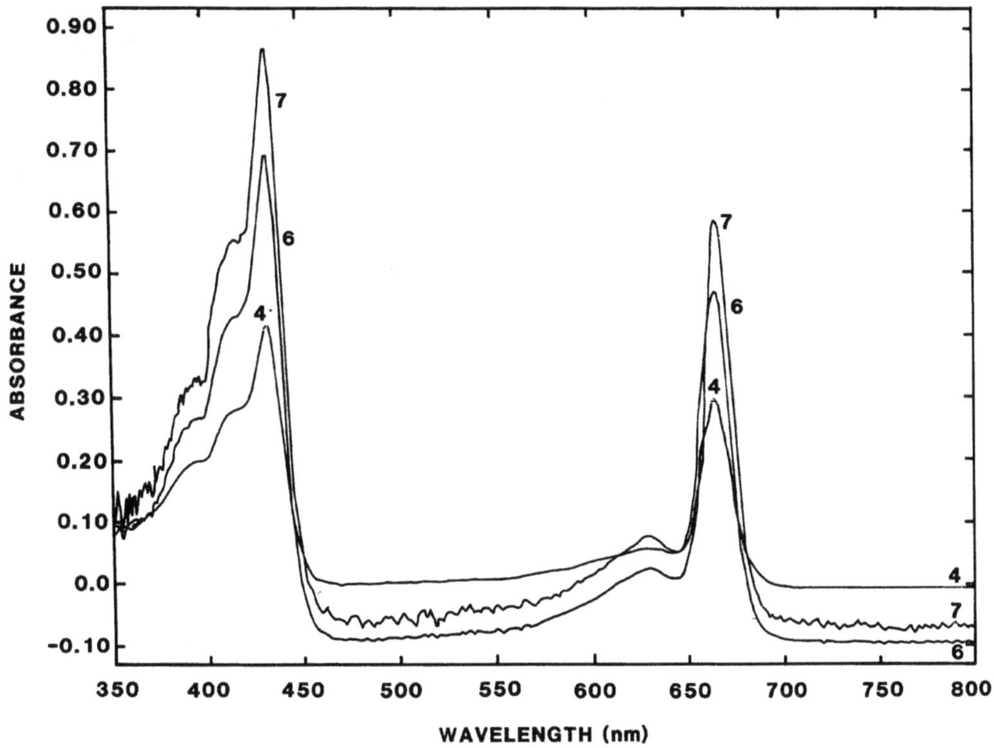

Fig. 3. Absorption spectra of three major bacteriochlorophyll *c* pigments in particulate matter from 11 m in Ace Lake. Possible structures are given in Fig. 4. Traces are offset to facilitate comparison.

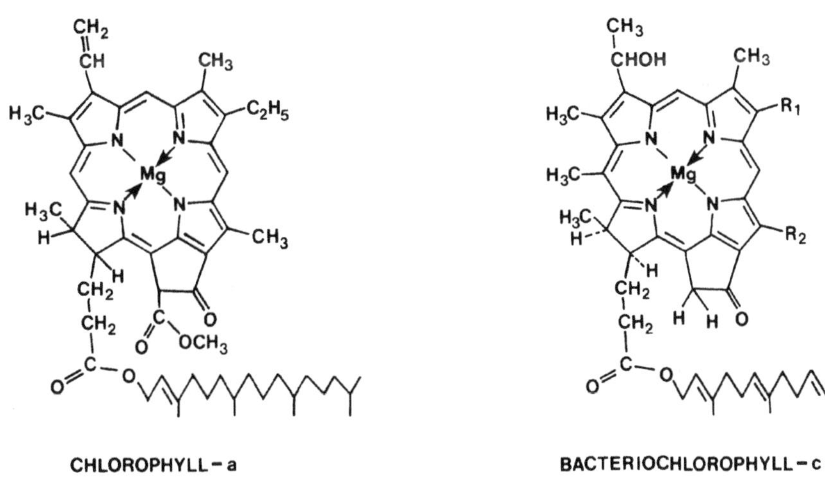

Fig. 4. Comparison of structures for chlorophyll *a* and those proposed by Holt *et al.* (1966), Jackson (1976) and Smith *et al.* (1983) for bacteriochlorophyll *c* isomers. I: R_1 = isobutyl, R_2 = ethyl; II: R_1 = n-propyl, R_2 = ethyl; III: R_1 = ethyl, R_2 = ethyl; IV: R_1 = ethyl, R_2 = methyl. The 2-(-1-hydroxyethyl) groups occur as R and S diasteriomers in structures I and II.

also present at 10 m and 23 m. Their presence at 10 m may be an artifact of the sampling method since the Kemmerer bottles sample over a depth of 50 cm and the sample at 10 m could include material from within the oxycline. Clearly the spatial separation of the bacterial and algal populations is much less than 50 cm indicating the need for more closely spaced sampling in future (e.g. Bayly, this volume).

One would not expect to find such high concentrations of bacteriochlorophylls at 23 m since there is no light penetration to this depth (Hand & Burton, 1981) and photosynthetic bacteria could not grow in such an environment. The bacterial cells probably settled out from higher in the water column and have been preserved in the relatively cold, dark and anoxic bottom waters. This agrees with work by Hand

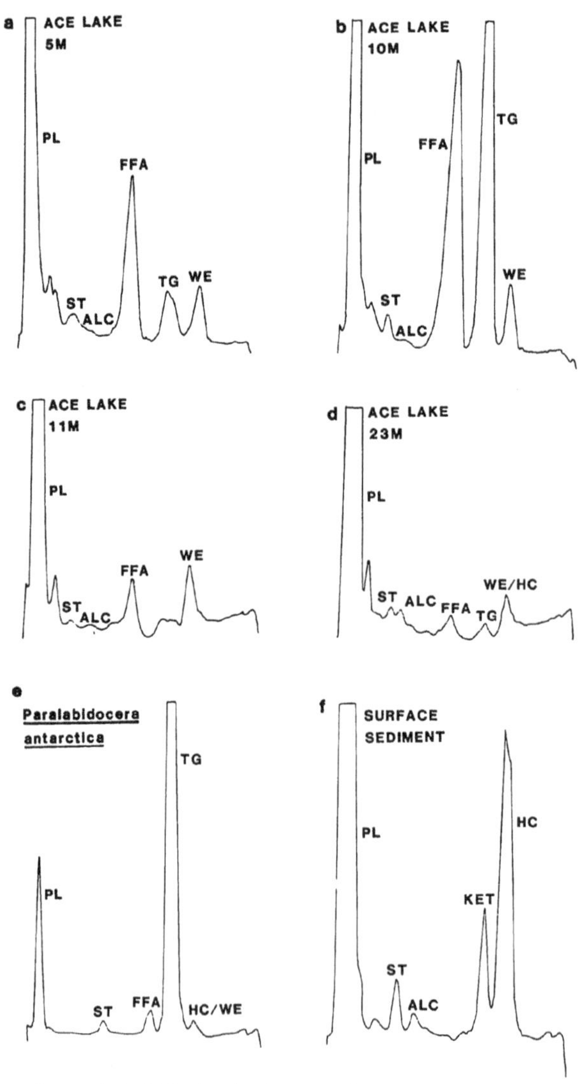

Fig. 5. Iatroscan chromatograms of non-polar lipid distributions in water column particulate matter samples (a – d), copepod *P. antarctica* (e), and surface sediment from Ace Lake (f). PL: polar lipids; ST: sterols; ALC: alcohols; FFA: free fatty acids; TG: triacylglycerols; HC: hydrocarbons; WE: wax esters. Concentration data are given in Table 2.

(1980) which showed that bacterial numbers in bottom waters of Ace Lake reached maximum values during the winter months as a result of sedimentation during the preceeding six months.

Lipid compositions of water column particulate matter

Chromatograms of non-polar lipid distributions in water column particulate matter, surface sediment and the copepod *P. antarctica* are shown in Fig. 5. Concentration data derived from these chromatograms, and others (not shown), are presented in Table 2. In all but the copepod sample, most of the lipid extracts consist of polar material which remains at the origin of the chromarod with the solvent system used. Re-examination using a more polar solvent system (not shown) indicated that little of this material corresponded to polar lipids such as phospholipids, glycolipids etc. In the 5 m and 10 m samples much of the polar material was chlorophyll pigments and carotenoids whereas in the deeper samples most of the polar material probably consisted of degraded lipid material and bacterial pigments.

There are major differences between the chromatograms obtained for the four water column samples, both in the relative abundances of the different lipid classes (Fig. 5), and their absolute concentrations (Table 2). These features result from the marked stratification of biological populations in the lake as already revealed by pigment analyses and electron microscopy. Matsumoto *et al.* (1984) also observed marked changes in the abundance of various lipid classes in another saline antarctic lake, Lake Vanda, and suggested that these reflected different vertical distributions of organisms. The low concentrations of unsaturated fatty acids and other labile lipids in their samples were cited as evidence for significant degradation of organic matter in the warm (20 °C) anoxic bottom waters of Lake Vanda. The same processes are likely to occur in the anoxic waters of Ace Lake but the lower bottom water temperature (1 °C cf. 20 °C) and shorter water column (23 m cf. 68 m), suggests that a greater proportion of the sedimenting material will reach the sediments of Ace Lake intact or only partially degraded.

Total lipid concentrations increase dramatically with depth in Ace Lake from a value of 31 $\mu g\,l^{-1}$ at 5 m to 295 $\mu g\,l^{-1}$ at 23 m (Table 2). Extractable organic carbon also increases with depth in Lake Vanda (Matsumoto *et al.*, 1984). In Ace Lake, most of this increase is accounted for by polar 'lipids' which comprise 38% of the extracted material at 10 m, and

Table 2. Concentrations of major lipid classes in water column particulate matter and surface sediment from Ace Lake.

Depth	Total lipids	Total[a] polars	Total non-polars	Sterols	Alcohols[b]	Free fatty acids	Triacyl-glycerols	HC/WE	Ketones & Esters
Water column: ($\mu g\,l^{-1}$)									
5 m	31	20	11	0.68	0.14	6.3	1.7	1.7[c]	ND[d]
10 m	124	47	77	1.2	0.25	34.0	36.7	3.6[c]	ND
11 m	149	131	18	0.22	0.25	4.5	1.1	8.4[c]	ND
23 m	295	287	8	1.2	0.15	1.7	1.7	2.2	ND
Sediment: ($mg\,g^{-1}$)									
0–5 cm	20	16.5	3.5	0.37	0.36	0.18	ND	0.20[e]	1.2

[a] Polar lipids are defined here as those lipids which remain at the chromarod origin using the hexane, diethylether, acetic acid (60/17/0.15) solvent system.

[b] Includes 4-methylsterols (minor components in water column).

[c] Includes β-carotene, wax esters and minor amounts of hydrocarbons.

[d] ND: None detected.

[e] Wax esters were minor constituents in sediment.

97% at 23 m. The concentrations of non-polar lipids from a very different profile with a maximum at 10 m, corresponding to the enhanced algal biomass (mainly *P. gelidicola*) at this depth.

The major non-polar lipids in the water column are free fatty acids, triacylglycerols and sterols together with small amounts of hydrocarbons, wax esters, alcohols and other uncharacterized material. Triacylglycerols are the major lipids at 10 m indicating that much of the organic matter at this depth is associated with living algal cells. The concentration of triacylglycerols at 5 m is about 20 times less reflecting the smaller algal biomass at that depth. The presence of triacylglycerols in the water samples from below the oxycline is of interest, since this lipid class is not common in bacteria. It seems likely that these triacylglycerols are at least partly associated with algal cells and copepods settling out from the upper water column.

Free fatty acids are abundant at each of the four depths with a maximum at 10 m, paralleling the concentration profile of triacylglycerols. Free fatty acids are not generally abundant in actively growing algae, but the high proportion of polyunsaturated free fatty acids (26.3%) at 10 m (Table 3) suggests that most are associated with intact cells. Polyunsaturated fatty acids are readily oxidized and unlikely to survive unaltered for more than a few days in oxygenated waters.

A detailed analysis of the free fatty acid distributions at 10 m and 11 m using capillary gas chromatography illustrates the sharp spatial differentiation between the algal bacterial populations at these two depths. At 10 m, polyunsaturated fatty acids as 16:4, 18:2, 18:3 and 18:4 (characteristic of green algae) are major constituents (Fig. 6), but at 11 m they collectively represent only 5.2% of the total fatty acids (Table 3), and in absolute concentration they are 38 times less abundant. Polyunsaturated fatty acids are rarely found in bacteria, with the possible exception of 18:2. These data therefore imply a biomass of green algae at 11 m which is less than 3% of that at 10 m.

The fatty acid distribution at 11 m is dominated by 16:0 and 16:1-Δ^9 which collectively comprise just over 50% of the total fatty acids. Less abundant constituents include 18:1-Δ^9, 18:0, 14:0 and 15:0 (Table 3). Various strains of *C. limicola* in culture have been shown to contain 16:0, 16:1-Δ^9 and 14:0 as major fatty acids with C_{18} constituents as minor components (reviewed by Kenyon, 1978). Most of the 16:0, 16:1-Δ^9 and 14:0 fatty acids at 11 m are therefore probably contributed by *Chlorobium* species, but other bacteria must also be important sources to account for the variety of fatty acids at this depth. Species of the Chromatiaceae contain 16:0, 16:1, 18:1 as major fatty acids, (Kenyon, 1978), and thus these bacteria could be a source of

Table 3. Relative percentage composition of fatty acid distributions in Ace Lake particulate matter from 10 m and 11 m, and surface sediment of Ace Lake.

Fatty acid	Particulate matter 10 m FFA[a]	Particulate matter 11 m FFA	Sediment FFA
Straight-chain saturates			
12:0	1.5	0.5	-
14:0	6.0	6.9	-
15:0	1.2	4.0	-
16:0	30.9	27.8	-
17:0	0.5	1.2	-
18:0	9.2	7.7	-
20:0	0.6	-	TR[b]
22:0	0.2	-	19.0
24:0	TR[b]	-	75.9
26:0	-	-	5.1
total:	50.4	48.1	100.0
Branched-chain saturates:			
iso 14:0	TR	0.8	-
iso 15:0	0.3	1.1	-
anteiso 15:0	0.6	3.6	-
iso 16:0	0.3	0.5	-
iso 17:0	0.2	0.2	-
anteiso 17:0	0.6	1.3	-
total:	2.0	7.5	TR
Monounsaturates:			
16:1 Δ 9	9.9	22.3	-
18:1 Δ 9	7.4	9.4	-
18:1 Δ 11	4.0	3.0	-
total:	21.3	34.7	TR
Polyunsaturates			
16:4 Δ 4,7,10,13	8.0	0.8	-
18:2 Δ 9,12	4.3	2.7	-
18:3 Δ 9,12,15	10.5	1.3	-
18:4 Δ 6,9,12,15	3.5	0.4	-
total:	26.3	5.2	TR

[a] FFA: free fatty acids isolated by column chromatography.
[b] TR: trace (less than 0.2%).

these acids, particularly 18:0 and 18:1. However, it is of interest that the major 18:1 isomer at 11 m is Δ^9 (i.e. oleic acid), and not Δ^{11} (vaccenic acid) which one would expect to predominate in bacteria using the anaerobic biosynthetic pathway. Branched-chain fatty acids are not common in photosynthetic bacteria (Kenyon, 1978), and these acids are probably derived from heterotrophic or sulphate reducing bacteria in the water column.

Copepods are another major respository of lipids in the oxygenated waters of Ace Lake. Analyses of 20-30 pooled individuals of *P. antarctica* gave an average value of 13 μg total lipid animal^{-1}. By comparison, the similar sized calanoid copepod *Calanus helgolandicus* Claus, caught off Plymouth had a total lipid content of 28 μg total lipid animal^{-1} (Sargent *et al.*, 1977) but these values will depend on season and growth stage. *P. antarctica* reaches a population density of 3 animals l^{-1} in Ace Lake (Hand & Burton, 1981), which corresponds to a lipid store of about 40 $\mu g\, l^{-1}$. This exceeds the lipid content of phytoplankton at 5 m and is one-third the concentration at 10 m (Table 1).

The present work presents the first lipid analysis of *P. antarctica*. The major storage lipid is triacylglycerol which comprises 77% of the total lipids. Analysis of other lipid classes (Fig. 5) gave values of: sterols (3%), free fatty acids (3.5%), pigments (4%) polar lipids (11%) and wax esters (1.5%). These values are similar to those reported for other calanoid copepods (Lee & Hirota, 1973; Lee, 1974, 1975), apart from the unusually low wax ester concentration. Copepods living in polar areas generally have high concentrations of wax esters which is thought to be a response to highly variable food supplies (Lee & Hirota, 1973; Sargent, 1978). The abundance of phytoplankton in Ace Lake may partially explain the predominance of triacylglycerols as the storage lipid of *P. antarctica*. It would be of interest to determine whether oceanic forms of *P. antarctica*, which have a much more varied diet, also use triacylglycerols as their major storage lipid.

Lipid composition of surface sediment

The distribution of lipids in the surface sediment is very different from any of the samples of water column particulate matter (Fig. 5). Polar lipids and pigments predominate but non-polar lipids are also present in substantial amounts (Table 2) compared with values reported for sediments from most lacustrine or marine environments.

Triacylglycerols were not detectable in the sediment even though large amounts are produced in the surface waters of the lake. Preliminary analyses of faecal pellets of *P. antarctica* indicated that triacylglycerols were present. Small amounts were also detected in the particulate matter at 23 m so we would expect some input from the settling of faecal pellets as well as small particles. The absence of these lipids in the sediment must reflect efficient utilization of organic matter by the microbial community there.

The abundance of free fatty acids in the sediment was much lower than would be expected from the amounts present in the lake waters (Table 2). Moreover, the C_{14}-C_{20} fatty acids found in the water column were not present in the sediment. It appears that fatty acids and other labile lipids, like triacylglycerols, are extensively broken down by the bacteria in the sediment. To our knowledge, this is the first report of a sediment which contains no C_{14}-C_{20} free fatty acids. In most sediments, some of these fatty acids would be contributed by the sediment bacteria (e.g. Perry *et al.*, 1979), but in Ace Lake the sediment microbial community is dominated by methanogenic bacteria which contain ether-linked lipids rather than the more common lipids based on ester-linked fatty acids.

The presence of 22:0, 24:0 and 26:0 fatty acids in the sediment was unexpected since these long-chain saturated fatty acids are commonly associated with higher plants. We can discount vascular plants as a possible source and it is also unlikely that these acids are derived from organisms in the water column since these acids were not detected in the water column particulate matter nor in *P. antarctica*. Moreover, it is difficult to envisage a mechanism by which these fatty acids would survive when shorter-chain acids are completely degraded. It may be that these long-chain acids are derived from bacteria in the sediment.

The Iatroscan chromatogram of the extractable lipids in the surface sediment contains a major peak

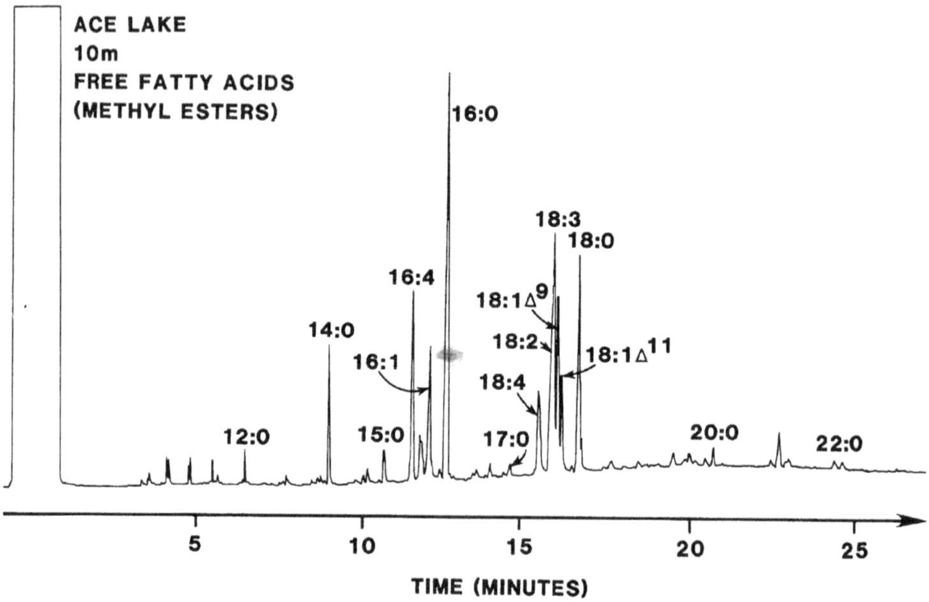

Fig. 6. Capillary gas chromatogram of free fatty acids (as methyl esters) in particulate matter from 10 m. Concentration data are given in Table 3. The abbreviation used is – no. of carbon atoms: no. of double bonds.

with a similar retention index to triacylglycerols (Fig. 5), but this peak was unaffected by transesterification. The compounds giving rise to this peak were identified as very high molecular weight straight-chain unsaturated methyl and ethyl ketones and fatty acid esters (Fig. 7; Table 4) similar to compounds isolated from the marine Prymnesiophyte

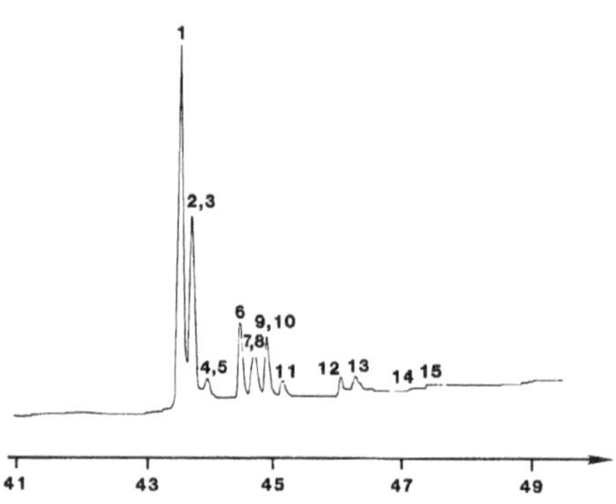

Fig. 7. Partial capillary gas chromatogram of very long-chain ketones and esters in surface sediment from Ace Lake. Peak identifications are given in Table 4.

Emiliania huxleyi (Lohm) Hay & Mohler (Volkman *et al.*, 1980a, b). The double bond positions were assigned on the basis of data presented by De Leeuw *et al.* (1980).

Approximately 1 μg of the 37:4 methyl ketone was isolated by thin layer chromatography and a fourier-transform infra-red spectrum obtained. This showed a major absorbance at 1713 cm^{-1}, which is typical of a carbonyl stretch in aliphatic ketones, as well as a strong absorbance at 962 cm^{-1} which is typical of a *trans*-double bond. This unexpected finding suggests that at least one, and possibly all, the double bonds in this molecule have the *trans* geometry and not the more common *cis* geometry found in most unsaturated lipids. Further work to confirm this finding is in progress.

Such long-chain ketones and esters have now been identified in six species of phytoplankton, all from the Prymnesiophyceae, order Isochrysidales, (Volkman *et al.*, 1980a, b; Marlowe *et al.*, 1984). As yet there are no reports of these unusual compounds in other algal classes suggesting that they may be specific biomarkers for Prymnesiophytes. The distribution of compounds in Ace Lake shows major

Table 4. Identification of very long straight-chain ketones and esters in surface sediments from Ace Lake.

Compound[a] No.	Type[b]	Carbon[c] No.	M.W.[d]	Chemical name
1	MK	37:4	526	Heptatriacontatetraen-2-one
2	MK	37:3	528	Heptatriacontatrien-2-one
3	ME[e]	37:4	542	Methyl hexatriacontatetraenoate
4	MK	37:2	530	Heptatriacontadien-2-one
5	ME	37:3	544	Methyl hexatriacontatrienoate
6	EE	38:4	556	Ethyl hexatriacontatetraenoate
7	EE	38:3	558	Ethyl hexatriacontatrienoate
8	EK	38:4	540	Octatriacontatetraen-3-one
9	EK	38:3	542	Octatriacontatrien-3-one
10	EE	38:2	560	Ethyl hexatriacontadienoate
11	EK	38:2	544	Octatriacontadien-3-one
12	MK[e]	39:4	554	Nonatriacontatetraen-2-one
13	MK[e]	39:3	556	Nonatriacontatrien-2-one
14	EK[e]	40:4	568	Tetracontatetraen-3-one
15	EK[e]	40:3	570	Tetracontatrien-3-one

[a] Compound numbers refer to Fig. 7.
[b] MK: methyl ketone; ME: methyl ester; EK: ethyl ketone; EE: ethyl ester.
[c] Number of carbon atoms: number of double bonds (includes esterifying alcohol for esters).
[d] MW: molecular weight.
[e] Compound not previously reported.

differences in the relative abundances of the various components and also includes several compounds not previously reported in either sediments or phytoplankton. These include a 37:4 methyl ester, 39:4 and 39:3 methyl ketones and 40:4 and 40:3 ethyl ketones (Table 4). These data indicate that the source of these compounds in Ace Lake is probably not one of the algae previously shown to synthesize these compounds. These long-chain compounds were not detectable in any of the water column particulate matter samples studied so it is improbable that they are derived from the green alga *P. gelidicola* or the *Cryptomonas* sp. Their most likely origin is the unidentified microflagellate (Fig. 1b), but this needs to be substantiated. Previous studies (Volkman *et al.*, 1980a) have shown that these long chain compounds are not readily degraded in aquatic environments. This could explain their high relative abundance in the sediments of Ace Lake, despite small populations of the source alga in the water column.

Long-chain unsaturated ketones have now been identified in a large number of marine sediments (e.g. Volkman *et al.*, 1980; De Leeuw *et al.*, 1980; Marlow *et al.*, 1984), and a few freshwater lakes (Cranwell, 1985) but this is the first report of their occurrence in an antarctic environment. In oceanic sediments their most likely source is the coccolithoporid *E. huxleyi*. This alga contains only small amounts of the tetra-unsaturated ketones (Marlow *et al.*, 1984) and thus in most sediments these particular ketones are not detectable, or are minor components. The distribution in Ace Lake is in marked contrast to this and represents the first report of a sediment in which these tetra-unsaturated ketones predominate. Although Marlowe *et al.* (1984) were unable to detect these ketones in a sediment from the Bransfield Strait in the Antarctic Peninsula, their occurrence in Ace Lake demonstrates that these unusual compounds are present in some antarctic environments, and confirms that algae other than coccolithophorids can be major sources.

Conclusions

This study has highlighted the potential of lipid and

pigment analyses for studying the spatial distribution of organisms in the water column and the recycling of organic carbon in aquatic environments. Ace Lake is particularly suited for such studies due to the small number of species present and the lack of allochthonous inputs, but the methods used are equally applicable to open ocean and other aquatic environments.

The combination of electron microscopy, lipid and pigment analyses shows that there are four main species of algae in Ace Lake which are confined to narrow depth intervals where conditions for growth are optimal. The most abundant alga is the Prasinophyte *P. gelidicola* which occurs mainly at 10 m, the base of the oxylimnion. A small biflagellate alga, thought to be a *Cryptomonas* sp. is more common at 5 m.

Bacteriochlorophylls *c* are the major pigments at 11 m, the top of the anoxylimnion, reflecting the abundance of green sulphur bacteria, *Chlorobium* spp., at this depth. Fatty acid distributions are very different at 10 m and 11 m, and indicate that the algal biomass at 11 m is less than 3% of that at 10 m. These data imply little settling out of algal cells, at least in summer. The presence of high concentrations of bacteriochlorophylls at 23 m provides convincing evidence for the settling out of bacterial cells from higher in the water column. These cells are only slowly degraded in the cold, anoxic bottom waters of Ace Lake.

The lipid content of the copepod *P. antarctica* has been determined for the first time and shown to consist mainly of triacylglycerols. The low abundance of wax esters is unusual since other studies have reported high concentrations of wax esters in antarctic copepod species.

Non-polar lipids are minor constituents of the lipids in the surface sediments of Ace Lake, and the absence of triacylglycerols implies extensive degradation of this lipid class by bacteria. The distribution of free fatty acids in the sediment is also quite different from those in the water column, which indicates almost complete turnover of labile lipids in the sediment. The major non-polar lipids in the sediment were found to be unusual very long chain unsaturated ketones and esters which previously have not been recorded from antarctic environments. Their presence suggests a wider occurrence in algae than previously thought, and provides circumstantial evidence that an unidentified microflagellate in Ace Lake may be a Prymnesiophycean alga.

Acknowledgements

We thank J. van den Hoff for technical assistance with electron microscopy, S. Wright for advice on pigment analyses and C. Burke and M. Burch for useful discussions on bacteria and phytoplankton in Ace Lake. We also thank D. P. Thomas and R. Croome for advice on algal identifications, and S. Wright and G. Hallegraeff for careful reviews of the manuscript.

References

Archibald, J. L., D. M. Walker, K. B. Shaw, A. Markovac & S. F. MacDonald, 1966. The synthesis of porphyrins derived from Chlorobium chlorophylls. Can. J. Chem. 44: 345–362.

Bayly, I. A. E., 1978. The occurrence of Paralabidocera antarctica (I. C. Thompson) (Copepoda: Calanoida: Acartiidae) in an antarctic saline lake. Aust. J. Mar. Freshwat. Res. 29: 817–824.

Bayly, I. A. E., this volume. Ecology of the zooplankton of a meromictic antarctic lagoon with special reference to Drepanopus bispinosus (Copepoda: Calanoida). Hydrobiologia.

Burch, M. D., this volume. Annual cycle of phytoplankton in Ace Lake; an ice covered, saline meromictic lake. Hydrobiologia.

Burke, C. M. & H. R. Burton, this volume. The ecology of photosynthetic bacteria in Burton Lake, Vestfold Hills, Antarctica. Hydrobiologia.

Burton, H. R. & R. J. Barker, 1979, Sulfur chemistry and microbiological fractionation of sulfur isotopes in a saline antarctic lake. Geomicrobiol. J. 1: 329–340.

Burton, H. R., 1980. Methane in a saline antarctic lake. In P. A. Trudinger & M. R. Walter (eds), Biogeochemistry of Ancient and Modern Environments. Aust. Acad. Sci., Canberra: 243–250.

Campbell, P. J., 1978. Primary productivity of a hypersaline antarctic lake. Aust. J. Mar. Freshwat. Res. 29: 717–724.

Cranwell, P. A., 1985. Long chain unsaturated ketones in recent lacustrine sediments. Geochim. Cosmochim. Acta. 49: 1545–1551.

De Leeuw, J. W., F. W. van der Meer, W. I. C. Rijpstra & P. A. Schenck, 1980. On the occurrence and structural identification of long chain unsaturated ketones and hydrocarbons in sediments. In A. G. Douglas & J. R. Maxwell (eds), Advances in Organic Geochemistry 1979. Pergamon Press, Oxford: 211–217.

Franzmann, P. D., G. W. Skyring, H. R. Burton & P. P. Deprez, this volume. Sulfate reduction rates and some aspects of the limnology of four lakes and a fjord in the Vestfold Hills, Antarctica. Hydrobiologia.

Gieskes, W. W. & G. W. Kraay, 1983. Unknown chlorophyll a

derivatives in the North Sea and the tropical Atlantic Ocean revealed by HPLC analysis. Limnol. Oceanogr. 28: 757–766.

Gillan, F. T., 1981. Lipids of aquatic ecosystems. Ph.D. thesis University of Melbourne.

Goedheer, J. C., 1966. Visible absorption and fluorescence of chlorophyll and its aggregates in solution. In L. P. Vernon & G. R. Seely (eds), The Chlorophylls. Academic Press: 147–184.

Hand, R. M., 1980. Bacterial populations of two saline antarctic Lakes. In P. A. Trudinger & M. R. Walter (eds), Biogeochemistry of Ancient and Modern Environments. Aust. Acad. Sci., Canberra: 123–129.

Hand, R. M. & H. R. Burton, 1981. Microbial ecology of an antarctic saline meromictic lake. Hydrobiologia 82: 363–374.

Holt, A. S., 1966. Recently characterized chlorophylls. In L. P. Vernon & G. R. Seely (eds), The Chlorophylls. Academic Press: 111–118.

Holt, A. S., J. W. Purdie & J. W. F. Wasley, 1966. Structures of Chlorobium chlorophylls (660). Can. J. Chem. 4: 88–93.

Jackson, A. H., 1976. Structure, properties and distribution of chlorophylls. In T. W. Goodwin (ed.), Chemistry and Biochemistry of Plant Pigments. Academic Press: 47–63.

Kaitaranta, J. K. & N. Nicolaides, 1981. Response and linearity of different lipid compounds when analysed by thin-layer chromatography with flame ionisation detector. J. Chromatogr. 205: 339–347.

Kenyon, C. N., 1978. Complex lipids and fatty acids of photosynthetic bacteria. In R. K. Clayton and W. R. Sistrom (eds), The Photosynthetic Bacteria. Plenum Press, N.Y.: 281–313.

Kerry, K. R., D. R. Grace, R. Williams & H. R. Burton, 1977. Studies on some saline lakes of the Vestfold Hills, Antarctica. In G. A. Llano (ed.), Adaptations Within Antarctic Ecosystems. Proceedings of the third SCAR Symposium on Antarctic Biology. Smithsonian Institution, Washington: 839–858.

Lee, R. F. & J. Hirota, 1973. Wax esters in tropical zooplankton and nekton and the geographical distribution of wax esters in marine copepods. Limnol. Oceanogr. 18: 227–239.

Lee, R. F., 1974, Lipid composition of the copepod Calanus hyperboreas from the Arctic Ocean. Changes with depth and season. Mar. Biol. 26: 313–318.

Lee, R. F., 1975. Lipids of arctic zooplankton. Comp. Biochem. Physiol. 51B: 263–266.

Mantoura, R. F. C. & C. A. Llewellyn, 1983. The rapid determination of algal chlorophyll and carotenoid pigments and their breakdown products in natural waters by reverse-phase high-performance liquid chromatography. Analyt. Chim. Acta. 151: 297–314.

Marchant, H. J. & D. P. Thomas, 1983. Polylysine as an adhesive for the attachment of nanoplankton to substrates for electron microscopy. J. Microscopy 131: 127–129.

Marlowe, I. T., S. C. Brassell, G. Eglinton and J. C. Green, 1984. Long-chain unsaturated ketones and esters in living algae and marine sediments. Org. Geochem 6: 135–141.

Matsumoto, G., T. Torii & T. Hanya, 1984. Vertical distributions of organic constituents in an antarctic Lake: Lake Vanda. Hydrobiologia 111: 119–126.

McIntyre, A. & A. W. H. Bé, 1967. Modern coccolithophoridae of the Atlantic Ocean. I. Placoliths and cyrtoliths. Deep-Sea Res. 14: 561–597.

Norgard, S., W. A. Svec, S. Liaaen-Jensen, A. Jensen & R. R. L. Guillard, 1974. Algal carotenoids and chemotaxonomy. Biochem. Syst. Ecol. 2: 7–9.

Parrish, C. C. & R. G. Ackman, 1983. Chromarod separations for the analysis of marine lipid classes by Iatroscan thin-layer chromatography-flame ionisation detection. J. Chromatogr. 262: 103–112.

Perry, G. J., J. K. Volkman, R. B. Johns & H. J. Bavor Jr., 1979. Fatty acids of bacterial origin in contemporary marine sediments. Geochim. Cosmochim. Acta. 43: 1715–1725.

Ricketts, T. R., 1970. The pigments of the Prasinophyceae and related organisms. Phytochemistry 9: 1835–1842.

Sargent, J. R., 1978. Marine wax esters. Sci. Prog. Oxf. 65: 437–458.

Sargent, J. R., R. R. Gatten, E. D. S. Corner & C. C. Kilvington, 1977. On the nutrition and metabolism of zooplankton XI: Lipids of Calanus helgolandicus grazing Biddulphia sinensis. J. Mar. Biol. Ass, U.K. 57: 525–533.

Sipos, J. C. & R. G. Ackman, 1978. Automated and rapid quantitative analysis of lipids with chromarods. J. Chromatogr. Sci. 16: 443–447.

Smith, K. M., G. W. Craig, L. A. Kehres & N. Pfennig, 1983. Reversed-phase high-performance liquid chromatography and structural assignments of the bacteriochlorophylls-c. J. Chromatogr. 281: 209–223.

Stanier, R. Y. & J. H. C. Smith, 1960. The chlorophylls of green bacteria. Biochim. Biophys. Acta. 41: 478–484.

Vincent, W. F. & C. L. Vincent, 1982. Factors controlling phytoplankton production in Lake Vanda (77 °S). Can. J. Fish. Aquat. Sci. 39: 1602–1609.

Volkman, J. K., G. Eglinton, E. D. S. Corner & J. R. Sargent, 1980a. Novel unsaturated straight-chain C_{37}-C_{39} methyl and ethyl ketones in marine sediments and a coccolithophore Emiliania huxleyi. In A. G. Douglas & J. R. Maxwell (eds), Advances in Organic Geochemistry, 1979. Pergamon Press: 219–227.

Volkman, J. K., G. Eglinton, E. D. S. Corner & T. E. V. Forsberg, 1980b Long-chain alkenes and alkenones in the marine coccolithophorid Emiliania huxeyi. Phytochemistry 19: 2619–2622.

Volkman, J. K., J. W. Farrington, R. B. Gagosian & S. G. Wakeham, 1983. Lipid composition of coastal marine sediments from the Peru upwelling region. In M. Bjorøy et al. (eds), Advances in Organic Geochemistry, 1981. Pergamon Press: 228–240.

Volkman, J. K., D. A. Everitt & D. I. Allen, 1986. Some analyses op lipid classes in marine organisms, sediments and seawater using thin layer chromatography-flame ionisation detection. J. Chromatogr. 356: 147–162.

Wright, S. W. & H. R. Burton, 1981. The biology of antarctic saline lakes. Hydrobiologia 82: 319–338.

Wright, S. W. & J. D. Shearer, 1984. Rapid extraction and high-performance liquid chromatography of chlorophylls and carotenoids from marine phytoplankton. J. Chromatogr. 294: 281–295.

Hydrobiologia 165: 59–75 (1988)
J. M. Ferris, H. R. Burton, G. W. Johnstone & I. A. E. Bayly (eds) Biology of the Vestfold Hills, Antarctica
Kluwer Academic Publishers

Annual cycle of phytoplankton in Ace Lake, an ice covered, saline meromictic lake

M. D. Burch[1]

Antarctic Division, Department of Science, Kingston, Tasmania, 7050, Australia; [1]Present address: State Water Laboratory, Private Mail Bag, Salisbury P.O., Salisbury, South Australia, 5108, Australia

Key words: Ace Lake, phytoplankton, meromictic, ice covered, annual cycle

Abstract

The annual cycle of phytoplankton in saline, meromictic Ace Lake (68°28.4′S, 78°11.1′E) in the Vestfold Hills, Antarctica, was studied from January, 1979 to January 1980. Ace Lake has permanent gradients of temperature, salinity, dissolved oxygen, and hydrogen sulphide, and is ice covered with up to 2 m of ice for 10–12 months each year. The phytoplankton community had low diversity, consisting of only four species, all flagellates – a prasinophyte *Pyramimonas gelidicola* McFadden *et al.*, a cryptophyte of the genus *Cryptomonas*, an unidentified colourless microflagellate, and an unarmoured dinoflagellate. These were restricted to the oxic zone of the lake from the surface to 10 m.

The phytoplankton had a cycle of seven months of active growth over spring and summer. Low numbers of cells survived in the water column over winter. Spring growth was initiated below the ice by increased light penetration through the ice into the lake, enhanced at the time by the removal of surface snow which accumulated on the ice over winter. Peak phytoplankton biomass production was by the shade adapted *P. gelidicola* and occurred at the interface of the oxic and anoxic zones where substantial available nitrogen as ammonia is found.

The three dominant phytoplankton species displayed distinct vertical stratification over the oxic zone. This stratification was not static and developed over spring as the flagellates migrated to preferred light climate zones. Mean cell volume of two of the flagellates varied significantly over the year. Minimum volumes were recorded in winter and volume increased progressively over spring to reach maximum mean cell volume in summer. Mean cell volume was positively correlated with light intensity (maximum ambient PAR at the respective depth for date of sample). Low cell volume in winter may be related to winter utilization of carbohydrate reserves by slow respiration, and may represent a survival mechanism.

Introduction

Ace Lake is a saline meromictic lake in the Vestfold Hills that has been extensively investigated since 1974 (H. R. Burton, Pers.commun.). Early studies examined sulphur chemistry (Burton & Barker, 1979) and microbial methane production (Burton, 1980) in the lake and revealed distinctive permanent gradients of temperature, salinity, dissolved oxygen, and hydrogen sulphide. Studies of the microbial ecology (Hand, 1980; Hand & Burton, 1981) described the phototrophic bacteria and sulphate-reducing bacteria which combined to form a sulphur cycle at the interface of the oxygenated and deoxygenated layers at 10 m depth. These studies suggested that Ace Lake was relatively productive. Hand (1980) further proposed that algal primary productivity in the upper waters contributed organic material for microbial activity at this interface. Burton (1980) proposed that this productivity ultimately fuels the significant methane production at greater depths. The present study was undertaken between January 1979 and January 1980 to investigate phytoplankton ecology in the upper layer of Ace Lake. This paper discusses the annual cycle of algal species and their vertical distribution in relation to the extreme seasonal variation in solar radiation and to the physical and chemical stratification in the lake. This paper is dedicated to the memory of the late Roger Barker, who contributed much to the early years of the Australian Antarctic Limnology Program and to our knowledge of Ace Lake.

Description of the lake

Ace Lake is located at 68°28.4′S, 78°11.1′E on a narrow section of Long Peninsula, the most northern of three peninsulas comprising the 400 km^2 ice-free Vestfold Hills area. The lake has a surface area of 1.32×10^5 m^2 and a maximum depth of 23 m. The catchment has an area of 5.11×10^5 m^2 and consists of low hills on which little snow accumulates due to wind exposure. The lake is a closed system, with no surface outflow and no seepage (Hand & Burton, 1981).

The lake had a marine origin (Johnstone *et al.*, 1973) and is currently 152 m from the sea at its nearest point. The complex evolution and genesis of its present meromixis is discussed by Burton & Barker (1979). Burton (1980) pointed out that Ace Lake differs from conventional meromictic lakes which have upper and lower layers separated by a salinity gradient or chemocline. In Ace Lake there is a steady increase in salinity with depth from 80% that of sea water at 3 m to 130% that of seawater at the bottom. Burton preferred to divide the lake into three zones based on oxygen concentrations – an upper oxylimnion separated from the lower anoxylimnion by a transition zone, or oxycline, between 7 m and 9 m. The maintenance of the stable meromictic condition in the lake in an area subject to strong winds is due to the protective ice cover of up to 2 m that usually exists for 9–10 months each year but may persist throughout summer (Burton & Barker, 1979).

Materials and methods

Sampling was conducted at approximately monthly intervals, although variations occurred due to inclement weather. No samples were taken in March and April as the lake was inaccessible due to the breakup of coastal sea-ice. Samples were taken from a boat on one open water occasion or through 12.5 cm holes cut through the ice with a SIPRE ice corer (Journal of Glaciology, 1958). The sample site was always over the deepest part of the lake.

Samples were taken with an acrylic non-metallic Kemmerer sampling bottle. Samples were immediately transferred to polyethylene bottles and transported in insulated boxes to Davis Station and analyses were commenced within four hours of collection. Samples for nutrient analysis were collected into acid washed polyethylene bottles and frozen. They were later stored at −20 °C and returned to Australia for analysis up to twelve months later.

Temperature was measured with a mercury thermometer clamped inside the transparent Kemmerer sampler. Photosynthetically available radiation (PAR) was measured with a Li-Cor underwater quantum sensor (LI 192 S). The extinction of PAR spectral components through the ice and water column was determined using Schott coloured glass filters fitted to the sensor. The filters used, with their peak transmission and glass thickness, were: blue: BG 12, 400 nm, 1 mm; green: VG 9, 520 nm, 1 mm; red: RG 610, 610 nm, 3 mm. All measurements of PAR were taken at solar midday. Vertical attenuation coefficients, K_d ($\log_e$ units m^{-1}), for the water column below the ice were determined as the linear regression coefficient of the plot of $\log_e$ PAR versus depth (Kirk, 1983). Inspection of the plot indicated whether the light penetration characteristics could be represented by a single value of K_d, or whether separate K_d values were required for different portions of the water column. An approximate value of K_d for the ice on each date was calculated from the formula.

$$K_{d\,(ice)} = \frac{1}{z} (\log_e I_o - \log_e I_z)$$

(i.e. Bouguer-Lambert Law), where z was ice thickness and I_0 and I_z were the values of downward irradiance above and below the ice respectively.

Chlorophyll concentration was determined spectrophotometrically after filtration onto glass fibre filters (Whatman GF/C) and extraction in 90% acetone (UNESCO, 1966). Pigment concentrations were calculated using the equations of Jeffrey & Humphrey (1975).

Phytoplankton samples were preserved shortly after collection with Lugol's iodine. Cells were sedimented and counted by inverted microscopy (Lund *et al.*, 1958). The phytoplankton were counted to meet minimum pre-set statistical criteria (Willen, 1976). Cell volumes (mean of 30 values) were determined for representative samples on each date from measurements of cell dimensions and calculation using volume formulae for equivalent geometric shapes (Willen, 1976).

Dissolved oxygen was determined by the Winkler method (Strickland & Parsons, 1972). Salinity was calculated from conductivity at 20 °C, measured with a Yeo-Kal inductively coupled salinometer (Model 601 Mk IV).

Total phosphorus and oxidized nitrogen (NO_2 + NO_3) were determined on samples stored frozen and returned to Australia. Analyses were performed spectrophotometrically. Phosphorus was determined by the ascorbic acid – molybdenum blue method after digestion with persulphate (APHA, 1981). Nitrite, and nitrate after cadmium reduction, were determined by the method of APHA (1981). Oxidized nitrogen was obtained by combining results for nitrite and nitrate.

Results

Physical and chemical features

The thermal cycle in Ace Lake for 1979 is shown in Fig. 1. The thermal profile changed little throughout the year, especially from April, when ice first covered the lake, to November, when the ice achieved its maximum thickness of 1.8 m. Increased water temperatures were not recorded until January 1980, and effective heating of the lake apparently did not commence until December. Seasonal increase in water temperature was therefore a truly summer phenomenon and did not coincide with increased solar radiation in spring. In contrast, hypersaline and ice-free Deep Lake in the Vestfold Hills begins to gain heat in spring (October/November), showing a more immediate response to the increasing solar radiation input (Ferris & Burton, this volume). Ace Lake's delayed response is probably related to the

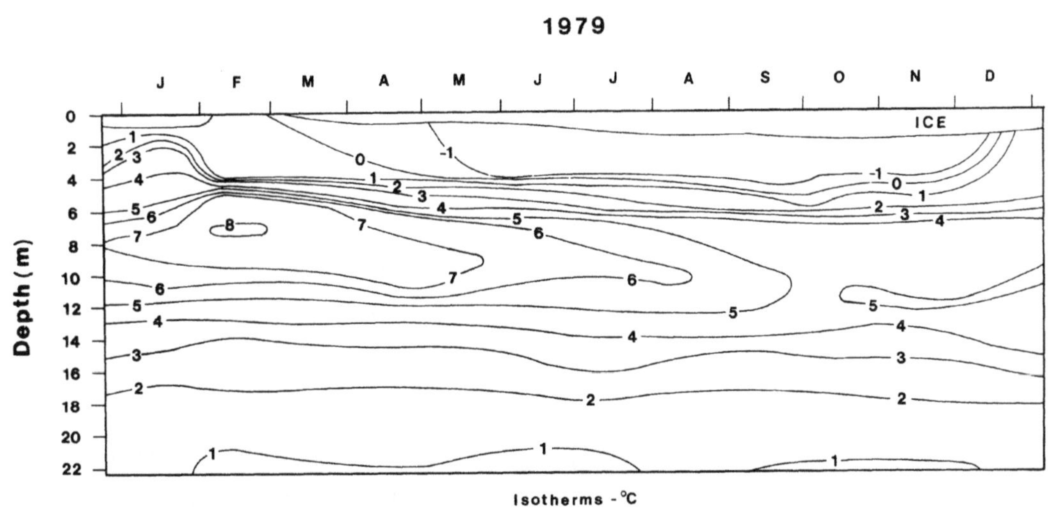

Fig. 1. Depth-time isotherms in Ace Lake during 1979.

presence of ice. Ice cover prevents wind-driven mixing and absorbs or reflects a significant amount of solar radiation, especially when snow covered as was the case for Ace Lake from August to early October 1979 (Fig. 4).

Ace Lake has four thermal zones. The first is a near-surface zone which remains for most of the year within the narrow temperature range of −1° to +1 °C. The second is the thermocline region with a 5−6 °C gradient, which usually lies between 4 m and 7 m but broadens and moves upwards towards the ice/water interface in January with summer heating. The third is an intermediate zone (7 m−12 m) which undergoes annual temperature variation of aproximately 4 °C. In this zone there was a steady decline in temperature from 8 °C in February to 4 °C in December. The fourth zone (12 m−23 m) is a region of virtually unchanging water temperature. The inverse thermal profile that exists in the upper 7 m−10 m of the water column demonstrates the comparative weakness of thermally induced density gradients. It is evident that salinity is the most important factor maintaining the long-term density stratification of Ace Lake.

Depth profiles of temperature, salinity and dissolved oxygen in the top 11 m, for ice-free (9 February 1979) and ice-covered (14 August 1979) conditions, are shown in Fig. 2. In summer, the melting of ice and freshwater inputs to the lake resulted in a layer of fresher water (43% of seawater salinity) overlying water of approximately seawater salinity. This salinity stratification enhanced the stability of the water column when the lake was ice-free and there was the potential for the meromixis to be destroyed by windy conditions. The maximum vertical penetration of wind-driven mixing occurred in February, when the upper 4 m of the water column was isothermal (0.5 °C) and probably isohaline (15‰; Fig. 2).

Comparison of the summer and winter temperature/salinity profiles (Fig. 2) shows that surface salinity increased in winter, weakening the salinity stratification and apparently deepening the pycnocline by 1 m−2 m.

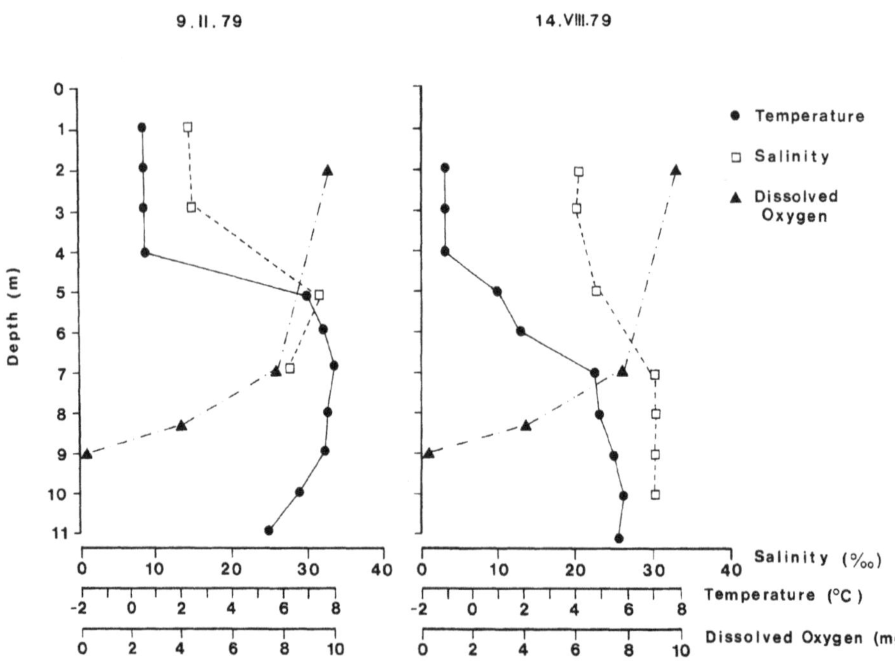

Fig. 2. Vertical profiles of temperature, salinity and dissolved oxygen from 0 m−11 m in Ace Lake under summer, ice-free conditions (9 February 1979), and winter ice cover (14 August 1979).

These effects probably resulted from salt exclusion by the developing ice cover (c.f. Kantha, 1979). Although it is evident that the water column was not vertically well mixed below 4 m, the profound cooling of the 5 m – 6 m region indicates that there was some mixing in this zone. The depth of the interface between oxic and anoxic water (c. 9 m) remained virtually unchanged between February and August of 1979 (Fig. 2), but deepened to about 10.5 m in October as the lake ice approached its maximum thickness. This may also be an effect of salt exclusion by newly forming ice, but the almost unchanging thermocline in the upper water layers (c. 5 m – 7 m; Fig. 1) tends to preclude any direct mixing from above. It may be speculated that saline density currents, originating at the shallow edges of the lake basin and flowing into the lake centre, provided the mechanism for mixing this deeper region without altering the thermal structure of the overlying water. Vertical displacement of the oxic/anoxic water interface was previously reported for Ace Lake by Hand (1980), and for another meromictic lake/lagoon in the Vestfold Hills by Burke & Burton (this volume). Burke & Burton ascribe the deepening of Burton Lake's oxic/anoxic water interface in 1983 to salt exclusion from the overlying ice. More recently, the formation of hypersaline brines in shallow, ice-covered margins has been observed and postulated as a mechanism for the formation and maintenance of saline stratification in two basins at the head of Ellis Fjord in the Vestfold Hills (J. B. Gallagher, pers. commun.).

Total phosphorus increases with depth and displays a stratification that corresponds to the prevailing thermal stratification (Fig. 3). Soluble reactive phosphorus (PO_4-P) was not determined separately because the samples had been frozen prior to analysis. Hand & Burton (1981) reported PO_4-P concentrations of around 100 $\mu g\ l^{-1}$ at 2 m and 5 m, and 546 $\mu g\ l^{-1}$ at 10 m in November, indicating that the PO_4-P forms a substantial component of the total-P. These figures match the profile measured on 23 September 1979 (Fig. 3). However in May (24 May 1979) total-P from 3 m – 6 m was appreciably higher than for the same depths in September. This could indicate the presence of more organic and particulate phosphorus in the upper waters at the end of the growing season which subsequently settles out over the winter months.

In contrast to phosphorus, oxidized nitrogen was

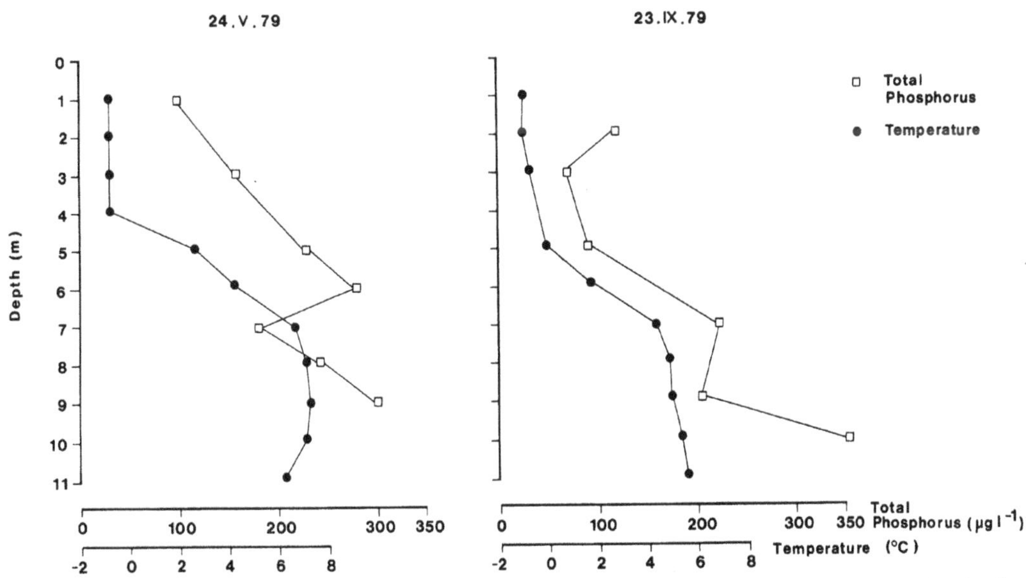

Fig. 3. Vertical profiles of total phosphorus and temperature from 0 m – 11 m in Ace Lake in late autumn (24 May 1979) and spring (23 September 1979).

below the level of detection (5 μg l^{-1}). Ammonia was not determined as subsantial ammonia has been found to be lost from samples stored frozen (H. R. Burton, pers. commun.). Hand & Burton (1981) found that inorganic nitrogen in Ace Lake was in the form of ammonia and was only found below the oxycline (7 m), a concentration of 466 μg l^{-1} NH_4-N being measured at 10 m.

Incident photosynthetic radition (I_o' sub-surface value of downward irradiance) and the depth of the conventional euphotic zone (1% I_o') on each sampling data are shown in Fig. 4. For ice-covered conditions I_o' was taken as the subsurface value at the water level in the sampling hole. This level was always 5 cm – 10 cm below the ice surface. Although the data is by no means complete it illustrates the wide seasonal variation in the daily maximum incident radiation received at this latitude, ranging from 1225 μE m^{-2} s^{-1} (9 February 1979) to 1.3 μE m^{-2} s^{-1} (3 July 1979). PAR climate within the lake reflects this range and by way of illustration the quanta value at 5 m (Fig. 4) varies over four orders of magnitude, from 245 μE m^{-2} s^{-1} for ice-free conditions (9 February 1979) to 0.03 μE m^{-2} s^{-1} shortly after midwinter with 1.2 m ice cover (3 July 1979).

A typical depth profile of PAR and spectral components for Ace Lake (Fig. 5) indicates that red light is attenuated most rapidly by the ice whereas blue and green are less strongly absorbed. Consequently 1% of incident red PAR does not penetrate below 6 m, while 1% of blue penetrates to between 8.5 m and 10 m, and 1% of green always reaches the opaque layer containing dense particulate matter at the top of the anoxylimnion. No PAR penetrates below this layer at 10 m – 11 m.

Optical properties of the ice cover and the water column varied over the year and are indicated by changes in vertical attenuation coefficients (K_d) (Table 1). The Bouguer-Lambert law (see Materials and methods) was used to characterize the attenuation of light through total lake cover (ice + snow)

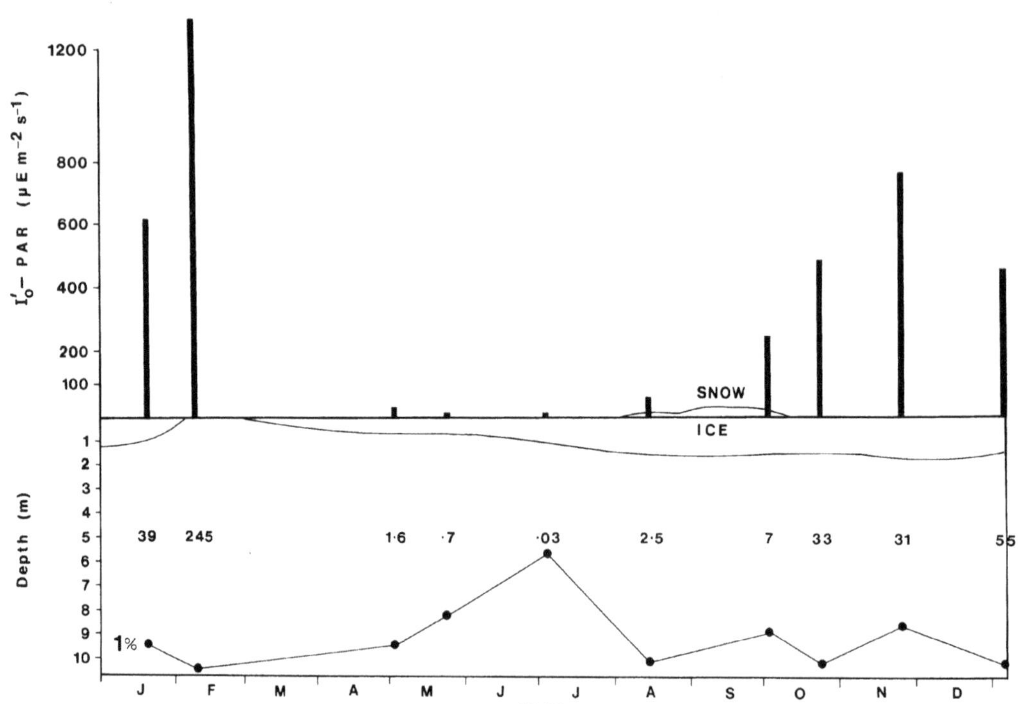

Fig. 4. Incident photosynthetically available radiation (I_o') and the depth of penetration of 1% of I_o' into Ace Lake during 1979. The value of PAR (μE m^{-2} s^{-1}) at 5 m on each sampling date is indicated. All measurements were made at solar midday.

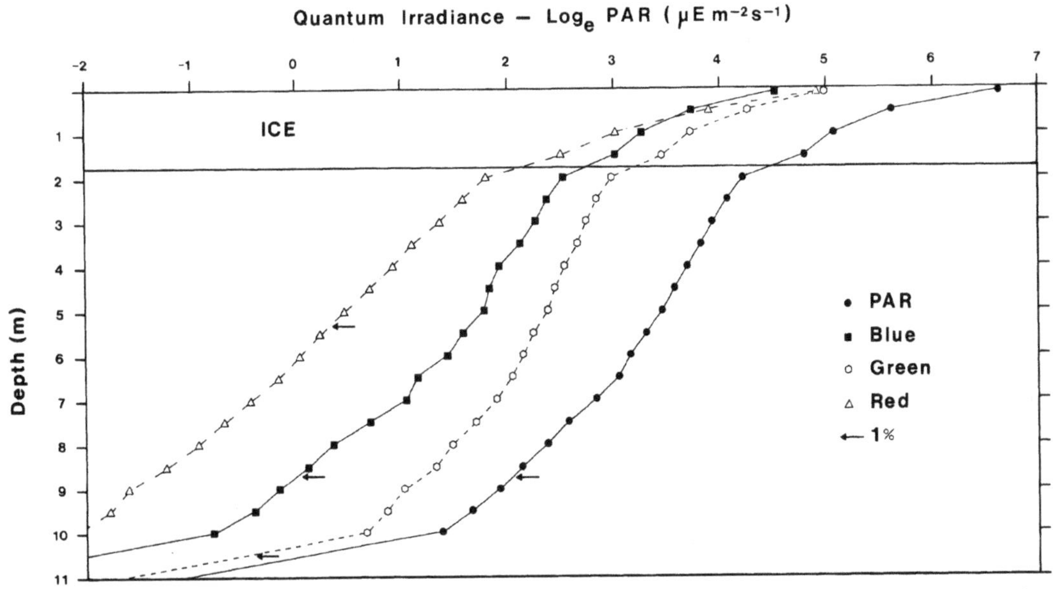

Fig. 5. Depth profile of downward quantum irradiance of PAR and its components measured with optical filters (Blue, BG 12; Green, VG 9; Red, RG 610) on 25 November 1979.

on different dates. However the attenuation did not always closely follow this law and the attenuation coefficient generally decreased with depth through the ice (Fig. 5). Similarly Welch & Kalff (1974) noted that light transmission through snow-pack that accumulated on top of lake ice did not follow the Bouguer-Lambert law, and that light extinction was nearer an exponential than a linear process. Strictly the Bouguer-Lambert law applies only to monochromatic light passing through a non-scattering medium (Kirk, 1977). However it was applied here to obtain a relative value of attenuation for the total cover on different dates. Adams (1978) also applied this law to calculate a single attenuation coefficient for an ice sheet, thereby averaging the diverse optical properties of different ice and snow strata within an ice sheet.

Attenuation by the ice varied with the age, from a K_d of 0.55 for 'old' rapidly deteriorating ice (19 January 1979) to 1.69 for 'new' ice (2 May 1979). The high attenuation coefficient of the ice around midwinter (3.03 on 24 May 1979) may indicate a compo-

Table 1. Vertical attenuation coefficients (K_d; log$_e$ units m^{-1}) of the ice cover, and the upper and lower zones of the euphotic water column in Ace Lake over 1979.

	Date								
	1979								1980
	19.I	9.II	2.V	24.V	14.VIII	2.X	23.X	25.XI	4.I
K_d (Ice)	0.55	–	1.69	3.03	1.34	1.52	0.97	1.2	0.97
K_d (Upper Zone)	0.38	0.33	0.25	0.42	0.31	0.17	0.23	0.26	0.19
K_d (Lower Zone)	0.39	0.38	0.42	0.42	0.26	0.27	0.34	0.47	0.52

nent due to the low angle of solar altitude. Low solar angle results in increased path-length of the solar beam and leads to an increase in vertical attenuation coefficient (Kirk, 1983). This effect is small in water, but is likely to be of greater significance to K_d in the more scattering medium of ice.

Snow accumulation on the ice increases attenuation by over 50% for a constant ice thickness, i.e. $K_d = 1.52$ on 2 October 1979 with 30 cm snow cover and $K_d = 0.97$ on 23 October 1979 without snow. Maykut & Grenfell (1975) demonstrated that first year arctic sea ice without snow transmitted an order of magnitude more energy than the same ice covered with a 25 cm layer of snow. In Ace Lake the difference was not quite so dramatic. Transmission of 21% of I_0' through 1.6 m of ice was reduced to 7% through the same ice with 30 cm snow cover.

Snow accumulation also has an influence on the relative spectral attenuation characteristics of the ice cover. Table 2 shows the K_d values for the spectral components of PAR with and without 30 cm snow for two dates in October. Snow cover caused dramatically more attenuation of blue and green light while not significantly affecting K_d for red light. Maykut & Grenfell (1975) noted that snow cover led to relatively more attenuation of energy in the blue band and a shift to a broad maximal energy peak in the blue-green (450–550 nm) from a narrow peak in the blue (470–480 nm) for ice alone.

The euphotic water column within the oxic zone (0 m–10 m) in Ace Lake can be divided into two optically distinct zones on the basis of light attenuation (Table 1; Fig. 5). The two zones roughly correspond to water above and below the thermocline, with significantly greater attenuation in the lower zone. Matched pairs comparison of K_d values shows a highly significant difference ($P < 0.01$) between the upper and lower zones.

Table 2. Vertical attenuation coefficients (K_d; $\log_e$ units m^{-1}) of various spectral components (Blue BG 12, Green VG 9, Red RG 610) of PAR through ice cover on Ace Lake on dates with and without snow cover respectively.

	Blue	Green	Red
2.X.79 (30 cm Snow)	1.42	1.37	1.30
23.X.79 (No Snow)	0.95	0.89	1.39

The boundary between the upper and lower zones shifts from 5 m in winter to 6.5 m in spring and summer and is accompanied by progressive increases in attenuation in the lower zone. This increased extinction will be shown to be associated with phytoplankton population increases over spring (see Biological features).

Biological features

The phytoplankton in Ace Lake consisted of only four species, all flagellates. These were the prasinophyte *Pyramimonas gelidicola* McFadden *et al.*, a cryptophyte of the genus *Cryptomonas* and two unidentified species, a microflagellate (2.5–7 μm diameter with two unequal flagella) and an unarmoured dinoflagellate which was the largest member of the phytoplankton (10–20 μm diameter) but was found in very low numbers. Occasionally, the diatoms *Navicula* sp. and *Pinnularia* sp. were found and were likely to have been washed in with meltwater, or suspended into the water column from the extensive sub-littoral benthic mats of cyanobacteria and diatoms.

The annual cycle of growth of the total phytoplankton population and the three numerically significant species can be seen from Fig. 6. The figure shows integral chlorophyll concentration and cell numbers over the oxic zone (0 m–10 m). No phytoplankton were found in the anoxylimnion.

The phytoplankton had a period of seven months of active growth during spring and summer. Maximum cell numbers were recorded in February with the dominants being *P. gelidicola* and the microflagellate. These species displayed maxima in mid-summer whereas the *Cryptomonas* sp. peaked in spring (November). Cell numbers of all species declined in autumn and stayed low up to the end of August. Spring growth commenced under ice in September with increased incoming radiation.

The chlorophyll *a* maximum (water column integral) for the period May 1979 to January 1980 occurred in November. Although chlorophyll appeared to follow the general trend of total cell numbers (Fig. 6), a simple linear regression of chlorophyll *a* on total cell numbers was not significant. Unfortunately, chlorophyll was not determined in January

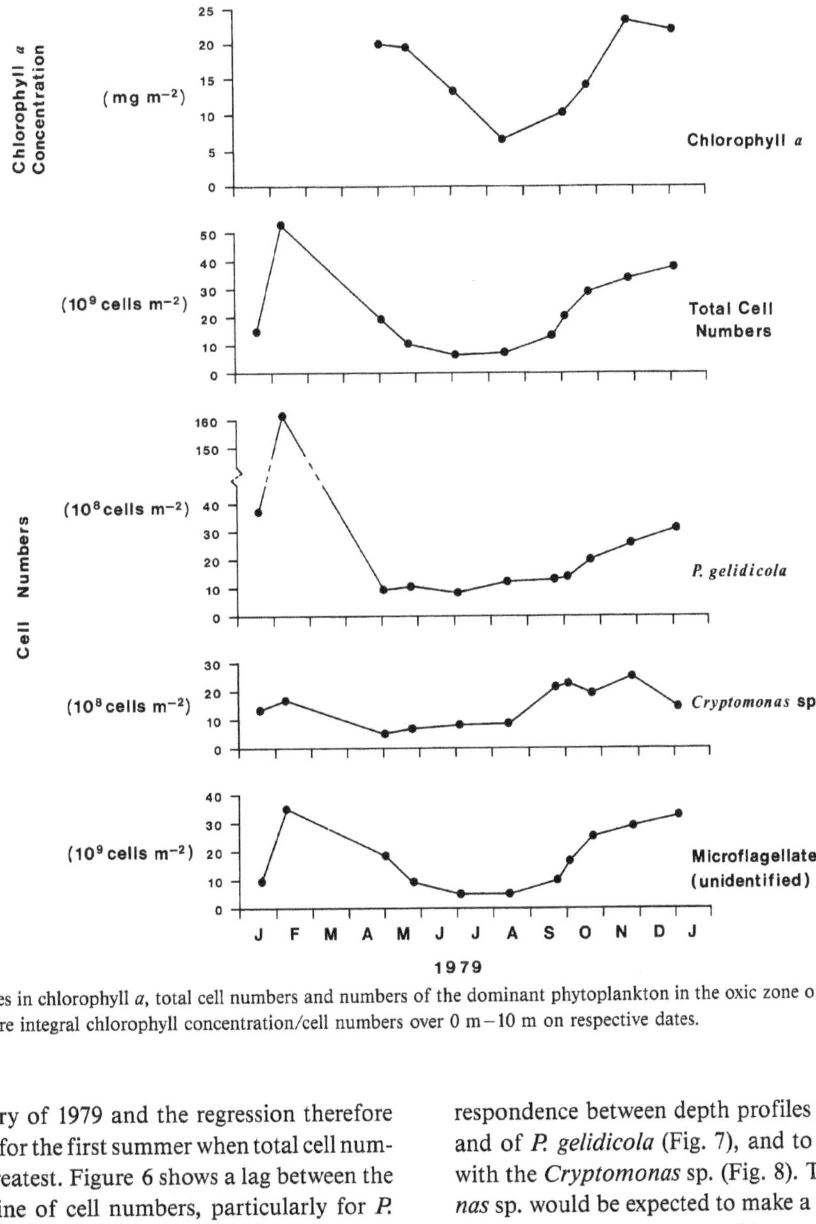

Fig. 6. Changes in chlorophyll *a*, total cell numbers and numbers of the dominant phytoplankton in the oxic zone of Ace Lake during 1979. Values are integral chlorophyll concentration/cell numbers over 0 m – 10 m on respective dates.

and February of 1979 and the regression therefore lacked data for the first summer when total cell numbers were greatest. Figure 6 shows a lag between the winter decline of cell numbers, particularly for *P. gelidicola* and *Cryptomonas* sp., and that of chlorophyll. This may have resulted from interference by degradation products in the chlorophyll determination, and certainly contributed to scatter in the regression.

The depth profile distributions of chlorophyll *a* and of the dominant phytoplankton species during 1979 are shown in Figs. 7 & 8. There is good correspondence between depth profiles of chlorophyll and of *P. gelidicola* (Fig. 7), and to a lesser extent with the *Cryptomonas* sp. (Fig. 8). The *Cryptomonas* sp. would be expected to make a smaller contribution to the total chlorophyll because of its relatively lower numbers. In contrast the distribution of the microflagellate was poorly related to chlorophyll. In the period October to January, the microflagellate was found mainly above 5 m at cell numbers of $6-9\times10^{3}$ cells ml^{-1} (Fig. 8), corresponding to chlorophyll concentrations in the range of $0.12-1.4\ \mu g\ l^{-1}$ (Fig. 7).

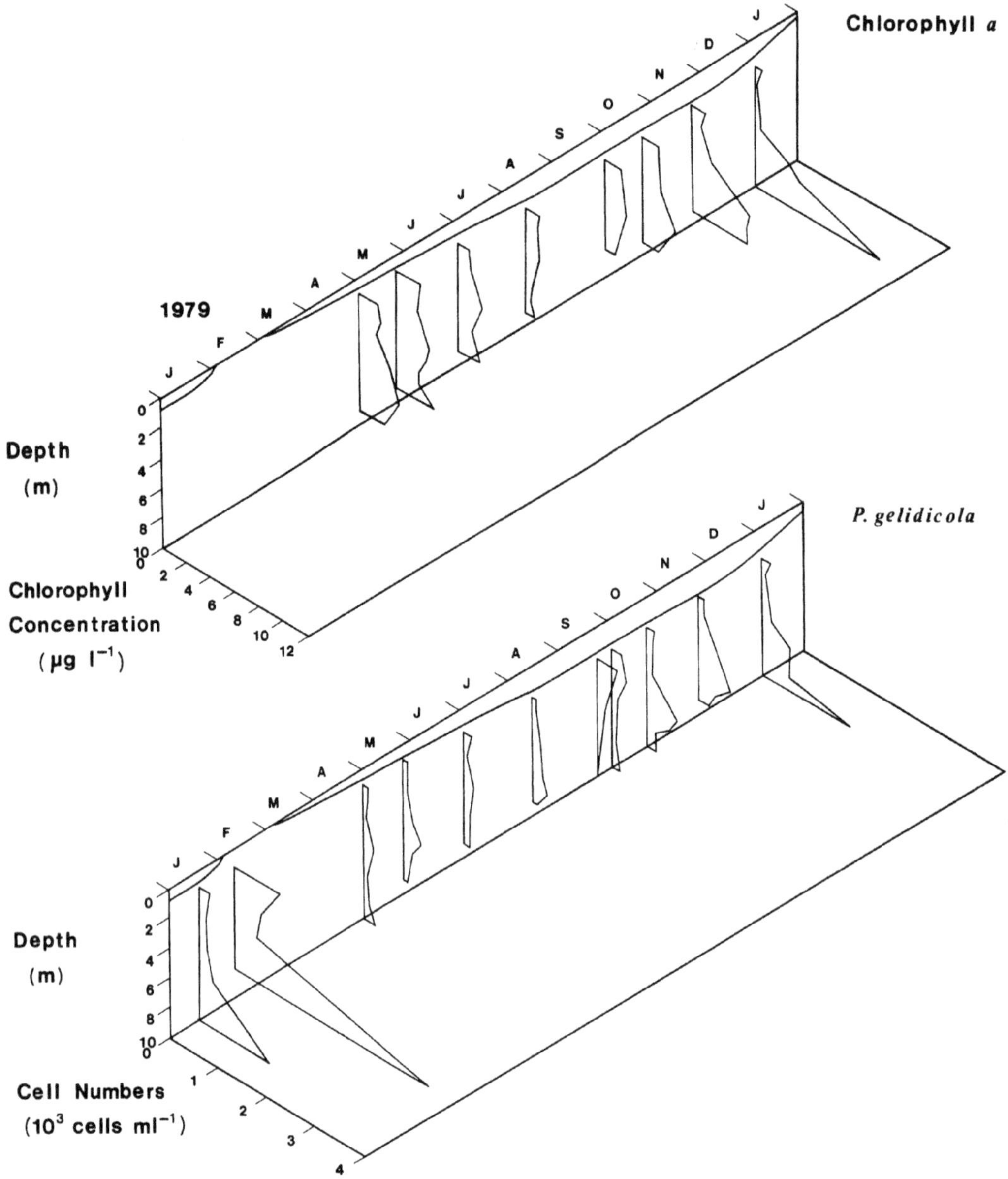

Fig. 7. Depth-time distributions of chlorophyll *a* and cell numbers of *Pyramimonas gelidicola* in Ace Lake during 1979. Note that chlorophyll was not measured in January or February of 1979.

Vertical stratification of the three dominant members of the phytoplankton occurred progressively as solar radiation penetration into the lake increased during spring. Initial spring growth of all three species occurred in September immediately below the ice (2 m – 5 m) where the water was coldest. The PAR below the ice at 2 m, where maximum cell numbers were recorded, was less than 10% of I_0', with a maximum PAR value of 10 μE m^{-2} s^{-1}. By the first week of October *P. gelidicola* and the *Crypto-*

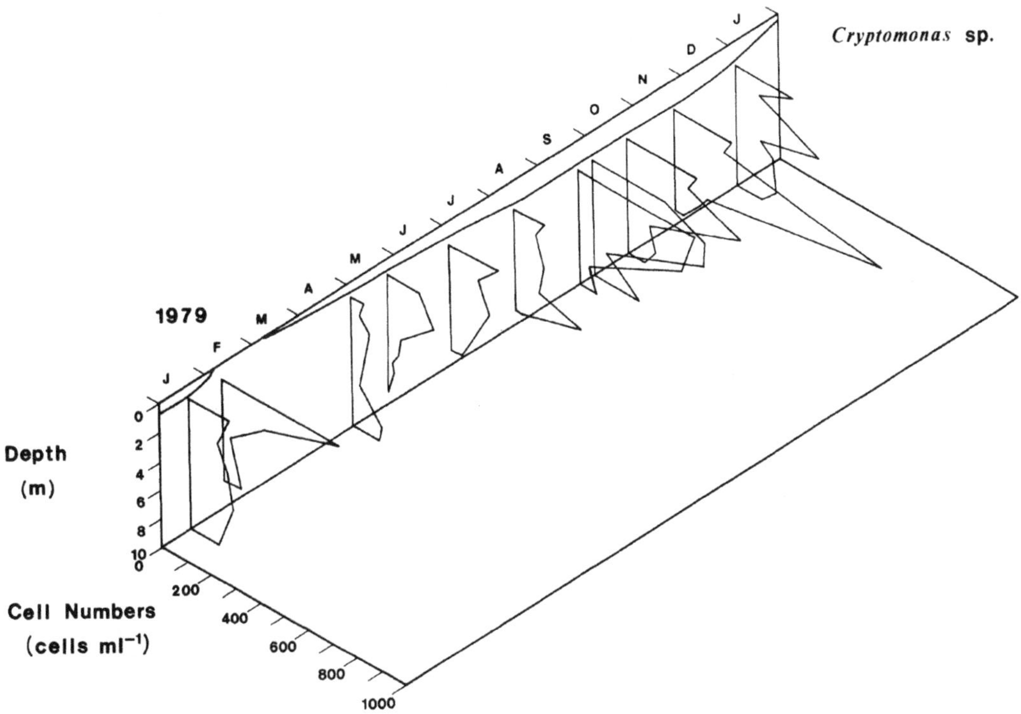

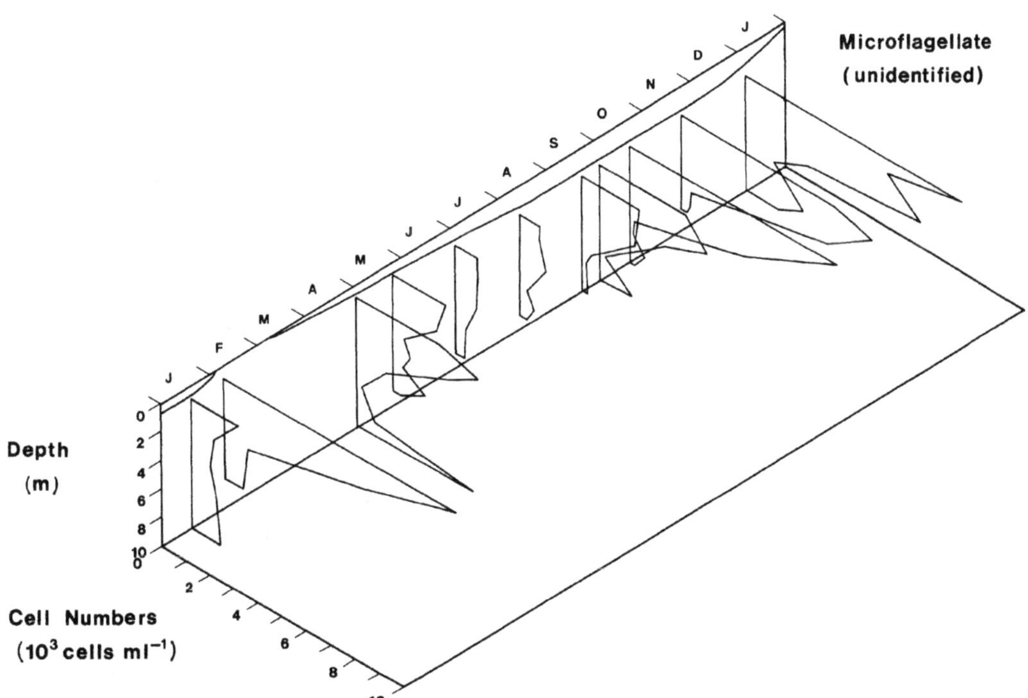

Fig. 8. Depth-time distributions of cell numbers of *Cryptomonas* sp. and the microflagellate (unidentified) in Ace Lake during 1979.

monas sp. had commenced downward migration, and their maximum numbers occurred at 3 m – 5 m on 2 October 1979.

By 23 October 1979 maximum numbers of the *Cryptomonas* sp. were at 5 m and stayed at this depth throughout the rest of the season, reaching highest cell numbers in late November. Maximum numbers of *P. gelidicola* were found at 7 m in October – November and had descended to the bottom of the oxic zone (10 m) by January 1980, as in the previous January and February.

The optimum depth of 7 m – 10 m for *P. gelidicola* had a light climate with a maximum PAR of less than 20 μE m^{-2} s^{-1} while the ice cover persisted. This depth coincided with the warmer water (5°–7 °C) below the thermocline. The combination of favourable temperature and the availability of nutrients at these depths is likely to have contributed to the high cell numbers and chlorophyll *a* concentration found at the bottom of the oxic zone (Fig. 7).

The microflagellate did not display any progressive depth migration and was restricted predominantly to 2 m – 5 m, i.e. always above the thermocline. The light climate below the ice did not cause apparent inhibition of growth of this species as its maximum number (9.5×10^3 cells ml^{-1}) was recorded at 1 m, under open water conditions in February.

The commencement of downward migration for the two larger flagellates coincided with the removal of snow from the surface of the ice in mid-October. This was a critical period in the acceleration of spring growth and has been previously shown to be associated with changes in light quantity and quality below the ice cover.

Analysis of the attenuation of spectral components of PAR (Table 3) over spring and summer provides evidence that specific absorption and scattering by algal cells contributed to overall extinction. Increase in phytoplankton numbers below the thermocline (6 m – 9.5 m) over spring, contributed to greater light attenuation in this zone. Attenuation coefficients (K_d) of all spectral components increased from early September through to January but increases were greater for blue and red components than for green. The increased attenuation corresponds to regions of the photosynthetic spectrum

Table 3. Vertical attenuation coefficients (K_d; $\log_e$ units m^{-1}) of the spectral components (Blue BG 12, Green VG 9, Red RG 610) of PAR in the lower zone (6 m – 9.5 m approx.) of the euphotic water column of Ace Lake on four dates over spring and summer.

Date	2.X.79	23.X.79	25.XI.79	4.I.80
K_d (Blue)	0.34	0.41	0.57	0.82
K_d (Green)	0.24	0.30	0.40	0.41
K_d (Red)	0.42	0.46	0.56	0.92

where the strongest specific absorption by phytoplankton occurs (Kirk, 1983).

Cell volumes of *P. gelidicola* and the microflagellate varied throughout the year (Fig. 9). The *Crypto-*

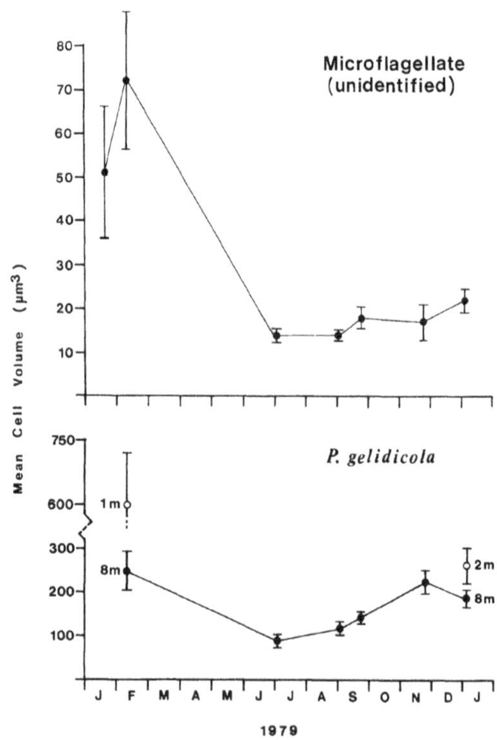

Fig. 9. Mean cell volumes of the microflagellate (unidentified) and *Pyramimonas gelidicola* on selected dates during 1979–1980. The volume indicated for each date is for a sample from a selected depth, except for *P. gelidicola* on 19 February 1979 and 4 January 1980 where volumes from different depth samples are given. Vertical bars are 95% confidence limits for the mean.

monas sp. underwent shape changes with preservation, and volume determinations were unreliable. Both *P. gelidicola* and the microflagellate had a minimum volume in winter and showed progressive increase in spring to reach a maximum volume in summer. For each species, mean cell volumes were compared between dates in each season, and the differences between winter (2 July 1979) and summer (9 February 1979) were highly significant ($P<0.001$). Significant differences also occurred from spring (25 November 1979) to summer (4 January 1980) ($P<0.05$). *P. gelidicola* also displayed volume differences with depth in summer. Cells at 2 m were significantly larger ($P<0.001$) than those found at 8 m in both summers.

Mean cell volume was positively correlated to light intensity (maximum ambient PAR at the depth of sample) for both *P. gelidicola* ($r=0.94$, $n=9$, $P<0.001$) and the microflagellate ($r=0.95$, $n=10$, $P<0.001$). However there was poor correlation between mean cell volume and temperature in the case of both species (*P. gelidicola*: $r=0.12$, $n=9$, $P>0.05$, not significant; microflagellate: $r=0.23$, $n=10$, $P>0.05$, not significant).

Discussion

Studies of phytoplankton ecology in antarctic lakes in the Dry Valleys of Southern Victoria Land (Vincent, 1981; Vincent & Vincent, 1982) and on Signy Island in the South Orkney Islands (Light, 1977; Light *et al.*, 1981) show some remarkable similarities and some significant differences to the patterns observed in Ace Lake. Lake Fryxell (77°35′S, 163°15′E) a permanently ice-covered meromitic lake in the Taylor Valley, had permanent gradients of temperature, salinity and dissolved oxygen over a 9.5 m depth similar to Ace Lake (Vincent, 1981). The phytoplankton were also all flagellates and were distributed in distinct layers over the water column in relation to these gradients, with *Ochromonas nannos* and *Chlamydomonas* sp. immediately beneath the ice and a deep maximum of *Chroomonas lacustris* and *Pyramimonas* sp. recorded at the bottom of the euphotic zone. The lower euphotic community was highly shade-adapted and not nutrient-limited (Vincent, 1981). Vincent suggested that these algae had high photosynthetic efficiency and were adapted to grow in the dimly lit environment utilizing nutrients mineralized by bacteria at the interface of the oxygenated and deoxygenated layers.

P. gelidicola in Ace Lake displayed shade adaptation and a seasonal downward migration in order to reside in a low light environment over the growing season. At the bottom of the oxic zone of Ace Lake temperature conditions and ammonia concentration provided optimal conditions for the development of substantial biomass over the summer season. Nitrogen limitation of primary production in the upper waters of Ace Lake was proposed by Hand & Burton (1981) and this may be the case. However, physiological assays (c.f. Vincent, 1981) are required to properly ascertain the nutrient status of Ace Lake phytoplankton.

The microflagellate in Ace Lake does not appear to have the growth requirements of *P. gelidicola* and the *Cryptomonas* sp. as it flourished in the nitrogen-poor surface waters just below the ice where high light conditions also prevail. Indeed the ecology and taxonomic status of this almost colourless microflagellate in Ace Lake are puzzling, especially as it makes an apparently minimal contribution to the chlorophyll concentration and has a vertical distribution inverse to that of the other flagellates.

Vincent & Vincent (1982) have also described a layered phytoplankton community in Lake Vanda (77°35′S, 161°40′E) with microflagellates below the ice and a deep production maximum by blue-green algae at a nutrient rich oxic/anoxic interface. These authors proposed that in both Lake Fryxell and Lake Vanda the maximum phytoplankton numbers occur early in the growing season. Their sampling program commenced after this period. They observed a decline in chlorophyll *a* level from mid-December to mid-January in both lakes, and suggested that the phytoplankton photosynthetic physiology may be adapted for higher production rates at the lower irradiance conditions in the spring rather than at the higher light regime of mid-summer. Individual species in Ace Lake had different times of peak abundance over the growing season. The *Cryptomonas* sp. peaked in November and *P. gelidicola* and the microflagellate were at their maximum numbers after mid-summer, in January and February. Without measurements of photosynthetic production it would be unwise to relate these peaks to the optimal production times of individual species. Population size at any one time is a balance of production and loss and a probable major loss factor in Ace Lake is zooplankton grazing by the Copepod *Paralabidocera antarctica* (I. C. Thompson) (Bayly & Burton, 1987). The extent of grazing is unknown but zooplankton are found in greatest numbers during summer (H. R. Burton, pers. commun.)

As a general observation ice-covered lakes in the Vestfold Hills show considerable phytoplankton growth in early spring (M.D. Burch, unpubl. data). Heath (this volume) in a study of annual primary productivity in freshwater Watts Lake (68°36′S, 78°13′E), found the phytoplankton had maximum photosynthetic assimilation rates in September. Monthly sampling of Watts Lake in 1979, in parallel with Ace Lake, showed it to have a maximum annual chlorophyll concentration peak in October followed by a progressive decline in concentration through to February 1980 (M. D. Burch, unpubl. data). An almost uni-algal phytoplankton population of *Dunaliella* sp. in hypersaline Organic Lake (nearby to Ace Lake) produced signficant increases in chlorophyll concentration from July to August 1979 (M. D. Burch, unpubl. data).

The only comparable complete annual study of phytoplankton ecology in an antarctic lake is that of Light *et al.* (1981) on Heywood Lake on Signy Island (60°43′S, 45°38′W), South Orkney Islands. Heywood Lake had a spring burst of growth (October–November) under the ice of both Chlorophytes and Prasinophytes, including two species of *Ankistrodesmus,* and the flagellates *Chlamydomonas* sp., *Pedinomonas minor* (Korshikov) and *Scourfieldia* sp. These declined in December as the ice disappeared and succession occurred during the summer open water period which gave rise to a population composed of the Cryptophytes *Rhodomonas lacustris* var. *nannoplanktica* and *Cryptomonas* sp.

Ace Lake differs from Heywood Lake in having no seasonal succession and also in its reduced community diversity. Collections of phytoplankton from most antarctic lakes, as well as arctic lakes and ice-covered lakes in general, are dominated by flagellates belonging to the divisions Chlorophyta, Prasinophyta, Chrysophyta and Cryptophyta (Vincent, 1981; Vincent & Vincent, 1982; Light *et al.,* 1981; Parker *et al.,* 1982; Hobbie, 1973; Pechlaner, 1971; Brettum, 1972). Members of the Chrysophyta often make a significant contribution but have not yet been found in any lakes in the Vestfold Hills (M. D. Burch, unpubl. data). The flagellates dominate as they can maintain their position in a water column not subject to wind induced turbulence (Goldman *et al.,* 1967; Rodhe *et al.,* 1966; Hobbie, 1973).

Heywood Lake is unusual amongst ice-covered lakes as the phytoplankton population is dominated by non-flagellate species (Light *et al.,* 1981). Hawes (1983) has found that under-ice turbulent mixing in Heywood Lake breaks down winter gradients of temperature, oxygen, chloride and nutrients at the start of spring and maintains the spring bloom of *Ankistrodesmus* sp. in suspension in the euphotic zone.

The phenomenon of vertical migration of flagellates in ice-covered lakes to depths of preferred light climate has been shown to occur both diurnally (Tilzer, 1973; Ilmavirta, 1974, 1975; Kalff & Welch, 1974) and seasonally (Tilzer, 1973; Brettum, 1972; Maeda & Ichimura, 1973; Kalff *et al.,* 1972). Brettum (1972) and Maeda & Ichimura (1973) specifically identified Cryptomonads as undergoing seasonal vertical shifts. Kalff *et al.* (1972), in a case closely resembling Ace Lake, found progressive downward migration of a flagellate dominated community through a 25 m water column over spring in Arctic Char Lake (74°42′N, 94°50′W). The progressive seasonal migration in Ace Lake has been shown to be differential for individual species, with *P. gelidicola* migrating to a lower preferred light zone than the *Cryptomonas* sp.

The question of the mechanism of overwinter survival of antarctic algae and algae living in ice-covered lakes has been reviewed extensively (Heywood, 1977; Wright & Burton, 1981; Light *et al.,* 1981). The arguments revolve around the early suggestion by Rodhe (1955) that heterotrophic or photoheterotrophic uptake of organic solutes by algae could contribute to survival under low light conditions. Later work (e.g. Morgan & Kalff, 1975). did not support this view and Pechlaner (1971) stated that low-light adaptation alone could account for winter survival in high-mountain lakes. Light *et al.* (1981) concluded that photoautotrophy by phytoplankton in winter in Heywood Lake was unlikely, and Priddle (1980) reported that heterotrophic uptake by phytoplankton was not detected in another lake on Signy Island.

The work on Ace Lake offers no physiological evidence to explain the survival of the few cells in the water column over the winter months. However the substantial decrease in cell size observed during winter may relate to slow respiration of carbohydrate reserves as the winter survival mechanism, as suggested by Pechlaner (1971) and Morgan & Kalff (1975). Morgan & Kalff (1975) investigated the survival of the flagellate *Cryptomonas erosa* (Skuja) in culture, under different temperature and light intensity conditions. They found that cells grown at 4 °C survived much longer in the dark than cells grown at 15 °C. Cell volume decreased during dark incubation and the cells stained progressively less with iodine, suggesting loss of carbohydrate reserves by slow respiration. Similarly, in Ace Lake, cells of *P. gelidicola* and the microflagellate were larger and stained more darkly with iodine in summer than in the winter when cells were smaller. Morgan & Kalff (1975) also found that the mean cell volume of *C. erosa* grown at 15 °C was positively correlated with light intensity which matches the ecological observations of *P. gelidicola* and the microflagellate in Ace Lake.

Conclusions

The annual cycle of phytoplankton in Ace Lake was one of active growth in spring and summer for a period of seven months, and overwinter survival by small numbers of cells which persisted throughout the winter below the ice. Spring growth commenced in September with increased solar radiation penetration into the lake which was enhanced by the removal of snow from the ice surface.

Overall production by the phytoplankton may be limited by the availability of inorganic nitrogen which was below the limit of detection in the upper 10 m of the lake. Peak biomass production by shade-adapted *Pyramimonas gelidicola* occurred at the interface of the oxic and anoxic zones where considerable ammonia was present.

The phytoplankton community has low diversity compared to similar antarctic lakes in other geographic areas and consisted of only four flagellates. Distinct vertical stratification of the three dominant phytoplankton species was observed. The pattern was similar to that observed in ice-covered, meromictic lakes in the Taylor Valley, Southern Victoria Land (Vincent, 1981; Vincent & Vincent, 1982). In Ace Lake this stratification was not static and was shown to undergo progressive development over spring as the three populations migrated to preferred light climate zones independent of both thermal and salinity stratification in the top 10 m of the lake.

Acknowledgements

I thank Mr H. R. Burton for initial inspiration to work on Ace Lake and for discussions and encouragement during the completion of this work. I also thank Dr P. A. Broady, Mr C. W. Heath, Dr J. M. Ferris and two anonymous referees for critically reviewing the manuscript, Dr G. G. Ganf for valuable advice on treatment of the solar radiation data, Mrs A. Holland and Mrs C. Sparnon for typing the drafts and Miss J. Burch for assistance with diagrams. Members of the 1979 ANARE to Davis Station provided field assistance, especially Dr W. Twycross who also assisted with some laboratory analyses.

References

Adams, W. A., 1978. Effects of ice cover on the solar radiation regime in Canadian lakes. Verh. int. Ver. Limnol. 20: 141–149.

APHA, 1981. Standard methods for the examination of water and wastewater. APHA-AWWA-WPCF, Washington, 1134 pp.

Bayly, I. A. E. & H. R. Burton, 1987. Vertical distribution of Paralabidocera antarctica (Copepoda: Calanoida) in Ace Lake, Antarctica, in summer. Aust. J. mar. Freshwat. Res. 38: 537–543.

Brettum, P., 1972. The phytoplankton of the Lake Övre Heimdalsvatn, central south Norway, 1969–70. Norw. J. Bot. 19: 79–90.

Burke, C. M. & H. R. Burton, this volume. The ecology of photosynthetic bacteria in Burton Lake, Vestfold Hills, Antarctica. Hydrobiologia.

Burton, H. R., 1980. Methane in a saline antarctic lake. In P. A. Trudinger & M. R. Walter (eds), Biogeochemistry of Ancient and Modern Environments. Proceedings of the Fourth International Symposium on Environmental Biogeochemistry (ISEB). Aust. Acad. Sci., Canberra: 243–251.

Burton, H. R. & R. J. Barker, 1979. Sulphur chemistry and microbiological fractionation of sulphur isotopes in a saline antarctic lake. Geomicrobiol. J. 1: 329–340.

Ferris, J. M. & H. R. Burton, this volume. The annual cycle of heat content and mechanical stability of hypersaline Deep Lake, Vestfold Hills, Antarctica. Hydrobiologia.

Goldman, C. R., D. T. Mason & J. E. Hobbie, 1967. Two antarctic desert lakes. Limnol. Oceanogr. 12: 295–310.

Hand, R. M., 1980. Bacterial populations of two saline antarctic lakes. In P. A. Trudinger & M. R. Walter (eds), Biogeochemistry of Ancient and Modern Environments. Proceedings of the Fourth International Symposium on Environmental Biogeochemistry (ISEB). Aust. Acad. Sci., Canberra: 123–129.

Hand, R. M. & H. R. Burton, 1981. Microbial ecology of an antarctic saline meromictic lake. Hydrobiologia 82: 363–374.

Hawes, I., 1983. Turbulence and its consequences for phytoplankton development in two ice covered antarctic lakes. Br. Antarct. Surv. Bull. 60: 69–81.

Heath, C. W., this volume. Annual primary productivity of an antarctic continental lake: phytoplankton and benthic algal mat production strategies. Hydrobiologia.

Heywood, R. B., 1977. Antarctic freshwater ecosystems: review and synthesis. In G. A. Llano (ed.), Adaptations within Antarctic Ecosystems. Smithsonian Institution, Wash. DC: 801–828.

Hobbie, J. E., 1973. Arctic limnology: A review. In M. E. Britton (ed.), Alaskan Arctic Tundra. Tech. paper No. 25, Arctic Inst. of N. Amer.: 127–168.

Ilmavirta, V., 1974. Diel periodicity in the phytoplankton community of the oligotrophic lake Pääjärvi, southern Finland. I. Phytoplanktonic primary production and related factors. Ann. bot. fenn. 11: 136–177.

Ilmavirta, V., 1975. Diel periodicity of the phytoplankton community of the oligotrophic lake Pääjärvi, southern Finland. II Late summer phytoplanktonic biomass. Ann. bot. fenn. 12: 37–44.

Jeffrey, S. W. & G. F. Humphrey, 1975. New spectrophotometric equations for determining chlorophylls a, b, c_1 and c_2 in higher plants, algae and natural phytoplankton. Biochem. Physiol. Pflanzen. 167: 191–194.

Johnstone, G. W., D. J. Lugg & D. A. Brown, 1973. The biology of the Vestfold Hills, Antarctica. ANARE Sci. Rep. Publ. 123. Antarct. Div., Dept. Sci., Melbourne, 62 pp.

Journal of Glaciology, 1958. Instruments and methods: Ice drills and corers. J. Glaciol. 3: 30.

Kalff, J., H. E., Welch & S. K. Holmgren, 1972. Pigment cycles in two high-arctic Canadian lakes. Verh. int. Ver. Limnol. 18: 250–256.

Kalff, J. & H. E. Welch, 1974. Phytoplankton production in Char Lake, a natural polar lake, and in Meretta Lake, a polluted polar lake, Cornwallis Island, Northwest Territories. J. Fish. Res. Bd Can. 31: 621–636.

Kantha, L. H., 1979. Turbulent entrainment at the bouyancy interface due to convective turbulence. In H. J. Freeland, D. M. Farmer & C. D. Levings (eds), Fjord Oceanography. Plenum Press, N.Y.: 205–213.

Kirk, J. T. O., 1977. Use of quanta meter to measure attenuation and underwater reflectance of photosynthetically active radiation in some inland and coastal south-eastern Australian waters. Aust. J. mar. Freshwat. Res. 28: 9–21.

Kirk, J. T. O., 1983. Light and photosynthesis in aquatic ecosystems. Cambridge University Press, Cambridge, 401 pp.

Light, J. J., 1977. Production and periodicity of antarctic freshwater phytoplankton. In G. A. Llano (ed.) Adaptations within Antarctic Ecosystems. Smithsonian Institution, Wash., DC: 829–837.

Light, J. J., J. C. Ellis-Evans & J. Priddle, 1981. Phytoplankton ecology in an antarctic lake. Freshwat. Biol. 11: 11–26.

Lund, J. W. G., C. Kipling & E. D. LeCren, 1958. The inverted microscope method of estimating algal numbers and the statistical basis of estimations by counting. Hydrobiologia 11: 143–170.

Maeda, O. & S. Ichimura, 1973. On the high density of a phytoplankton population found in a lake under ice. Int. Revue ges. Hydrobiol. 58: 673–689.

Maykut, G. A. & T. C. Grenfell, 1975. The spectral distribution of light beneath first-year sea ice in the Arctic Ocean. Limnol. Oceanogr. 20: 554–563.

Morgan, K. & J. Kalff, 1975. The winter dark survival of an algal flagellate Cryptomonas erosa (Skuja). Verh. int. Ver. Limnol. 19: 2734–2740.

Parker, B. C., G. M. Simmons Jr., K. G. Seaburg, D. D. Cathey & F. C. T. Allnutt, 1982. Comparative ecology of plankton communities in seven antarctic oasis lakes. J. Plankton Res. 4: 271–286.

Pechlaner, R., 1971. Factors that control the production rate and biomass of phytoplankton in high-mountain lakes. Verh. int. Ver. Limnol. 19: 125–145.

Priddle J., 1980. The production ecology of benthic plants in some antarctic lakes. II. Laboratory physiology studies. J. Ecol. 68: 155–166.

Rodhe, W., 1955. Can plankton production proceed during winter darkness in subarctic lakes? Verh. int. Ver. Limnol. 12: 117–122.

Rodhe, W., J. E. Hobbie & R. T. Wright, 1966. Phototrophy and heterotrophy in high mountain lakes. Verh. int. Ver. Limnol. 16: 302–313.

Strickland, J. D. H. & T. R. Parsons, 1972. A practical handbook of seawater analysis. 2nd. Edition. Bull. Fish. Res. Bd. Can. 167: 310 pp.

Tilzer, M. M., 1973. Diurnal periodicity in the phytoplankton assemblage of a high mountain lake. Limnol. Oceanogr. 18: 15–30.

UNESCO, 1966. Determination of photosynthetic pigments in seawater. Report of SCOR-UNESCO Working group 17. Monographs on Oceanographic Methodology 1. UNESCO, Paris, 69 pp.

Vincent, W. F., 1981. Production strategies in antarctic inland waters: phytoplankton eco-physiology in a permanently ice-covered lake. Ecology 62: 1215–1224.

Vincent, W. F. & C. L. Vincent, 1982. Factors controlling phytoplankton production in Lake Vanda (77°S). Can. J. Fish. aquat. Sci. 39: 1602–1609.

Welch, H. E. & J. Kalff, 1974. Benthic photosynthesis and respiration in Char Lake. J. Fish Res. Bd Can. 31: 609–620.

Willen, E., 1976. A simplified method of phytoplankton counting. Br. phycol. J. 11: 265–278.

Wright, S. W. & H. R. Burton, 1981. The biology of antarctic saline lakes. Hydrobiologia 82: 319–338.

Hydrobiologia 165: 77–87 (1988)
J. M. Ferris, H. R. Burton, G. W. Johnstone & I. A. E. Bayly (eds) Biology of the Vestfold Hills, Antarctica
Kluwer Academic Publishers

Annual primary productivity of an antarctic continental lake: Phytoplankton and benthic algal mat production strategies

C. W. Heath[1]
Antarctic Division, Department of Science, Kingston, Tasmania, 7050, Australia; [1]Present address: Metropolitan Water Sewerage & Drainage Board, Scientific Laboratories, P.O. Box 73, West Ryde, Sydney, 2114, Australia

Key words: Antarctica, annual production, phytoplankton, benthic algae

Abstract

Primary production in Watts Lake, Vestfold Hills, Antarctica (68°36′S, 78°13′E), was measured from March 1981 to February 1982. Phytoplankton production peaked in autumn and spring, with a September maximum (340 mgC m^{-2} d^{-1}), then declined in summer and was not detectable in winter. Benthic algal production peaked in summer at 74 mgC m^{-2} d^{-1}. Production strategies differed, with the more efficient phytoplankton adapted to growth at low light, while benthic production increased with increasing light in summer. Estimation of annual production was 10.1 gC m^{-2} and 5.5 gC m^{-2} for the phytoplankton and benthos respectively.

Introduction

Primary production in antarctic lakes has been almost exclusively measured during the austral summer. Studies have been undertaken on offshore or subantarctic islands (Goldman, 1964, 1970; Goldman *et al.*, 1963, 1972; Light, 1977; Light *et al.*, 1981; Priddle, 1980 and Samsel & Parker, 1971, 1972), and on continental antarctic water bodies. These latter include the Dry Valley lakes of Victoria Land (Goldman, 1964; Koob & Leister, 1972; Parker *et al.*, 1977; Vincent, 1981 and Vincent & Vincent, 1982); the Molodezhnaya Oasis lakes (Kaup, 1975; in Light *et al.*, 1981) and pools around Syowa Station, Prince Olav Coast (Tominaga, 1977). In the Vestfold Hills, Campbell (1978) attempted to measure primary production in hypersaline Deep Lake, with limited success, over the summer of 1975–1976. He successfully measured primary production in Watts Lake once in 1975. Measurement of annual production has only recently been attempted, notably by Light (1977), Light *et al.* (1981) and Priddle (1980), all of whom conducted their studies in lakes on Signy Island. There is, however, a need for more annual studies to compare seasonal variations and provide an insight into the growth strategies of the algae in response to the environmental extremes encountered in Antarctica.

Algal production is the sum of that contributed by the phytoplankton and the benthic and epiphytic algae. In high latitude lakes, however, only the studies of Fogg & Horne (1970), Goldman *et al.* (1972), Kairesala (1980), Parker *et al.* (1977) and Tominaga (1977) have considered total lake production. This paper presents a study of the annual cycle of production of an antarctic continental lake and attempts to compare limnetic and benthic production strategies.

Study Site

For a period in the recent past (c. 4000 a BP), Watts Lake (68°36′S, 78°13′E), was a coastal embayment in the Vestfold Hills. Subsequent to icecap retreat and crustal uplift, meltwater inflows have largely flushed out the saline water. The ionic balance and concentration of remaining salts (2–3 g l^{-1}) are evidence of its marine origin.

The Lake consists of a shallow (10 m) Western basin connected by a narrow stretch of water with a minimum depth of 4 m to a deeper (35 m) Eastern basin. The lake presently lies 4.4 m below mean sea level, has a surface area of 0.38 km^2 and volume of 4.5×10^6 m^3. Ice cover, up to 1.9 m thick, persists for 10–12 months. Summer thaw can result in fully open water conditions (1974), but more often is incomplete (1981, 1982) with the residual ice remaining on the deeper basin.

Materials and methods

Water samples were collected with a darkened 2.25 l Kemmerer bottle, through an 11 cm hole in the ice, bored with a SIPRE ice auger (Journal of Glaciology, 1958). Benthic samples were collected by means of a grab constructed to fit down the ice hole (Heath & Singleton, this volume). Samples for total inorganic carbon and biomass (chlorophyll *a*) were taken concurrently with the samples for productivity measurements.

Photosynthetically active radiation (PAR) was measured with a LAMBDA Li-cor submersible quantum sensor (LI 192S). Water samples were taken from depths corresponding approximately to 1%, 5%, 10%, 20% and where possible, 50% of the value measured with the sensor immersed below the water surface in the ice hole. On most occasions less than 50% of the light penetrated the ice and an under-ice depth (usually 2 m) was substituted for the 50% light level depth. With insufficient light to register on the meter prior to the experiments in May and July, the depths sampled were based on light profiles taken a few days previously at noon.

Total inorganic carbon was measured titrimetrically according to the method of Golterman (1969). Chlorophyll *a* was measured as an estimate of biomass. Water was filtered through 47 mm diameter Whatman GF/C filters which were immediately frozen and analysed within 3 d of collection. Filters were ground in 90% acetone for less than 2 min with a Teflon pestle in glass, stored in the dark at 4 °C for not less than 1 h and centrifuged at 500 G for 10 min. Absorbance of the supernatant was measured in a Varian 635DB spectrophotometer with 1 cm or 4 cm pathlength cuvettes. Pigments were determined according to the equation of Parsons & Strickland (1963), with phaeopigment correction for chlorophyll after acidification (Strickland & Parsons, 1972).

Photosynthetic carbon fixation was assayed by the ^{14}C technique outlined by Strickland & Parsons (1972) and Vollenweider (1974). Samples were incubated for 4 h (10:00–14:00 h) in 300 ml light and dark bottles at the depths of collection, with 10 μCi ^{14}C sodium bicarbonate per bottle. Four hour incubations were necessary for reliable measurement of uptake but sufficiently brief to minimize bottle effects (Vollenweider & Nauwerck, 1961). Benthic samples were placed into a darkened bucket containing water from a similar depth, and subsamples dispensed into the production bottles, previously filled with water from immediately above the mat. All manipulations were carried out in an on-ice shelter over or adjacent to the ice hole. At each depth, two light bottles, two dark bottles and one light bottle pre-fixed with 5 mls of 40% formalin were suspended horizontally in racks. In the water column, shading was prevented by placing the darkened bottles centrally in each rack and offsetting the racks radially with respect to each other. Benthic samples were suspended immediately above the mat to avoid shading. In winter, there was significant refreezing of the ice hole during the incubation period. A reaming device, described in Heath *et al.* (this volume), was used to return the ice hole to its original dimensions and permit the recovery of the production bottles.

As soon as practicable after retrieval, the contents of the bottles were filtered through 47 mm Whatman GF/C filters and the bottles rinsed twice with clean water to flush inorganic ^{14}C from the filters. Detritus in some benthic samples greatly increased filtration times. In such cases, only one rinse was used.

On return to the laboratory, the filters were purged over fuming HCl for 4 h to remove residual inorganic ^{14}C and placed in glass scintillation vials to which

NSC tissue solubilizer (1 ml) and Aquasol scintillant (10 ml) were added. Activity was measured in a Packard 101 liquid scintillation spectrometer and adjusted for quenching by the external standard method.

Results

General features of the water column

Throughout the study period, conductivity profiles remained stable (Fig. 1), increasing with depth. The lake remained stratified thermally, with the exception of a period of approximately one week in mid-March 1981. The maximum oxygen concentration recorded at 5 m in December 1980 (13.5 g m^{-3}) represented 100% saturation. While little change in oxygen concentration occurred at any depth

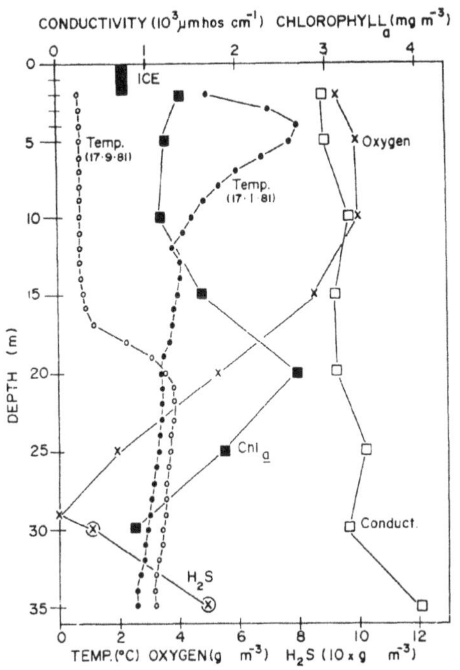

Fig. 1. Characteristics of the water column of Watts Lake. Dissolved oxygen (×), hydrogen sulphide (⊗), conductivity (□) and chlorophyll *a* (■) shown as means from December 1980 to January 1982. Temperature profiles show extremes in summer (●, 17 January 1981) and winter (○, 17 September 1981).

throughout the year, absolute levels declined with depth to 29 m, below which an anoxic hydrogen sulphide layer developed. The depth to which 1% PAR penetrated, varied from 5.5 m to 27.5 m (Table 6). Chlorophyll content was low (range 0.1–10 mg m^{-3}), peaking in spring and late summer at depth. Inputs to the lake occur in summer as a result of wind, snow melt, and leachate from an adjacent relict marine deposit of algae and serpulid worm tubes. Seasonal overflows from a nearby oligotrophic freshwater lake also contributed to Watts Lake.

The phytoplankton consisted mainly of a *Rhodomonas* like flagellate (10 µm diameter) and an unidentified ovoid Chlorophyte (5 µm diameter). *Cosmarium* spp. dominated at depth in late summer (20–25 m). Pennate diatoms were also present in the water column and the benthic mat. This mat principally consisted of the filamentous algae *Vaucheria* sp. with *Oedogonium* sp. and *Mougeotia* spp. as minor components. Rising in places 1 m above the substrate as an entangled mass, it extended to a depth in excess of 12 m.

Limnetic production

Primary production of Watts Lake phytoplankton varied from peaks in March and September, to a minimum in July when no net uptake was measurable (Table 1). The maximum rate measured was 6.4 mgC m^{-3} h^{-1} at 4 m in September, corresponding to an areal production rate of 340 mgC m^{-2} d^{-1} integrated over the euphotic zone. This active growth phase in spring was reflected in the assimilation number (ratio of photosynthesis to chlorophyll *a*; Table 2). Peak assimilation rates occurred in September, with lesser levels corresponding to the autumn peak in production. Isolated high assimilation numbers were observed at 1 m in May and 15 m in September (11.23 and 17.31 mgC mgChl a^{-1} h^{-1} respectively) without obvious cause.

Photosynthetic efficiency (photosynthesis per unit light intensity) was maximal in autumn and spring (Table 3), declining in summer during peak illumination. Similarly, greater efficiency was achieved at lower light intensities with depth in all but the April and May assays. On these two occa-

Table 1. Carbon fixation rate (mgC m^{-3} h^{-1}) of Watts Lake phytoplankton over the period 27 March 1981 to 9 January 1982 and 19 February 1975 (Campbell, 1978).

Depth (m)	27.III 1981	25.IV 1981	30.V 1981	11.VII 1981	18.VIII 1981	11.IX 1981	23.X 1981	14.XI 1981	20.XII 1981	9.I 1982	19.II 1975
0	0.10										
1			1.46								0.65
2	2.46	1.17		0	0.18	5.30	1.38	1.47	0.35	0.61	
3		0.66									
4	0.36	0.38	0.16	0	0.21	6.44					
5								1.85	0.63		0.94
6			0.12	0	0.18	6.39	2.40			0.61	
7	0.54										
8							1.54	1.68	0.66		
9											
10										0.48	
11											
12											
13											
14											
15					0.11	2.25					0.20
16	0.51										
17											
18											
19											
20							0.34	1.22	1.34	0.17	
21											
22											
23											
24											
25										0.53	

sions, underwater illumination was markedly attenuated by snow and ice (Table 6).

Changes in production, relative to percentage surface irradiance, from midwinter darkness to continuous sunlight in midsummer, are shown in Fig. 2. During this period incident irradiance at noon increased from 6 μE m^{-2} s^{-1} in July, to a measured maximum of 1 590 μE m^{-2} s^{-1} on a clear day in December.

By grouping production results into bands covering a percentage of surface irradiance, production increased then declined within each band as absolute irradiance increased towards midsummer. The peak of production occurred progressively later in the year with decreasing percentage light availability, as absolute intensity increased. A similar trend occurred in assimilation numbers (Table 2), indicating that the shift was not wholly due to changes in chlorophyll concentration seasonally or with depth.

Benthic production

Due to the nature of the benthic mat and the difficulty of sampling benthos through ice, assimilation number was considered a more accurate means of assessing production than area or volume measures. Benthic algal mat assimilation rates (Table 4), increased during spring and, with the exception of January, remained at elevated levels throughout summer. Assimilation was greater at shallower depths during the period of increasing light intensity (August to October) but similar at both depths in summer. This trend was also observed in photosyn-

Table 2. The ratio of carbon fixation rate to chlorophyll *a* content (assimilation number) of Watts Lake phytoplankton over the period 27 March 1981 to 9 January 1982 (mgC mgChl a^{-1} h^{-1}).

Depth (m)	27.III 1981	25.IV 1981	30.V 1981	11.VII 1981	18.VIII 1981	11.IX 1981	23.X 1981	14.XI 1981	20.XII 1981	9.I 1982
0	0.18									
1			11.23							
2	3.51	2.85		0	2.11	5.20	0.41	0.39	0.11	0.29
3		1.43								
4	0.49	0.72	1.00	0	2.63	7.49				
5								0.67	0.41	
6			0.75	0	1.50	7.26	0.75			0.41
7	0.79									
8							0.58	0.62	0.48	
9										
10										0.39
11										
12										
13										
14										
15					5.50	17.31				
16	1.11									
17										
18										
19										
20							0.43	1.17	0.71	0.11
21										
22										
23										
24										
25										0.11

thetic efficiency (Table 4). A marked increase in pigment content in the deeper mat was observed between October and November. A second assay was conducted in November to confirm this observation (Table 5). Part of the pigment variation lay in differences in sample size, but the carotenoid:chlorophyll *a* ratio (Table 5) demonstrated a shift in pigment content in favour of chlorophyll in the deeper mat. At that time, photosynthetic activity of the mat at greater depth increased to levels similar to those of the shallower mat.

The contribution to lake production by the benthic mat growing near the lower limit of its extent was assessed in February (Table 4). Although assimilation at 5 m was greater than at 12 m, in response to higher light intensity, the deeper mat was more efficiently utilizing the available light.

Factors influencing production

Throughout the study, the 1% PAR light transmission depth varied from 5.5 m to 27.5 m (100% PAR was that value measured with the sensor immediately below the water surface in the ice hole). Although it was difficult to apportion their relative importance, ice and snow cover influenced the 1% PAR depth. When the ice was thinnest in March (0.5 m), areal production was 53 mgC m^{-2} d^{-1}, considerably less than the spring peak of 340 mgC m^{-2} d^{-1}, which occurred when the ice was thickest (1.5 m). Ice conditions could, however, influence light penetration, as was the case in May when opaque ice greatly reduced the 1% PAR depth (Table 6).

A comparison of conditions existing in April and August (Table 6) demonstrates the role snow cover

Table 3. Photosynthetic efficiency (assimilation per unit irradiance) of Watts Lake phytoplankton over the period 27 March 1981 to 9 January 1982 (mgC mgChl a^{-1} h^{-1} μE m^{-2} s^{-1}). Irradiances were averages at each depth over the period of incubation.

Depth (m)	27.III 1981	25.IV 1981	30.V 1981	11.VII 1981	18.VIII 1981	11.IX 1981	23.X 1981	14.XI 1981	20.XII 1981	9.I 1982
0	0.004									
1			4.76							
2	0.12	0.18		0	0.05	0.04	0.004	0.001	0.001	0.002
3		0.15								
4	0.03	0.12	1.58	0	0.10	0.10				
5								0.004	0.004	
6			3.29	0	0.09	0.15	0.02			0.006
7	0.08									
8							0.03	0.008	0.01	
9										
10										0.01
11										
12										
13										
14										
15					1.32	1.49				
16	0.58									
17										
18										
19										
20							0.27	0.17	0.15	0.02
21										
22										
23										
24										
25										0.06

Table 4. Carbon fixation of Watts Lake benthic algal mat over the period 19 August 1981 to 16 February 1982. Fixation expressed in terms of (A) assimilation number (μgC mgChl a^{-1} h^{-1}) and (E) photosynthetic efficiency (μgC mgChl a^{-1} h^{-1} μE m^{-2} s^{-1}).

Date	19.VIII.81		12.IX.81		24.X.81		15.XI.81		21.XI.81		21.XII.81		10.I.82		16.II.82	
Depth (m)	A	E	A	E	A	E	A	E	A	E	A	E	A	E	A	E
3.0	40	NL[a]														
3.5																
4.0																
4.5			10	0.14	280	3.75	50	0.40	170	1.13	250	1.43				
5.0	2	NL[a]											20	0.11	200	2.68
5.5					20	0.86	130	1.57	160	1.71	230	2.26	30	0.17		
6.0			1	0.01												
12.0															90	5.87

[a] No light profile.

Table 5. Changes in chlorophyll *a* content of Watts Lake benthic algal mat between 24 October and 21 November 1981 (μg Chl *a* per equivalent size mat sample). The ratio of carotenoids : chlorophyll *a*, also shown, removes variation due to different mat sample size.

Depth (m)	24.X.81		15.XI.81		21.XI.81	
	chl *a*	*carot.* chl *a*	chl *a*	*carot.* chl *a*	chl *a*	*carot.* chl *a*
4	102.3	0.45	94.5	0.53	36.7	0.51
6	66.4	1.00	82.3	0.77	103.3	0.47

played in determining euphotic depth. Day-length (7 h 46 min) and noon incident light intensities were similar on both occasions (187 and 157 $\mu E\ m^{-2}\ s^{-1}$ for April and August respectively). A blizzard in April deposited deep snow banks on the ice. By August, ablation and wind scour had removed much of this, producing large areas of ice free from any snow cover. Only 11% of incident light penetrated below the ice in April, compared with 23% in August. Euphotic depths (1% PAR) correspondingly were, 5.5 m in April and 27.5 m in August. While light attenuation by the denser phytoplankton crop in April (0.56 mgChl *a* m^{-3}), would have reduced the euphotic depth to a greater extent than in August (0.12 mgChl *a* m^{-3}), the albedo of the snow markedly reduced the amount of light entering the water column, more directly affecting euphotic depth.

No change in water column temperature was observed during the spring peak in limnetic production. Water temperature at that time was 1 °C, increasing only 1 °C from August to December. Summer warming (January 1981, 8 °C; January 1982, 4.5 °C) was not reflected in increased production in the water column (Table 1). The development of a moat around the lake margin in summer indicated warming in the shallows, in the vicinity of the benthic mat.

Discussion

The extreme seasonal light regimes in high latitudes elicit distinct production strategies in the algae of the region (Fogg & Horne, 1970; Goldman, 1970). Within Watts Lake, the phytoplankton and benthic algae responded in different ways to these conditions (Fig. 3). Carbon fixation rates in the phytoplankton varied from no detectable uptake in midwinter darkness, to spring and autumn peaks. Maximum benthic algal photosynthesis occurred in summer, which coincided with maximum light availability, increased water temperature and the insolation effect of the mat itself, which was demonstrated by Fogg & Horne (1970).

Although the assessment of benthic production

Table 6. Comparison of areal primary production and selected physical characteristics of Watts Lake from March 1981 to January 1982.

Month	Primary production mg C $m^{-2}\ d^{-1}$	Snow cm	Ice cm	Under ice light intensity $\mu E\ m^{-2}\ s^{-1}$	1% PAR Depth m
III.81	53	5	50	34	15.6
IV.81	8	50	75	21	5.5
V.81	9	6	95	2.7	5.6
VII.81	0	1	110	2.7	19.0
VIII.81	6	0	135	36	27.5
IX.81	340	0	145	102	27.5
X.81	166	4	150	107	18.7
XI.81	171	2	150	337	19.1
XII.81	91	0	145	153	20.0
I.82	151	0	140	84	23.4

Table 7. Comparison of selected physical and biological parameters in two antarctic lakes.

	Watts L.[a]	Heywood L.[b]
Location	68°36′S, 78°13′E	60°43′S 45°38′W
PO_4-P (mg m^{-3})	1 – 20[c]	10 – 60
NO_3-N (mg m^{-3})	10[c]	ND[d]
Chl *a* (mg m^{-3})	0.1 – 10	2 – 38
P. max.[e] (mgC m^{-2} d^{-1})	340	4000
Max. assimilation no. (mgC mgChl *a* h^{-1})	17.3	10.5

[a] This study; [b] Light, 1977; Light *et al.*, 1981; [c] Unpublished ANARE results; [d] Not determined; [e] Maximum photosynthetic rate.

was difficult and could only be considered an approximation, some comparison of the contribution to total lake production from both components was possible. On the directly comparable basis of assimilation rates, the phytoplankton were found to be approximately thirty times more productive than the benthic mat at maximum rates (Tables 2 & 4). Integrating the discrete results throughout the study, over the euphotic zone and the entire mat surface, produced an estimated annual net production of 10.1 gC m^{-2} a^{-1} by the phytoplankton and 5.5 gC m^{-2} a^{-1} by the benthic algal mat.

The phytoplankton in Watts Lake may have shown light saturated photosynthesis, or photosynthetic inhibition by high light intensity in summer. This was indirectly indicated, first, by the fact that volumetric and areal production peaked in spring and then declined in summer (Tables 1 & 6). Second, that maxima of volumetric production and assimilation number were often found some metres below the depth of maximum light availability (i.e. the ice/water interface; Tables 1 & 2). Third, by the downward shift of maximum volumetric production in summer (Fig. 2). Laboratory tests were not conducted to determine photosynthesis versus light intensity curves for Watts Lake phytoplankton. Nevertheless, the lack of increased production in response to warm summer water temperatures, suggested that the algae may have been inhibited by the high light intensities.

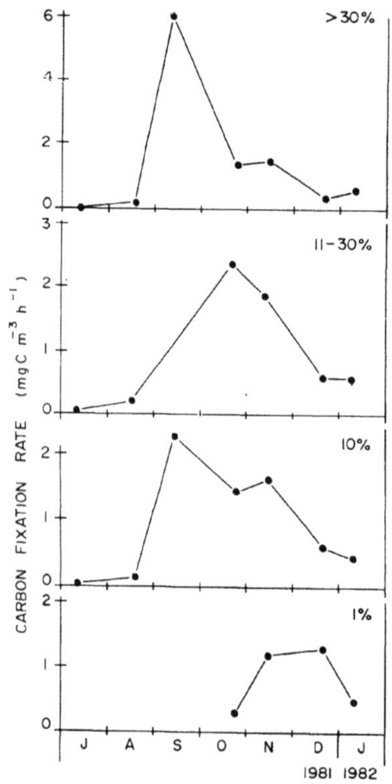

Fig. 2. Carbon fixation rates (mgC m^{-3} h^{-1}) of Watts Lake phytoplankton from midwinter darkness to summer continuous light. Rates were grouped according to percentage of surface light intensity shown on the figures. Values are either individual results or means at similar intensities.

Increased photosynthetic efficiency at lower light intensity in winter and at depth (Table 3), suggest that the phytoplankton were adapted to low light intensity. This would have enabled saturation of photosynthesis to occur at lower intensities than for the benthic mat, where self-shading was a significant factor. Self-shading by the phytoplankton would appear unlikely at the low biomass levels observed (0.1–10 mgChl *a* m^{-3}). Similar low light adaptation has been demonstrated in the phytoplankton of Lake Fryxell (Vincent, 1981) and Lake Vanda (Vincent & Vincent, 1982).

The results reported here, while in general agreement with the annual trend in production reported from Heywood Lake on Signy Island (Light, 1977; Light *et al.*, 1981), differed in some respects, due

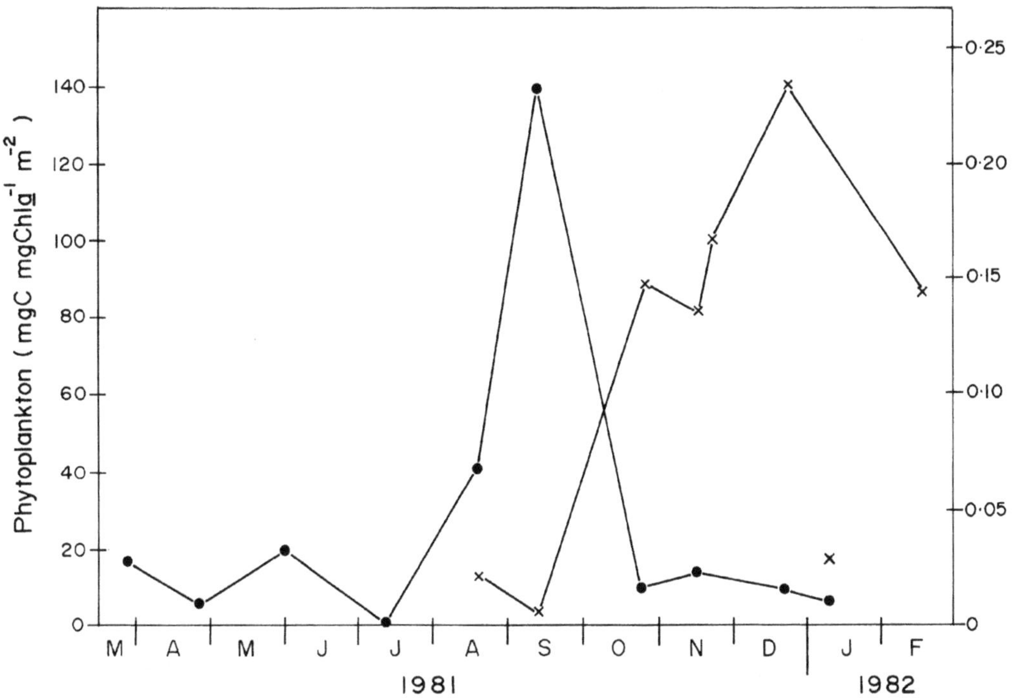

Fig. 3. Comparison of primary productivity of phytoplankton (●) and the benthic algal mat (×) in Watts lake on the basis of assimilation number (carbon fixation:chlorophyll). Values plotted are integrated area under the curve for phytoplankton (mgC mgChl a^{-1} m^{-2}) and means for benthic mat (mgC mgChl a^{-1} h^{-1}). Mean values for phytoplankton yield a plot of similar form with a peak of 9.4 mgC mgChl a h^{-1}.

most likely to the unusual nature and eutrophic status of Heywood Lake (Light *et al.*, 1981). The maximum areal photosynthetic rate measured in Watts Lake was considerably less than in Heywood Lake (Table 7), and occurred in spring with a lesser peak in autumn. Summer production declined to a level similar to that measured in February 1975 (Campbell, 1978). Heywood Lake phytoplankton production, however, increased during spring to a summer maximum (Light, 1977; Light *et al.*, 1981).

Maximum assimilation numbers measured in Watts Lake exceeded those measured by Light *et al.* (1981) in Heywood Lake (Table 7), although both were high for reported freshwater systems (Talling, 1975). The spring maximum in Watts Lake contrasted with a summer peak in Heywood Lake at the onset of open water. As ice remained on Watts Lake over the 1981 and 1982 summers, the effect of open water could not be studied. However, the maximum assimilation rate of 0.11 mgC mgChl a^{-1} h^{-1} measured in permanently ice covered Lake Fryxell (Vincent, 1981), was similar to the assimilation rates measured in Watts Lake in summer (Table 2). Thus, ice cover may have played a role in determining assimilation rate, in a fashion similar to that whereby snow cover regulated underwater light levels. The Watts Lake study and those of Light (1977), Light *et al.* (1981) and Vincent (1981), all point to the importance of snow cover in determining levels of primary production.

Photosynthetic efficiency increased with depth in Watts Lake (Table 3), Lake Fryxell (Vincent, 1981) and Lake Vanda (Vincent & Vincent, 1982), suggestive of low-light adapted algae. Seasonal comparison was difficult as light attenuation by the 4.5 m thick ice sheet on Lake Fryxell reduced light penetration to levels only encountered at or below 20 m in Watts Lake at a similar time. Nevertheless, at comparable light intensities, Watts Lake phytoplankton were nine times more efficient than those in Lake Fryxell (170 and 18.5 mgC mgChl a^{-1} unit light intensity^{-1}, respectively). Watts Lake phytoplankton were also three to five times more photosynthetically efficient than Lake Vanda algae at similar light intensities (Vincent & Vincent, 1982).

Lake morphometry played a major role in determining relative contribution to total production by the phytoplankton and benthos in the few studies of comparative production attempted in antarctic lakes. Tominaga (1977) measured greater production in the benthos than the phytoplankton in shallow lakes near Syowa station. Similarly, Fogg & Horne (1970) measured production by the *Phormidium* felt in Heywood lake at a rate 25 times greater than that of the phytoplankton, however, Light *et al.* (1981) measured phytoplankton production two orders of magnitude greater than Fogg & Horne (1970) in the same lake.

These studies were based on thin algal felts, growing as a carpet over the mud surface of shallow lakes. Kairesalo (1980) found that production was confined to the surface layer of the epiphytic and epipelic algae in a similar felt growing in a shallow Finnish lake, and that the upper layer shaded lower strata. On a total biomass basis, planktonic production was three times that of the epiphytes and twenty times that of the epipelon (Kairesalo, 1980).

In contrast to these lakes, the depth of Watts Lake enabled the development of a significant phytoplankton component. The benthic algal layer, as distinct from a thin felt, was more restricted in its extent and prone to self shading. Thus, production by the phytoplankton exceeded benthic production in Watts Lake.

Comparison of primary production in Watts Lake with studies reported elsewhere (McLaren, 1969; Goldman, 1970; Kalff, 1970; Light, 1977; Priddle, 1980; Light *et al.*, 1981; Vincent, 1981; Vincent & Vincent, 1982) placed Watts Lake amongst the most productive of high latitude lakes. Phytoplankton production exceeded that of other oligotrophic and several mesotrophic and eutrophic lakes (Light *et al.*, 1981: Table 2). Benthic algal mat production was similar to that measured in cyanophyte felts in three lakes on Signy Island (3–9 gC m^{-2} a^{-1}) and was similar to other rates reported in that study (Priddle, 1980).

Year-round measurements of production need to be undertaken in other antarctic lakes to investigate the extent to which the Watts Lake spring peak in phytoplankton production occurs elsewhere. Vincent (1981) suggests the most probable period of maximum metabolism in antarctic lakes is early spring when nutrients are plentiful and ambient irradiance has risen above compensation point. This appears to be the case in Watts Lake. If so, previously reported studies of antarctic lake primary productivity (conducted in summer) may have underestimated production.

Acknowledgements

I thank the Antarctic Division of the Australian Department of Science for financial and logistic support.

I also thank the personnel of the Australian National Antarctic Research Expedition (1981) at Davis Station for their assistance. Secondment to the Department of Science, from the Metropolitan Water, Sewerage and Drainage Board, Sydney, Australia, to undertake the study is gratefully acknowledged.

References

Campbell, P. J., 1978. Primary productivity of a hypersaline antarctic lake. Aust. J. Mar. Freshwat. Res. 29: 717–724.

Fogg, G. E. & A. J. Horne, 1970. The physiology of antarctic freshwater algae. In M.W. Holdgate (ed.), Antarctic Ecology, 2. Academic Press, Lond.: 632–638.

Goldman, C. R., 1964. Primary productivity studies in antarctic lakes. In R. Carrick, M. Holdgate & J. Prevost, (eds), Biologie Antarctique. Hermann, Paris: 291–299.

Goldman, C. R., 1970. Antarctic freshwater ecosystems. In M. W. Holdgate (ed.), Antarctic Ecology, 2. Academic Press, Lond.: 609–627.

Goldman, C. R., D. T. Mason & B. J. B. Wood, 1963. Light injury and inhibition in antarctic freshwater phytoplankton. Limnol. Oceanogr. 3: 313–322.

Goldman, C. R., D. T. Mason & B. J. B. Wood, 1972. Comparative study of the limnology of two small lakes on Ross Island, Antarctica. In G. A. Llano (ed.), Antarctic Terrestrial Biology. Antarct. Res. Ser. 20, Wash., D.C.: 1–50.

Golterman, H. L., 1969. Methods for Chemical Analysis of Fresh Waters. I.B.P., 8. Blackwell, Oxford, 166 pp.

Heath, C. W. & W. J. Singleton, this volume. A device for remote sampling of benthic algae under ice. Hydrobiologia.

Heath, C. W., R. Sidebottom & W. J. Singleton, this volume. A device for reaming holes in ice. Hydrobiologia.

Journal of Glaciology, 1958. Instruments and methods: Ice drills and corers. J. Glaciol. 3: 30.

Kairesalo, T., 1980. Comparison of *in situ* photosynthetic activity of epiphytic, epipelic and planktonic algal communities in an oligotrophic lake, Southern Finland. J. Phycol. 16: 57–62.

Kalff, J., 1970. Arctic lake ecosystems. In M. W. Holdgate (ed.), Antarctic Ecology, 2. Academic Press, Lond.: 651–661.

Koob, D. D. & G. L. Leister, 1972. Primary productivity and associated physical, chemical and biological characteristics of Lake Bonney: A perennially ice-covered lake in Antarctica. In G. A. Llano (ed.), Antarctic Terrestrial Biology. Antarct. Res. Ser. 20, Wash., D.C.: 51–68.

Light, J. J., 1977. Production and periodicity of antarctic freshwater phytoplankton. In G. A. Llano (ed.), Adaptations within Antarctic Ecosystems. Smithsonian Institution, Washington, D.C.: 829–837.

Light, J. J., J. C. Ellis-Evans & J. Priddle, 1981. Phytoplankton ecology in an antarctic lake. Freshwat. Biol. 11: 11–26.

McLaren, I. A., 1969. Primary production and nutrients in Ogac Lake, a landlocked fjord on Baffin Island. J. Fish. Res. Bd. Can. 26: 1561–1576.

Parker, B. C., R. C. Hoehn, R. A. Paterson, J. A. Craft, L. S. Lane, R. W. Stavros, H. G. Sugg, Jr., J. T. Whitehurst, R. D. Fortner & B. L. Weand, 1977. Changes in dissolved organic matter, photosynthetic production, and microbial community composition in Lake Bonney. Southern Victoria Land, Antarctica. In G. A. Llano (ed.), Adaptations within Antarctic Ecosystems. Smithsonian Institution, Wash. D.C.: 873–889.

Parsons, T. R. & J. D. H. Strickland, 1963. Discussion of spectrophotometric determination of marine plant pigments, with revised equations for ascertaining chlorophylls and carotenoids. J. Mar. Res. 21: 155–163.

Priddle, J., 1980. The production ecology of benthic plants in some antarctic lakes. I. *In situ* production studies. J. Ecol. 68: 141–153.

Samsel, G. L. & B. C. Parker, 1971. Comparison of two antarctic lakes with different trophic states. Va. J. Sci. 22: 177–182.

Samsel, G. L. & B. C. Parker, 1972. Limnological investigations in the area of Anvers Island, Antarctica. Hydrobiologia 40: 505–511.

Strickland, J. D. H. & T. R. Parsons, 1972. A practical handbook of seawater analysis, 2nd ed. Bull. Fish. Res. Bd Can., 310 pp.

Talling, J. F., 1975. Primary production of aquatic plants – conclusions. In J. P. Cooper (ed.), Photosynthesis and Productivity in Different Environments. I.B.P., 3. Cambridge University Press, Cambridge: 281–294.

Tominaga, H., 1977. Photosynthetic nature and primary productivity of antarctic freshwater phytoplankton. Jap. J. Limnol. 38: 122–130.

Vincent, W. F., 1981. Production strategies in antarctic inland waters: Phytoplankton eco-physiology in a permanently ice-covered lake. Ecology 62: 1215–1224.

Vincent, W. F. & C. L. Vincent, 1982. Factors controlling phytoplankton production in Lake Vanda (77°S). Can. J. Fish. Aquat. Sci. 39: 1602–1609.

Vollenweider, R. A., 1974. A manual on methods for measuring primary production in aquatic environments 2nd ed., I.B.P., 12. Blackwell, Oxford, 225 pp.

Vollenweider, R. A. & A. Nauwerck, 1961. Some observations on the C-14 method for measuring primary production. Verh. int. Ver. Limnol. 14: 134–139.

Hydrobiologia 165: 89–96 (1988)
J. M. Ferris, H. R. Burton, G. W. Johnstone & I. A. E. Bayly (eds) Biology of the Vestfold Hills, Antarctica
Kluwer Academic Publishers

Comments on the antarctic Rotifera

M. Sudzuki
Biological Laboratory, Nihon Daigaku University, Omiya-Shi, Saitama-Ken, Japan 330

Key words: Rotifera, Antarctica, taxonomy, free-swimming

Abstract

This paper reconsiders the taxonomy and distributional ecology of the antarctic free-swimming Rotifera with special reference to *Notholca* and *Keratella*. *Notholca verae* is a valid species separate from the *squamula* group. *N. salina* may be different from *N. squamula* in that the former has a smooth pectoral margin and the shortest occipital spines located at the lateral region. *N.* sp. nov. of Dartnall (1983) is included within those of *N. marina*. A species of *Keratella* from Kerguelen, identified as *K. sancta*, is clearly different in the dorsal sculpturing on the lorica from the original description by Russell (1944). The occurrence of *Filinia* species with large numbers of nuclei in the vitellarium and relatively long bristles is noteworthy.

Introduction

This symposium has provided the opportunity to consider problems related to the taxonomy and distributional ecology of antarctic Rotifera on the basis of older material collected by both Australian and Japanese workers, and new material from near lakes Vanda, Bonny and Fryxell collected by H. Yusa and H. Nakaya. In doing so, I have reviewed most of the relevant literature. I have restricted consideration to some freely swimming rotifers only.

Results and discussion

Ecological characteristics of antarctic Rotifera

All species recorded from Antarctica are listed in Table 1. Three general points emerge from this table:

1. Some rotifers live among the littoral vegetation of lakes or ponds.
2. Some rotifers tend to dwell in algal felts or moss carpets near lakes or ponds if there is sufficient water to support them.
3. Some rotifers live in the interstitial water of mosses, lichens, or occasionally of soils and sand.

Most rotifers found in aquatic habitats of the Antarctic belong to 'the pseudo-periphyton' in the sense of Sladecekova & Sladecek (1977) or to the bottom creeper category. Truly planktonic rotifers living in the open water are very rare. This is perhaps due to the severe antarctic climate. The occurrence of such free-swimming rotifers as *Keratella, Notholca* and *Filinia,* and of the seldom studied *Rhinoglena*, is thus noteworthy.

The genus Notholca

Everitt (1981) mentioned that *N. verae* was observed occasionally from 21 January 1981 in Deep Lake Tarn. It has been observed in abundance in deeper lakes in the Vestfold Hills (D. R. Grace, H. R. Burton, pers. commun.). These are the only records since the species was discovered in the Bunger Hills by Kutikova (1958a). The following species have also been reported from the Antarctic or subantarctic: *N. squamula* by Dartnall (1977), *N. salina* and *N. walterkostei* by De Paggi (1982), *N.* sp. nov. Dartnall

Table 1. List of Rotifera reported from the Antarctic. Rih. = Richters (1908); Mur = Murray (1910); Kut = Kutikova (1958a, 1958b); Rus = Russell (1959); Sud = Sudzuki (1979; & Shimoizumi, 1967); Don = Donner (1972); Dat = Dartnall (1977, 1983); Evr = Everitt (1981); Pag = De Paggi (1982); Lai & Kos = Lair & Koste (1984). F = Freshwater; M = Mixohaline. I = Interstitial; PP = Pseudo-periphytic; C = Creeping. Brackets indicate a taxon found nearby but outside continental Antarctica.

Order Ploimida

Family Brachionidae

Keratella americana Pag, F, PP; *K. cochlearis* Pag, F, PP; *K. sancta* (Russ, M, PP): (Lai & Kos, M, PP); *K.* sp. Pag, F, PP; *Notholca jugosa* (Lai & Kos, M, PP); *N. salina* Pag, F, PP; *N. squamula* (Dat, F, PP); *N. verae* Kut, F, PP: Evr. F, PP; *N. walterkostei* Pag, F, PP; *N.* sp. (Dat, F, PP)

Family Coluridae

Lepadella acuminata (Rus, F, PP, M, PP); *L. elliptica* (Rus, F, PP); *L. patella* (Rus, F, PP, M, PP): Kut, F, PP: (Dat, F, PP): Pag, F, PP; *L. p. matsudai* Sud, I, C; *Colurella adriatica* (Rus, M, PP); *C. colurus* (Rus, F, PP)

Family Lecanidae

Lecane closterocerca (Dat, F, PP); *L. lunaris* Pag, F, PP: (Dat, F, PP); *L. mawsoni* (Rus, F, PP, M, PP); *Proales reinhardti* Kut, F, PP

Family Epiphanidae

Rhinoglena fertoensis Kut, F, PP; *Lophocharis oxysternon* (Rus, F, PP, M, PP); *Euchlanis*? Rih, F, PP; *Epiphanes senta* (Mur, F, PP): Kut, F, PP, M, PP: Don, F, PP: Pag, F, PP: (Dat, F, PP)

Family Notommatidae

Scaridium bostjani (Dat, F, PP); *S. longicaudum* (Dat, F, PP); *Eosphora najas* (Dat, F, PP); *Resticula gelida* (Dat, F, PP); *Notommata cyrtopus* (Rus, M, PP); *Cephalodella catellina* (Dat, F, PP): (Rus, M, PP); *C. gibba* Pag, F, PP: (Dat, F, PP); *C. sterea* Kut, F, PP; *C. tenuior* ♀, ♂ (Mur, F, PP); *C.* sp. (Rus, F, PP)

Family Trichocercidae

Diurella bidens (Rus, F, PP); *D. tigris* (Dat, F, PP)

Family Dicranophoridae

Encentrum antarctica Sud, I, C; *E. bryocolum* Sud, I, C; *Dicranophorus uncinatus* Pag, F, PP; *Pleurotrocha* sp. (Mur, F, PP)

Order Gnesiotrocha

Family Testudinellidae

F. terminalis = *(Filinia major)* (Rus, M, PP); *F. terminalis kergueleniensis* (Lai & Kos, M, PP)

Family Flosculariidae

Ptygura melicerta (Dat, F, PP); *P.* sp. Evr, F, PP

Family Collothecidae

Collotheca ornata cornuta Kut, F, PP: (Dat, F, PP); *C.* sp. (Mur, F, PP)

Order Bdelloidea

Family Habrotrochidae

Habrotrocha angularis (Mur, F, PP); *H. angusticollis* (Rih, I, C); *H. constricta* (Mur, F, PP): (Don, F, PP): Evr, I, C; *H. elusa* (Don, F, PP); *H. glosa* Sud, I, C; *H. tridens* (Mur, I, C); *H.* spp. Kut, F, PP: Sud, I, C

Family Philodinidae

Macrotrachela habita (Mur, F, PP); *M. insolita* (Don, I, C): Sud, I, C; *M. quadricornifera* Evr, F, PP; *M. nixa* Sud, I, C; *M. tridens* (Mur, F, PP); *M.* sp. Sud, I, C; *Rotaria longirostris* (Rih, F, PP); *R.* sp. Sud I, C; *Philodina alata* (Mur, F, PP): Kut, F, PP; *Ph. antarctica* (Mur, F, PP); *Ph. gregarina* (Mur, F, PP, M, PP): (Don, F, PP): Evr, F, PP: (Dat, F, PP); *Ph.* sp. (Mur, F, PP): Kut, F, PP, M, PP: Evr, F, PP: Sud, I, C; *Mniobia russeola* Evr, F, PP; *M.* spp. Kut, F, PP: Sud, I, C

Family Adinetidae

Adineta barbata? (Mur, F, PP); *A. grandis* (Mur, F, PP, M, PP): (Don, F, PP): (Dat, F, PP, M, PP); *A. longicornis*? (Mur, F, PP): (Rih, I, C); *A. vaga* (Mur, F, PP); *A.* sp. Sud, I, C; *A. gracilis* (Mur, F, PP): (Don, F, PP): Sud, I, C

(1983), and *N.* cf. *jugosa* by Lair & Koste (1984).

Comparisons of these antarctic species with others of the *labis-marina* and *salina-squamula* groups in the genus *Notholca* are made in Tables 2 & 3 and they may be dealt with as follows:

a) *N. squamula* (after various authors; note that in most papers on *Notholca* the trophi are not shown). From Table 2 the following may be noted for the squamula group: there are only two good criteria for classification, viz. the features of the pectoral margin and the position of the highest part (G in Table 2; +, posterior; −, anterior) although the latter is rarely observed by authors. This is because all species but one (*verae*) have an oval shaped lorica (A in Table 2) and the ratio of lorica length/width (B in Table 2) varies between 1.2−1.6 (cf. Pejler,

Table 2. The genus *Notholca:* comparison of the antarctic species *verae, salina, squamula* and *jugosa,* with others of the *salina-squamula* group.

Characters[a]	A	B	C	D	E	F	G
	verae Kutikova						
Kutikova (1958a, b)	cylind.	2.1 − 2.5	sm.	wav.	1-3-2 1-2-2	0.5	+
	pacifica Russell						
Russell (1962)	oval	1.2	?	st.	1-1-2	post.	+
	salina (Focke)						
Focke (1961)	oval	1.2			1-2-3	post.	
Björklund (1972)	oval	1.3 − 1.4	sm.		1-2-3 1-2-2	0.5	
De Paggi (1982)	oval	1.3 − 1.5	sm.	st.	1-2-3 1-3-2	post.	−
	laurentiae Stemberger						
Stemberger (1976)	oval	1.3 − 1.5	und.	st.	1-2-3	0.5	−
	kostei Chengalath						
Chengalath (1978)	oval	1.2 − 1.4	und., ptd.	st.	1-2-2 1-2-3	0.5	+
	cristata Grese						
Grese (1955) (cited in Kutikova, 1970)	oval	1.3		st.	1-2-3	0.5	
	squamula (O.F. Muller)						
Carlin (1943)	oval	1.3	?		1-3-2 1-2-2	0.5	
Gillard (1948)	oval	1.3	?		1-3-2	0.5	
Hauer (1952)	oval	1.3	ptd.		1-2-2 1-2-1	0.5	
Althaus (1957)	oval	1.2 − 1.3	sm., und.		1-2-1	post.	
Pejler (1957)	oval	1.2 − 1.4			1-3-2 1-2-1 2-3-1	0.5	+
Klement (1960)	oval	1.3	und.	st.	1-3-2	post.	
Koste & Wulfert (1964)	oval	1.4	und.	st.	1-3-2	post.	
Kutikova (1970)	oval	1.2 − 1.3	und.	st.	1-3-2 1-2-1	post.	
Stemberger (1976)	oval	1.5	und.	st.	1-2-2 2-3-1	post.	
Pejler (1977)	oval	1.3 − 1.4	und.		1-3-2 1-2-2	0.5	
	mulleri Focke						

Table 2. (Continued).

Characters[a]	A	B	C	D	E	F	G
Focke (1961)	oval	1.3 – 1.6			1-2-1 1-3-2		
	jugosa Hudson & Gosse						
Hudson & Gosse (1889)	oval	1.5	?	st.	2-3-1	post.	
					1-2-1		
Kutikova (1970)	oval	1.5 – 1.6	und.	st.	1-3-2	0.5	+
Lair & Koste (1984)	oval	1.6	und.	st.	1-3-2	0.5	±
	frigida Jaschnov						
Carlin (1943)	oval	1.2	und.		1-3-2	0.5	
Kutikova (1970)	oval	1.3	und.		1-3-2	0.5	
Pejler (1977)	oval	1.3 – 1.4	und.	st.	1-3-2 1-2-3	0.5	
	michiganensis Stemberger						
Stemberger (1976)	oval	1.2 – 1.3	und.	st.	1-3-2	0.5	±

[a] See text for explanation.

1957; Focke, 1961), except for *verae* (2.1 – 2.5). Further, the widest part of the lorica (F in Table 2) is at about half the body length (0.5 in Table 2), or slightly posteriorly (post. in Table 2), and the striation on the dorsal surface of the lorica (D in Table 2) is straight (st. in Table 2), except for *verae* where it is wavy (wav. in Table 2). Particularly important is the fact that the order of length in the occipital spines (E in Table 2) is variable. For example, Pejler (1957: Figs. 59 – 64) shows this feature as follows: median spines longest, intermedian spines shortest, and lateral spines second longest (conveniently abbreviated to the formula 1-3-2); or median and lateral spines the longest, intermedians shortest (1-2-1); and median spines second longest, intermedian spines shortest, lateral spines longest (2-3-1) in the type for *squamula* among the Swedish samples. But in this case it is clear that the intermedian spines are always the shortest. For *salina,* Björklund (1972: Figs. H & I) shows 1-2-3 and 1-2-2. So here the lateral spines are always shortest. While De Paggi (1982: Fig. 5, B & C) shows 1-2-3 and 1-3-2, indicating that the laterals or intermedians are shortest. For *kostei,* Chengalath (1978: Figs. 1 & 2) shows 1-2-2 and 1-2-3, i.e. the laterals are shortest (as *salina* of Björklund, above). In summary, the order of length in the occipital spines is not an adequate taxonomic criterion (if the frequency of the order is not examined) except for one case, viz. 1-1-2 in *pacifica* (Russell, 1962: Figs. 3 – 5) where medians and intermedians are equal in length and the order is unique.

b) *N. verae* is a good species lying between the *salina* and *marina* groups. But, if some individuals in a population of *verae* possessed a posterior extension they would be indistinguishable from the *marina* group (*walterkostei, marina,* etc., see Table 3) without close examination of other characters. This situation arises from the work of Björklund (1972: Fig. 2, G – X) who reported the existence of specimens without any posterior extensions even in *marina.*

c) *N. salina,* as described by De Paggi (1982), is identical with *N. salina* of Björklund (1972: Fig. 1, H & I). Further, it closely resembles *N. squamula* of Klement (1960: Fig. 14) in the general shape of the lorica. In my opinion, *N. salina* is different from *N. squamula* in the following points:

i) The pectoral margin of *salina* (C in Table 2) is usually smooth (sm. in Table 2), but that of *squamula* is ribbed. The margin can also be described as undulated (und.) or pointed (ptd.).
ii) The order of length in the occipital spines of *salina* is 1-2-3 (Focke, 1961; Björklund, 1972; De Paggi, 1982); 1-2-2 (Björklund, 1972) or 1-3-2 (De Paggi, 1982); in these cases the median spines are always the longest and the lateral spines the short-

est (or equal to the intermedians), except for De Paggi's (1982) single example, i.e. 1-3-2 (see *N. squamula*).

In *squamula*, however, the condition is 1-3-2 (Carlin, 1943; Gillard, 1948; Pejler, 1957; Klement, 1960; Koste & Wulfert, 1964; Kutikova, 1970; Pejler, 1977), or 1-2-1 (Hauer, 1952; Althaus, 1957; Pejler, 1957; Kutikova, 1970) or occasionally 2-3-1 (Pejler, 1957; Stemberger, 1976), or even 1-2-2 (Carlin, 1943; Hauer, 1952; Stemberger, 1976; Pejler, 1977). Thus the occipital spine formula 1-2-3 dominates in *salina*, while the 1-3-2 condition is dominant and the 1-2-3 condition never occurs in *squamula*. Consequently there are morphological grounds for regarding *N. salina* as a subspecies of *squamula*.

d) *N.* cf. *jugosa* of Lair & Koste (1984: Fig. 4) is identical with *jugosa* of Kutikova (1970: Fig. 1009). However, both nomina differ from the specimens of Hudson & Gosse (1889 for Gosse, 1887: Fig. 59) in the order of length in the occipital spines (E in Table 2), namely 2-3-1 (or 1-2-1) in Hudson & Gosse (1889) but 1-3-2 in both Kutikova (1970), and Lair & Koste (1984). In other words, both nomina have a common character only in the position of the shortest spines, but differ from each other in the position of both the longest spines and second longest spines. For this reason, the specimens reported from the subantarctic could be treated as *squamula* together with that of Kutikova (1970).

Concerning Table 3, it should be noted that characters for *walterkostei* and sp. nov. of Dartnall (1983)

Table 3. The genus *Notholca:* comparison of the Antarctic species, *walterkostei* and sp. nov. of Dartnall, with others of the *labismarina* group.

Characters[a]	I	II	III	IV	V	VI	VII	VIII
	walterkostei De Paggi							
De Paggi (1982)	el.-oval cylind.	1.9 – 2.2 (2.0)	10.8 – 14.8	27.8 – 37.4 (30.5)	0.7 – 0.9 (0.8)	und.	1-3-2	+
	new species Dartnall							
Dartnall (1983)	oval	1.8?	17.3	38.1	0.8	?	1-2-3 1-3-2	±
	marina Focke							
Björklund (1972)	cylind.	2.0 – 2.4	0.0 – 28.0	0.0 – 16.9	0.0 – 5.8	und.	1-3-2 1-2-1 1-2-2	±
Donner (1978)	cylind.	1.9 – 2.2	6.5 – 18.4	6.6 – 16.4	0.5 – 3.1		1-1-1 1-3-2	±
	latistyla (Oloffson)							
Oloffson (1918) (cited in Voigt, 1956)	el.-oval	1.7	12.3	5.8	3.6	und.?	2-3-1 2-2-1	±
	complexa Kutikova							
Kutikova (1970)	el.-oval	1.6 – 2.0	12.7 – 43.3	7.1 – 26.8	1.2 – 10.8	und.	1-3-2	±
	labis Gosse							
Gosse (1887) cited in Hudson & Gosse, 1889)	oval	1.4	19.3	17.4	1.5	?	1-3-2	+
Hudson & Gosse (1889)	oval	1.4	15.6	13.6	1.5	?	1-3-2	
Dief.-Sach. (1911) (cited in Pejler, 1957)	oval	1.5	12.9	14.5	1.4	?	1-3-2	+
Daday (1913) (cited in Pejler, 1957)	oval	1.3	21.8	18.9	1.5	?	1-3-2	+
Jaschnov (1922) (cited in Pejler, 1957)	oval	1.4	14.0	8.5	2.3	?	1-3-2 1-2-3	+

Table 3. (Continued).

Characters[a]	I	II	III	IV	V	VI	VII	VIII
Carlin (1943)	el.-oval	1.8	11.3	11.2	1.8	?	1-3-2	±
Gillard (1948)	el.-oval	1.8	11.7	11.5	1.8	?	1-3-2	+
Pejler (1957)	oval el.-oval	1.3–1.6	12.9–37.7	6.2–10.5	1.6–6.1	?	1-3-2 1-2-1	±
Althaus (1957)	oval el.-oval	1.5–1.6	18.3–18.9	18.2–22.0	1.2–1.7	und.	1-2-1	+
Kutikova (1959)	el.-oval	1.7	13.1	7.0	3.2	und.	1-2-1	
Sudzuki (1964)	oval	1.3–1.4	4.3–9.2	10.4–11.5	0.5–1.3	sm.	2-3-1 1-2-3	+
Kutikova (1970)	oval el.-oval	1.6–1.9	11.2–14.1	6.9–14.1	1.3–3.4	und.	1-2-1 1-1-1	+
Björklund (1972)	el.-oval	1.5–1.7	4.5–18.1	8.7–18.1	0.4–5.4	und.	1-3-2	+
Donner (1978)	oval	1.3–1.8	8.1–12.1	11.9–13.9	0.8–3.2	?	1-3-2	+
	species (?) Kutikova							
Kutikova (1970)	oval	1.4	13.0	11.6	1.6	und.	1-3-2	+
	lapponica Ruttner-Kolisko							
Ruttner-Kolisko (1966)	el.-oval	1.7	3.3	15.3	0.0–0.4	und.	1-1-2	+
Pejler (1977)	oval	1.4–1.5	10.4–12.5	11.0–17.1	0.9–1.1	und.	1-3-2 2-3-1	±
	limnetica (Levander)							
Levander (1901) (cited in Pejler, 1957)	el.-oval	1.7–1.8	23.3–24.3	15.8–16.5	2.5–2.6	ptd.	1-3-2	±
Levander (1905) (cited in Pejler, 1957)	el.-oval	1.6–1.7	24.8–40.0	5.6–7.6	9.2	ptd.	1-3-2	±
Hofsten (1923) (cited in Pejler, 1957)	oval	1.7	27.3	5.6	8.2	und.	1-3-2	±
Carlin (1943)	oval	1.4–1.5	13.2–24.9	8.5–13.2	1.3–3.0	?	1-3-2 2-3-1 1-2-1	+
Klement (1957)	cylind.	1.9	37.0	7.9	8.9	sm.	1-2-3	
Fadeev (1926) (cited in Pejler, 1957)	el.-oval	1.3	29.2	10.8	3.5	ptd.	2-3-1	
Kutikova (1970)	el.-oval cylind.	1.9	26.4	7.4	6.6	und.	2-3-1	±
	rectospina Kutikova							
Kutikova (1970)	el.-oval	1.6–1.7	31.1–36.6	6.0–7.7	7.0–10.0	und.	2-3-1	±

[a] See text for explanation.

are included within the variation of *marina*; e.g. the ratio of length/width of lorica (II in Table 3) is 1.9–2.2 (average 2.0 in *walterkostei,* 1.8 (?) in sp. nov. of Dartnall (1983), 2.0–2.4 in *marina* of Björklund (1972) and 1.9–2.2 in *marina* of Donner (1978). The percentage of extension length/lorica length (III in Table 3) 10.8–14.8 in *walterkostei,* 17.3 in sp. nov. of Dartnall (1983), 0.0–28.0 in *marina* of Björklund (1972), and 6.5–18.4 in *marina* of Donner (1978). The ratio of extension length/extension width (V in Table 3) is 0.8 (average) in *walterkostei,* 0.8 in sp. nov. of Dartnall (1983), 0.0–5.8 in *marina* of Björklund (1972), and 0.5 – 3.1 in *marina* of Donner (1978). The general shape of the lorica (I in Table 3) and the ornamentation on the lorica (VI in Table 3; undulated (und.); pointed (ptd.); smooth (sm.)) are not very useful characters. For *walterkostei* there are two significant points: the ratio of extension width/lorica width (IV in Table 3) and position of the widest part of the lorica (VIII in Table 3). Concerning sp. nov. of Dartnall (1983) there are also two significant characters: the % of extension width/lorica width (IV in Table 3), and the order of length of the occipital spines (VII in Table 3). I can say nothing about the affinity of sp. nov. of Dartnall (1983) with *walterkostei,* since the description of Dartnall's species is still in press.

N. marina differs from the *labis* group of the same genus in the general shape of the lorica (I in Table 3), the ratio length/width of the lorica (II in Table 3), the ratio extension length/body length (III in Table 3), and the ratio extension length/extension width (V in Table 3). All these features are related to the posterior extension of the lorica.

In general, *Notholca* is regarded as a cold stenothermal genus by various authors. I have found *N. labis* exclusively in winter, both in Tokyo and Urawa (Sudzuki, 1964, 1981). A lot of species have also been recorded as endemic to a single lake – Lake Baikal. The genus *Notholca* may be capable of diversifying under antarctic conditions.

Comments on antarctic species of Keratella

The antarctic species of the genus *Keratella* have hitherto been recorded only from the Potter Peninsula, South Shetland, by De Paggi (1982). They are *K. americana, K. cochlearis* and *K.* sp. belonging to the *valga* group. On the other hand, a subantarctic species, *K. sancta,* has been recorded by Russell (1959) and Lair & Koste (1984), both from the Kerguelen archipelago. Not a single specimen has been detected from aquatic habitats in the Vestfold Hills.

Of these *Keratella* species, only *sancta* is compared here with related species, for it is not certain whether the species from Kerguelen is really identical with that from New Zealand. However, it is safe to say that *K. sancta* is a valid species, even if this species has for a long time been overlooked by most of the rotifer workers. *K. sancta* has the following characteristics differentiating it from other related species:

1. The carinal plaques are big and there is one pair.
2. A posterio-carinal area is present.
3. There are two median plaques, of which the anterior is smaller than the posterior.
4. A posterior median area is clearly present.
5. The anterio-median plaques are quadrilateral.
6. There are four pairs of lateral plaques.
7. There are two pairs of marginal areas.
8. The lateral plaque is angular in shape and equal in size.

Of these characters, the specimen from Kerguelen has three points in common with *sancta* from New Zealand, viz. 1, 2 and 6, but lacks the remaining five features. For this reason I consider that the species from Kerguelen should be treated as a separate one from *sancta.* No species of *Keratella* has hitherto been treated as sufficiently polymorphic to encompass the many morphological differences recorded in this case.

Comments on Filinia *species from the subantarctic*

No *Filinia* species have been discovered from Antarctica but the genus has been recorded from the Kerguelen archipelago – *Filinia major* by Russell (1959), and *F. terminalis kergueleniensis* by Lair & Koste (1984: Fig. 1). Both belong to the *longiseta* group, the taxonomy of which seems to have become rather complicated by the studies of Pejler (1957) and Hofmann (1974) who separated *terminalis* from *longiseta* by biometric methods. Formerly, Sudzuki (1964: Fig. 79, pl. 27) used the number of vitellarium nuclei and relative length of the blade as taxonomic features. On this point it is noteworthy that *Filinia* from Kerguelen has a vitellarium with many nuclei like *mystacina* in Japan, and relatively long bristles like *limnetica* in Japan, in one and the same individual.

Acknowledgements

I would like to express my cordial thanks to the Organizing Committee of the Vestfold Hills Symposium for their kind invitation to the Antarctic Division, Hobart, in August 1984.

References

Althaus, B., 1957. Faunistisch-ökologische Studien an Rotatorien salzhaltiger Gewässer Mitteldeutschlands. Wiss. Z. Martin-Luther-Univ. Halle-Wittenb. 6: 17–157.

Björklund, B. G., 1972. Taxonomic and ecological studies of species of Notholca (Rotatoria) found in sea- and brackish-water, with description of a new species. Sarsia 51: 25–65.

Carlin, B., 1943. Die Planktonrotatorien des Motalaström. Zur Taxonomie und Ökologie der Planktonrotatorien. Meddn Lunds Univ. limnol. Instn 5: 1–256.

Chengalath, R., 1978. A new species of the genus Notholca Gosse, 1886 (Brachionidae: Rotifera) from great Slave Lake, N.W.T. Can. J. Zool. 56: 363–364.

Dartnall, H. J. G., 1977. Antarctic freshwater rotifers. Arch. Hydrobiol. Beih. Ergebn. Limnol. 8: 240–242.

Dartnall, H. J. G., 1983. Rotifers of the Antarctic and subantarctic. Hydrobiologia 104: 57–60.

De Paggi, S. B. J., 1982. Notholca walterkostei sp. nov. y otros rotiferos dulceacuicolas de la peninsula Potter, Isla 25 de Mayo (Shetland) del Sur, Antartida. Revta Asoc. Cienc. nat. Litoral 13: 81–95.

Donner, J., 1972. Bericht über Funde von Rädertieren (Rotatoria) aus der Antarktis. Pol. Arch. Hydrobiol. 19: 251–252.

Donner, J., 1978. Material zur saprobiologischen Beurteilung mehrerer Gewässer des Donau-Systems bei Wallsee und in der Lobau, Österreich, mit besonderer Berücksichtigung der litoralen Rotatorien. Arch. Hydrobiol. Suppl. 52: 117–228.

Everitt, D. A., 1981. An ecological study of an antarctic freshwater pool with particular reference to Tardigrada and Rotifera. Hydrobiologia 83: 225–237.

Focke, E., 1961. Die Rotatoriengattung Notholca und ihr Verhalten im Salzwasser. Kieler Meeresforsch. 17: 190–205.

Gillard, A. A. M., 1948. De Brachionidae (Rotatoria) van België met Beschouwingen over de Taxonomie van de Familie. Natuurwet. Tijdschr. 30: 159–218.

Hauer, J., 1952. Pelagische Rotatorien aus dem Windegefällweiher, Schuluchsee and Titisee im südlichen Schwarzwald. Arch. Hydrobiol. Suppl. 20: 212–237.

Hofmann, W., 1974. Zur Taxonomie und Verbreitung von Filinia-Arten (Rotatoria) in holsteinischen Gewässern. Faunistisch-ökol. Mitt. 4: 437–444.

Hudson, C. T. & P. H. Gosse, 1889. The Rotifera or wheel animalcules, both British and foreign. Supplement. Longmans Green, Lond., 64 pp.

Klement, V., 1957. Zur Rotatorienfauna des Monrepos-Teiches bei Ludwigsburg. Jh. Ver. vaterl. Naturk. Württ. 112: 238–263.

Klement, V., 1960. Zur Rotatorienfauna ephemerer Gewässer. Jh. Ver. vaterl. Naturk. Württ. 115: 338–349.

Koste, W. & K. Wulfert, 1964. Rotatorien aus der Wüste Gobi. Limnologica 2: 483–490.

Kutikova, L. A., 1958a. O novoy kolovratkye v Antarktidye. Inf. Byull. sov. antarkt. Eksped. 1955–58 (2): 45–46.

Kutikova, L. A., 1958b. K faunye kolovratok s poberzh'ya Vostochnoy Antarktidye. Inf. Byull. sov. antarkt. Eksped. 1955–58 (3): 7.

Kutikova, L. A., 1959. K izucheniyu fauni kolovratok latbii. Fauna Latviiskoi (II): 211–231.

Kutikova, L. A., 1970. Kolovratki fauni SSSR (Rotatoria). Leningrad Akademiya Nauk 104, 744 pp.

Lair, N. & W. Koste, 1984. The rotifer fauna and population dynamics of Lake Studer 2 (Kerguelen Archipelago) with description of Filinia terminalis kerguelenensis n. sp. and a new record of Keratella sancta Russell 1944. Hydrobiologia 108: 57–64.

Murray, J., 1910. Antarctic Rotifera. In J. Murray (ed.) British Antarctic Expedition, 1907–09. Reports on the Scientific Investigations: Biology 1(3). Heinemann, Lond.: 41–73.

Pejler, B., 1957. On variation and evolution in planktonic Rotatoria. Zool. Bidr. Upps. 32: 1–66.

Pejler, B., 1977. On the global distribution of the family Brachionidae (Rotatoria). Arch. Hydrobiol. Suppl. 53: 255–306.

Russell, C. R., 1944. A new rotifer from New Zealand. J. r. microsc. Soc. 64: 121–123.

Russell, C. R., 1959. Rotifera. Br., Aust, N.Z. Antarct. Res. Exped. 1929–31. Rep. Ser. B. 8(3): 81-87.

Russell, C. R., 1962. Additions to the Rotatoria of New Zealand, IX. Trans. r. Soc. N.Z. Ser. Zool. 1: 337–341.

Ruttner-Kolisko, A., 1966. Notholca lapponica, nov. spec., eine psammophile Rotatorienart aus Nord-Skandinavien. Anz. öst. Akad. Wiss. 9: 1–4.

Sladecekova, A. & V. Sladecek, 1977. Periphyton as indicator of the reservoir water quality, II. Pseudoperiphyton. Arch. Hydrobiol. Beih. Ergebn. Limnol. 9: 177–191.

Stemberger, R., 1976. Notholca laurentiae and N. michiganensis, new rotifers from the Laurentian Great Lakes region. J. Fish. Res. Bd Can. 33: 2814–2818.

Sudzuki, M., 1964. New systematical approach to the Japanese planktonic Rotatoria. Hydrobiologia 23: 1–124.

Sudzuki, M., 1979. On the microfauna of the antarctic region, III. Microbiota of the terrestrial interstices. Mem. natn Inst. Polar Res. Spec. Issue 11: 104–126.

Sudzuki, M., 1981. Periphytic biota (Aufwuchs) in sewages and park waters. In M. Sudzuki (ed.), Some Approaches to Saprobiological Problems. Sanseido: 69–107.

Sudzuki, M. & J. Shimoizumi, 1967. On the freshwater microfauna of the antarctic region, II. Stability of faunistic composition of antarctic microorganisms. Jap. antarct. Res. Exped. Sci. Rep. Spec. Issue 1: 216–235.

Voigt, M., 1956. Rotatoria. Gebrüder Borntraeger, Berlin & Stuttgart, 508 pp.

Hydrobiologia 165: 97–101 (1988)
J. M. Ferris, H. R. Burton, G. W. Johnstone & I. A. E. Bayly (eds) Biology of the Vestfold Hills, Antarctica
Kluwer Academic Publishers

Iodine distribution in an antarctic meromictic saline lake

E. C. V. Butler[1,3] H. R. Burton[2] & J. D. Smith[1]
[1]*Marine Chemistry Laboratory, School of Chemistry, University of Melbourne, Parkville, Victoria, 3052, Australia;* [2]*Antarctic Division, Department of Science, Kingston, Tasmania, 7050, Australia;* [3]*Present address: CSIRO Marine Laboratories, Division of Oceanography, GPO Box 1538, Hobart, Tasmania, 7001, Australia*

Key words: iodine, iodide, iodate, lakewaters, meromictic, biological cycling

Abstract

Water samples collected from a range of depths in Ace Lake in the Vestfold Hills, Antarctica were analysed for iodine species by a polarographic method. Contrary to thermodynamic prediction, iodide was the predominant species in oxic surface waters. Total dissolved iodine was closely correlated with bacterial density down the water column. This observation and other data are used to postulate a scheme for the biological mediation of iodine distribution in the lake.

Introduction

Iodine can be a useful element in characterizing a water mass (Lloyd *et al.*, 1982; Butler & Smith, 1985). Its distribution is influenced by biological activity as observed in seawater (Elderfield & Truesdale, 1980), but its speciation also reflects the redox chemistry of the surrounding water.

Although chemical investigations of antarctic saline lakes have been numerous, iodine analyses were reported on only two previous occasions. Iodate was detected but not quantified in Lake Vashka, Victoria Land (Johannesson & Gibson, 1962), and iodine concentration was below the detection limit of 1 mg kg^{-1} in a McMurdo saline lake (Torii *et al.*, 1975). We report here the concentration of iodine and its speciation in Ace Lake, Princess Elizabeth Land (Australian Antarctic Territory). Use is made of this data and previously reported nutrient concentrations to postulate a biological mechanism responsible for the distribution of these species with depth.

Description of the lake

Ace Lake (surface area 1.32×10^5 m^2, maximum depth 23 m – at time of sampling) is a saline meromictic lake in the low-lying ice-free oasis known as the Vestfold Hills (Fig. 1) on the Princess Elizabeth Land coast. Permanent stratification is the result of density and chemical differences between water layers. Wind-driven mixing is inhibited for most of the year by ice cover up to 2 m thick. The physical, chemical and biological characteristics of the lake have been extensively studied (Hand & Burton, 1981 and other reports in this volume). It is in effect a closed system; there is no surface outflow and no apparent underground seepage. Low, rocky and barren hills make up the surrounding catchment (5.11×10^5 m^2); as well as being ice-free, snow rarely accumulates on account of strong winds. Accordingly, the input of fresh water to Ace Lake is minimal.

Isostatic uplift of the coast following the last Ice Age (about 5000–8000 a B.P.) is believed to have isolated the system from the sea (Adamson & Pickard, 1983). The lake waters, quite different now from the

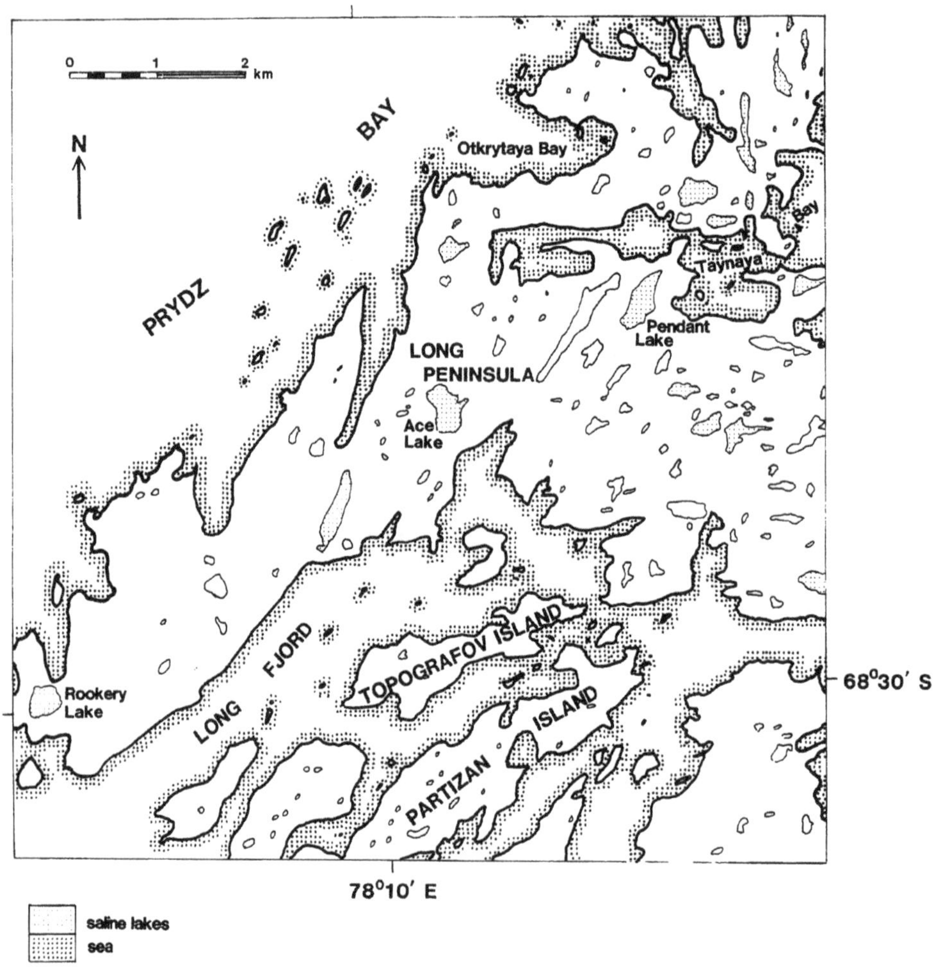

Fig. 1. Location of Ace Lake, Princess Elizabeth Land, Australian Antarctic Territory.

original seawater have been classified with respect to oxygen content (Burton, 1980; see Fig. 2). The top 7 m of the lake is supersaturated with oxygen (oxylimnion). Between 7 m and 9 m dissolved oxygen concentration falls to zero (oxycline). At depths in excess of 10 m, hydrogen sulphide and methane are detected (anoxylimnion); both increase with depth. Chlorinities increase down the water column (Fig. 2), spanning that of seawater.

The lake supports an abundant range of microorganisms, which are present at all depths (Hand & Burton, 1981; Burke & Burton, this volume). The only input of carbon into the lake is through algal fixation of carbon dioxide in the oxylimnion. Fallout of copepod faecal pellets transports organic carbon to the anoxylimnion and the lake floor. Phototrophic and heterotrophic bacteria are particularly active in the upper anoxylimnion, while methane-producing bacteria are found below 20 m.

Materials and methods

Sample collection

Water samples were collected by passing a 2 l transparent plastic Kemmerer water sampler through a hole cut in the surface ice sheet by a SIPRE ice corer

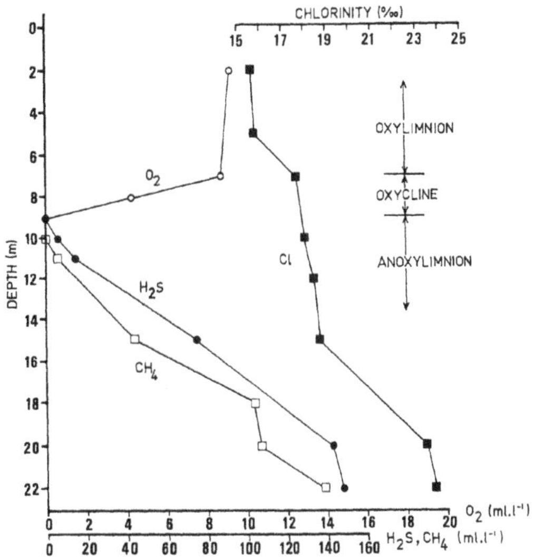

Fig. 2. Vertical profiles of dissolved gases (O_2, H_2S and CH_4) and chlorinity in Ace Lake (after Hand & Burton, 1981).

(Journal of Glaciology, 1958). A depth profile (Profile 1) taken on 24 May 1979 included waters from oxylimnion and oxycline. A later suite of samples (Profile 2) collected on 12 September 1979 comprised samples from the complete water column. All were stored frozen and unfiltered in low density polyethylene bottles. Under these conditions there might have been some degree of cell lysis.

Experimental

Iodine species were determined by differential pulse polarography (Butler & Smith, 1980). Iodate was measured directly in 20 ml of filtered (0.45 μm) water sample by the method of standard additions, with zinc interference masked by ethylenediaminetetraacetic acid (EDTA). Dissolved inorganic iodine was estimated after chlorine water oxidation; and total dissolved iodine was determined following preliminary ultra-violet (u.v.) photo-oxidation. Evaluation of the difference between (i) iodate and inorganic, and (ii) inorganic and total dissolved iodine gives a measure of the concentration of iodide and 'organo'-iodine, respectively.

Inorganic iodine determinations were not attempted in the reducing waters of the anoxylimnion because of the expected high chlorine demand of such samples. To ensure sufficient dissolved oxygen during photo-oxidation for total iodine measurement, anoxic water samples were both diluted with twice-distilled water and treated with 30% hydrogen peroxide before u.v.-irradiation. Dilution of these waters was feasible because of their high iodine concentrations; quantitative recovery of known additions of iodide at different sample dilutions also confirmed that it was a reliable procedure.

Results and discussion

Iodine analyses of profiles 1 and 2 are presented in Table 1. The iodate is below the method detection limit (<1 μg I l^{-1}) in Ace Lake waters with the exception of the surface layer (0–3 m) of Profile 1. There was interference in the polarographic determination of iodate in the anoxy-limnetic waters, possibly from sulphide (Davison, 1977). However, no io-

Table 1. Concentration of iodine species (μg l^{-1}) in Ace Lake profiles 1 and 2.

Profile	Depth (m)	Iodate (IO_3^-)	Inorganic Iodine	Σ I[a]	Σ I'[b]
1	1	1	9	10	-
	3	2	9	12	-
	5	<1	12	15	-
	6	<1	10	13	-
	7	<1	13	16	-
	8	<1	17	30	-
	9	<1	24	28	-
2	2	<1	-	6	7
	3	<1	-	5	6
	5	<1	-	6	7
	7	<1	-	18	20
	9	<1	-	21	23
	10	<1	-	59	63
	12	<1	-	96	101
	15	ND[c]	-	47	49
	18	ND[c]	-	71	64
	20	ND[c]	-	145	120
	22	ND[c]	-	165	134

[a] Σ I = total iodine; [b] Σ I' = specific total iodine concentrations normalized to 19.364‰ Cl; [c] ND = not determined, because of interference.

date should exist in these waters given its absence from the less reducing waters above.

The prevalence of iodide in the oxylimnion is contrary to thermodynamic prediction in a system where the O_2/H_2O couple dominates (Sillen, 1961; Dyrssen & Wedborg, 1980). Doubtless, the redox potential is perturbed by biological activity, and possibly by small scale mixing processes, like turbulent diffusion and edge effects caused by seiching, which transfer reducing agents from the anoxylimnion. Iodide can persist in oxygenated waters, even if thermodynamically unfavoured, because of the notoriously slow kinetics of its oxidation (Riley & Chester, 1971; Emersen *et al.*, 1979).

The total iodine concentration increases with depth (Table 1) with a peak in Profile 2 at the top of the anoxylimnion. There is a nearly constant difference of 3 μg I l^{-1} between total iodine and inorganic iodine in the oxylimnion and oxycline (Profile 1), which implies the existence of a small fraction of 'organo'-iodine.

When total iodine concentrations are normalized to a seawater chlorinity of 19.374 g kg^{-1}, (Table 1), it is seen that there is an uneven distribution of iodine in the water column. Relative to the mean seawater iodine concentration of 60 μg I l^{-1} (Riley & Chester, 1971) there has been an increase in iodine

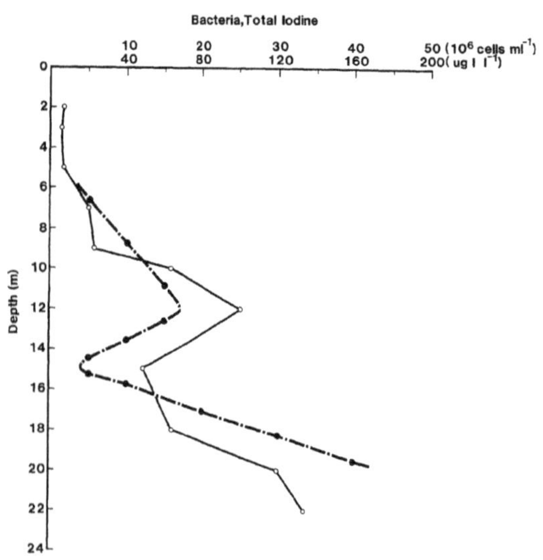

Fig. 3. Vertical profiles of total dissolved iodine, 12 September 1979 (o) and bacterial population density (•) in Ace Lake. Bacterial data selected from Hand & Burton (1981).

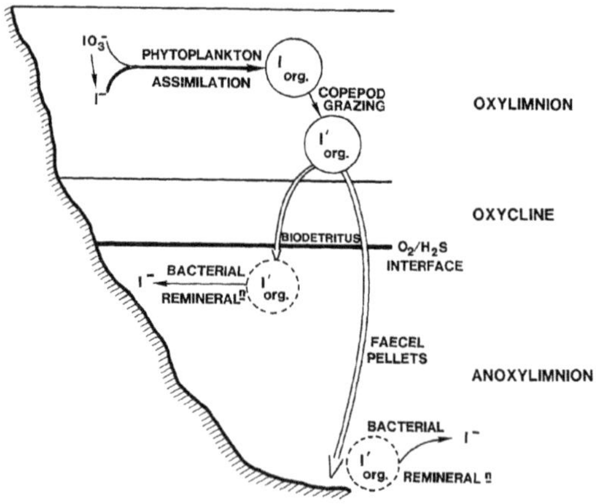

Fig. 4. A scheme postulated to explain the biological mediation of iodine distribution in Ace Lake.

content of the bottom waters at the expense of the surface layers.

A close correlation is noted between total iodine concentration and bacterial numbers (Fig. 3). The distribution of iodine might be the result of a long term interaction with micro-organisms, where the element is assimilated by phytoplankton, which are grazed by zooplankton (copepods). Iodine then sinks in faecal pellets, and other biodetritus, to be remineralized by heterotrophic bacteria at the top of the anoxylimnion or in near-bottom and sediment pore-waters (see Fig. 4).

Some support for this scheme is provided by previous nutrient analyses of Ace Lake waters (Burton, 1980; see Fig. 5). Nitrogen is severely depleted by biological activity in surface oxic waters, while phosphate concentrations are also low. From vertical profiles of phosphate and ammonia-plus-amino acids, microbial degradation of particulate or organically-bound nutrients takes place in the anoxylimnion or sediment. Phosphate is regenerated in the sediment; but the maximum dissolved concentration of ammonia-plus-amino acids occurs in the upper anoxylimnion, which suggests this is an important site for nitrogen remineralization. Iodine is seemingly a suitable substrate at both zones of bacterial activity. This might reflect a widespread distribution of iodine among particulate phases, while phosphate regeneration and ammonia production are res-

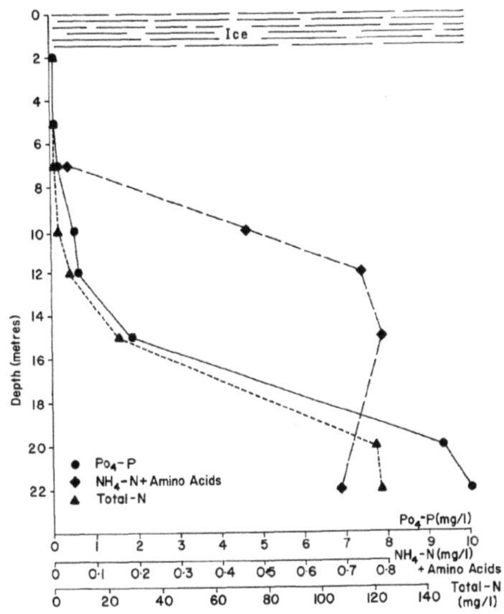

Fig. 5. Vertical profiles of phosphate and ammonia-plus-amino acids in Ace Lake (after Burton, 1980).

tricted to mutually-exclusive particulate fractions. Alternatively, bacterial population differences between the two zones could produce the same chemical partitioning.

Acknowledgement

We are grateful to Mike Burch for collecting the water samples used in this study.

References

Adamson, D. & J. Pickard, 1983. Late Quaternary ice movement across the Vestfold Hills, East Antarctica. In R. L. Oliver, P. R. James & J. B. Jago (eds), Antarctic Earth Science. Aust. Acad. Sci., Canberra: 465–469.

Burke, C. M. & H. R. Burton, this volume. Photosynthetic bacteria in meromictic lakes and stratified fjords of the Vestfold Hills, Antarctica. Hydrobiologia.

Burton, H. R., 1980. Methane in a saline antarctic lake. In P. A. Trudinger & M. R. Walters (eds), Biogeochemistry of Ancient and Modern Environmnents. Proc. 4th International Symposium on Environmental Biogeochemistry (ISEB). Australian Academy of Science, Canberra: 243–251.

Butler, E. C. V. & J. D. Smith, 1980. Iodine speciation in seawaters – the analytical use of ultra-violet photo-oxidation and differential pulse polarography. Deep-Sea Res. 27: 489–493.

Butler, E. C. V. & J. D. Smith, 1985. Iodine and arsenic redox species in oxygen-deficient estuarine waters. Aust. J. Mar. Freshwat. Res. 36: 301–309.

Davison, W., 1977. The polarographic measurement of O_2, Fe^{2+}, Mn^{2+} and S^{2-} in hypolimnetic water. Limnol. Oceanogr. 22: 746–753.

Dyrssen, D. & M. Wedborg, 1980. Chemical speciation in estuarine waters. In E. Olausson & I. Cato (eds), Chemistry and Biogeochemistry of Estuaries. John Wiley & Sons, Chichester: 71–119.

Elderfield, H. & V. W. Truesdale, 1980. On the biophilic nature of iodine in seawater. Earth planet. Sci. Lett. 50: 105–114.

Emerson, S., R. E. Cranston & P. S. Liss, 1979. Redox species in a reducing fjord: equilibrium and kinetic considerations. Deep-Sea Res. 26: 859–878.

Hand, R. M. & H. R. Burton, 1981. Microbial ecology of an antarctic saline meromictic lake. Hydrobiologia 82: 363–374.

Johannesson, J. K. & G. W. Gibson, 1962. Nitrate and iodate in antarctic salt deposits. Nature 194: 567–568.

Journal of Glaciology, 1958. Instruments and methods: Ice drills and corers. J. Glaciol. 3: 30.

Lloyd, J. W., K. W. F. Howard, N. R. Pacey & J. H. Tellman, 1982. The value of iodide as a parameter in the chemical characterization of groundwaters. J. Hydrol. 57: 247–265.

Riley, J. P. & R. Chester, 1971. Introduction to Marine Chemistry. Academic Press, Lond., N.Y., 465 pp.

Sillen, L. G., 1961. The physical chemistry of seawater. In M. Sears (ed.), Oceanography. American Association for the Advancement of Science, Washington (D.C.): 549–581.

Torii, T., N. Yamagata, S. Nakaya, S. Murata, T. Hashimoto, O. Matsubaya & H. Sakai, 1975. Geochemical aspects of the McMurdo saline lakes with special emphasis on the distribution of nutrient matters. Mem. Nat. Inst. Polar Res., Spec. Issue (Jpn) 4 (Geochem. Geophys. Stud. Dry Val., Victoria Land Antarct.): 5–29.

Hydrobiologia 165: 103–114 (1988)
J. M. Ferris, H. R. Burton, G. W. Johnstone & I. A. E. Bayly (eds) Biology of the Vestfold Hills, Antarctica
Kluwer Academic Publishers

Trace element distributions in some saline lakes of the Vestfold Hills, Antarctica

N. Masuda[1], S. Nakaya[2], H. R. Burton[3] & T. Torii[4]
[1]*Department of Chemistry, Faculty of Fisheries Science, Hokkaido University, Hakodate-city, Japan 041;* [2]*Department of Earth Science, Faculty of Science, Hirosaki University, Hirosaki-city, Japan 036;* [3]*Antarctic Division, Department of Science, Kingston, Tasmania, 7050, Australia;* [4]*Chiba Institute of Technology, Tsudanuma, Narashino-city, Japan 275*

Key words: trace elements, saline lakes, Antarctica

Abstract

The distributions of trace elements in Shield, Ace and Burton Lakes of the Vestfold Hills were investigated. Three aspects are discussed as follows: (1) the vertical distribution of 18 trace elements in the three lakes, (2) the behaviour of trace elements in the lakes, especially that of manganese in Shield Lake, and (3) the origin of trace elements in antarctic saline lakes.

High concentrations of trace elements were found in these coastal saline lakes, when compared to open ocean water.

We suggest that the peak of total extractable manganese, found at 20 m in Shield Lake, was related to the oxic/anoxic water interface brought about by microbiological activity. Solid phase manganese at the upper oxic layer may have precipitated and then reached the anoxic boundary to be there reduced to manganese ion. This dissolved manganese may then have diffused upwards to be reoxidized to a solid form. This cycle, repeated many times, may have produced the Mn profile.

The alkali, alkaline earth elements and Cl were probably derived from relict seawater. Other elements were present in similar concentration ratios to those of South Polar aerosols. Residence time calculations indicate that fallout of aerosol particles, themselves derived from various sources, is capable of accounting for the measured concentrations of some trace elements in Shield Lake. This source of trace elements may be significant for other antarctic saline lakes.

Introduction

A large ice-free area, the Vestfold Hills, exists on the coast of Princess Elizabeth Land, Antarctica. This has an area of some 400 km^2 and contains several hundred lakes and ponds. Davis Station, the Australian base, lies centrally in the area (68°34′S, 77°58′E).

Some chemical characteristics of Dingle, Stinear, Deep and Club Lakes were reported by McLeod (1964). Kerry *et al.* (1977) described major cation and nutrient concentrations of four saline lakes in the area and Burton & Barker (1979) reported on the sulfur chemistry at Ace Lake by using isotopic information. Horiuchi *et al.* (1977) reported on the radium content of lake deposits of four Vestfold saline lakes. Matsubaya *et al.* (1979) studied natural isotopes and reported on δ D and δ ^{18}O in 53 Vestfold lakes. Matsumoto *et al.* (1983) studied aspects of lake biogeochemistry and reported on the stenols and phytol content of the sediment from some Vestfold lakes.

Trace elements can be both essential elements for biota and toxic in higher concentration (Anderson & Morel, 1978; Jackson & Morgan, 1978). However, there have been few geochemical studies conducted on the trace elements of antarctic lakes.

Boswell *et al.* (1967a, b) reported on the concentrations of six trace elements in the bottom waters of six lakes of the Dry Valleys area, Victoria Land, and concluded that some of these trace elements were not derived directly from sea-water.

Masuda *et al.* (1982a) reported on the vertical distribution of Fe, Cu, Al and Ni from Lake Vanda in the Dry Valleys of South Victoria Land. They demonstrated that an iron cycle existed in the water column and suggested that at least Cu originated from airborne particles brought into the lake via glacial melt water. Masuda *et al.* (1982b) described the vertical distributions of 14 trace elements from both Lake Vanda itself and its inflow water and concluded that the pathway of these trace elements was from tropospheric aerosol particles – precipitation – glacier – glacial melt water – Lake Vanda.

This study presents the vertical distribution of 23 chemical elements in Shield, Ace and Burton Lakes. The origin of chemical elements from other antarctic saline lakes is discussed with respect to the results of the Vestfold Hills' and the Dry Valleys studies.

Materials and methods

Water samples were taken in the austral summer (February, 1984) and the sampling locations are presented in Fig. 1. The water samples were taken with a polyvinylchloride Van Dorn water sampler which was previously soaked in 4 M nitric acid for 1 day. The depths which were sampled are tabulated in Table 1. Unfiltered samples for trace metal analyses were immediately poured into 1 l polyethylene bottles which had been thoroughly washed, conditioned in distilled deionized water for a week and soaked for one day in 4 M nitric acid. Nitric acid (B.D.H. Aristar grade, 5 ml) was added to keep the samples at pH 1.0–1.2. Samples for specific gravity, sulfide and sulfate were collected at the same sites and on the same day from successive profiles.

Chloride was analysed by the Mohr titration method (Strickland & Parsons, 1972). The specific gravity was measured in an Anton Paar DMA 55 density meter. The Eh was measured with a platinum electrode, a calomel reference electrode and a millivoltmeter in the laboratory on the same day as sample collection. Samples for Eh measurement were kept in fully filled McCartney bottles until analysis. Sulfide was measured by the DTNB (5,5-dithiobis-2-nitrobenzoic acid) colorimetric method (Skyring & Chambers, 1976) and sulfate was measured gravimetrically after precipitation with barium. The samples for sulfate analyses had added to them, immediately after collection, crystals of cadmium chloride to precipitate the sulfide. The concentrations of Na, K, Mg and Ca were determined by atomic absorption spectrometry.

For trace element analyses, basically two types of neutron activation methods were used. Cu, Mn and Al, which produce short-lived radioactive nuclides after neutron irradiation, were analysed with a combination of solvent extraction and neutron activation methods.

Fig. 1. Location of Ace Lake, Shield Lake and Burton Lake in the Vestfold Hills.

The samples contained high concentrations of chloride, up to four times that of sea-water. Chlorine-38 was produced from chlorine-37 at the same irradiation as Cu, Mn and Al and this Cl-38 strongly interfered with the other short lived nuclide countings. The neutron activation analyses of Cu, Mn and Al were therefore performed after solvent extraction (see below) to avoid chloride interference.

50 ml aliquots of water were adjusted to pH 9.0 and then extracted with 0.01 M Oxine (8-hydroxyquinoline) containing chloroform. Back extraction of the metals was performed with 6 M nitric acid (super special grade). The acid layers were separated and dried onto filter papers under an infra-red lamp in a clean cabinet. The filter papers were then analysed for concentrations of trace elements by neutron activation analysis. The extraction efficiency for Mn was reported as 97.6% (Landing & Bruland, 1980). Total blank concentrations from the pH buffer, oxine, chloroform, nitric acid, filter paper and from the drying procedure as well as general laboratory environment contamination were determined by following the same routine.

The reactor which we used was a TRIGA II reactor at Rikkyo University, operating with a thermal neutron flux of 1.0×10^{11} n cm^{-2} s. The conditions of neutron activation analyses are presented in Fig. 2. The Cu, Mn and Al data therefore refers to total extractable metal concentrations and not total concentrations.

The other trace elements (Sc, Cr, Fe, Co, Ni, Zn, Se, Rb, Sr, Cs, Ce, Eu, Hf and Th) produce long-lived radioactive nuclides after irradiation. The analytical procedure was as follows. Each 100 ml of the non-filtered samples was dried in a Teflon beaker under an infra-red lamp, the residue was weighed (1–3 g) and then put into polyethylene bags (Fig. 2). The coal fly ash (SRM 1633) of NBS was used as a standard reference material (Ondov *et al.*, 1975). The absolute blank levels of this method were as follows:

Al 2.5 ± 0.2 μg, Mn < 3×10^{-2} μg, Fe < 1.0 μg, Co < 1.0×10^{-2} μg, Cr 5 ± 1 × 10^{-2} μg, Sc 6 ± 1 × 10^{-4} μg, Zn < 5.0×10^{-2} μg, Rb < 7×10^{-2} μg, Sr < 5×10^{-1} μg, Cs 1.6 ± 0.9 × 10^{-3} μg, Cu < 5.0×10^{-2} μg, Ni < 1.0×10^{-1} μg, and Sb, Se, Ce, Eu, Hf and Th were < 1.0×10^{-3} μg.

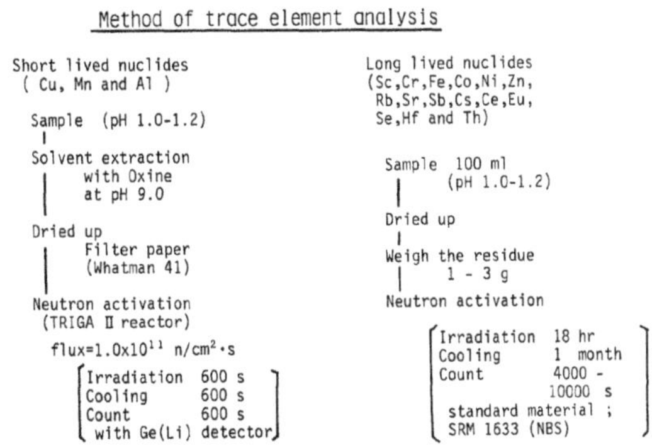

Fig. 2. Method of trace element analysis. Two types of instrumental neutron activation analysis were performed.

Results and discussion

The vertical distributions of 23 chemical elements in Shield, Ace and Burton Lakes were determined (Table 1). In general, the concentrations of Fe, Cr, Zn, Cu, Co and Sc increased towards the bottom of the lakes in the anoxic zone, whereas the maximum Mn value lay just below the oxic/anoxic boundary layer in Shield Lake.

The values of trace metals found in this study are, in general, several orders of magnitude greater than those found in open oceanic waters (Bruland, 1983). However they are in overall accord with values from Lake Vanda, a saline lake in the Dry Valleys region of Antarctica (Masuda *et al.*, 1982b); and continental lakes do have far greater terrigenous inputs (see below).

The very real dangers of contamination during sampling (Bruland, 1983) were recognized and the greatest care possible was taken to avoid them. However sampling from a small row-boat in Antarctica did prove difficult and polyethylene gloves were not used.

The high values for trace metals in the surface waters of Burton Lake, relative to oceanic values, are at first sight difficult to explain as there is some exchange between these waters and the sea during late summer and autumn when a shallow ice-plug, separating the two environments, melts (Burke & Burton, this volume). However it may be that this exchange is limited and that higher concentrations of trace metals are nonetheless maintained in the lake. The lake plankton certainly maintain a very distinct

Table 1. Chemical composition of Vestfold Hills saline lakes. The error denotes the analytical standard deviation of the uncertainties due to counting statistics.

	Sp. gr. (20 °C)	Cl ($g\ kg^{-1}$)	Sulfate ($g\ kg^{-1}$)	Na ($g\ kg^{-1}$)	K ($g\ kg^{-1}$)	Mg ($g\ kg^{-1}$)	Ca ($g\ kg^{-1}$)
Ace lake							
5 m	1.0139	12.2	0.5378	6.16	0.24	0.86	0.12
10	1.0204	13.4	0.5656	7.35	0.30	0.96	0.14
18	1.0234	19.7	0.1450	11.3	0.39	1.44	0.21
23	1.0286	22.8	0.0268	12.2	0.45	2.06	0.29
Shield L.							
1 m	1.0163	17.6	1.346	11.8	0.43	1.23	0.27
4	1.0570	56.5	4.624	36.9	1.20	3.80	0.75
8	1.0711	63.7	4.387	36.4	1.20	3.85	0.84
12	1.0814	67.7	5.331	38.8	1.23	3.99	0.81
16	1.0847	67.7	4.996	39.3	1.27	4.26	0.82
20	1.0892	68.9	5.995	40.2	1.27	4.24	0.82
24	1.1268	84.7	10.57	52.2	1.61	5.72	0.83
28	1.1159	83.8	9.933	51.5	1.74	5.46	0.78
37	1.1280	88.2	12.81	52.7	1.83	6.12	0.72
Burton L.							
2 m	1.0120	10.2	1.07	5.25	0.08	0.64	0.22
5	1.0253	18.3	2.28	9.80	0.28	1.22	0.45
10	1.0282	22.5	2.80	11.5	0.34	1.43	0.45
12	1.0300	22.9	2.74	12.2	0.40	1.53	0.48
15	1.0308	24.5	1.86	12.5	0.45	1.63	0.49
16	1.0309	24.5	1.75	12.9	0.45	1.63	0.49

identity, in comparison to the adjacent fjord, throughout summer (Bayly, 1986). Annual additions of trace metal rich anoxic waters from Burton Lake are mixed into the upper oxygenated waters in late winter when the chemocline is eroded due to descending waters of increased salinity excluded from the growing surface ice (Burke & Burton, this volume).

Behaviour of trace elements

In Shield Lake, the concentration of Mn at the surface was low, but at 4 m the concentration increased to 130 $\mu g\ l^{-1}$ and then lower values persisted until 20 m (Fig. 3). At 20 m, Mn had a maximum value of 260 $\mu g\ l^{-1}$ and decreased again to 30 $\mu g\ l^{-1}$ at the bottom. The high concentration at 4 m can, perhaps, be explained by two effects: one is biological activity and the other is the surface ice formation process. When the ice formed, metals could have been left behind to enrich the water just below the ice. This phenomenon has been reported from Lake Vanda in the Dry Valleys, Victoria land (Masuda *et al.*, 1982a).

The maximum of total extractable Mn at 20 m was strongly related to the oxic and anoxic water conditions. Both negative Eh values and sulfide appeared from a depth of 18 m onward to the bottom. These results demonstrate that below 20 m Shield Lake was anoxic. This was due to sulfate reduction in these long term stratified waters (Franzman *et al.*, this volume). We suggest that the solid phase Mn in the upper oxic layer precipitated downward to the boundary layer where it was reduced to manganous ion, dissolved phase. This dissolved Mn then slowly diffused upward to be reoxidized to the solid phase in the upper oxic layer. This cycle, repeated many times, could have produced the Mn profile we measured. The downward diffusion of Mn ions at the

Table 1. (Continued).

Al ($\mu g\ kg^{-1}$)	Sc ($\mu g\ kg^{-1}$)	Cr ($\mu g\ kg^{-1}$)	Mn ($\mu g\ kg^{-1}$)	Fe ($\mu g\ kg^{-1}$)	Co ($\mu g\ kg^{-1}$)
74.4 ± 2.2	0.0075 ± 0.0013	0.60 ± 0.49	2.1 ± 0.2	37.9 ± 8.3	0.166 ± 0.020
91.1 ± 2.3	0.0099 ± 0.0009	< 0.27	4.1 ± 0.3	52.7 ± 9.1	0.154 ± 0.015
129 ± 3	0.019 ± 0.001	3.8 ± 0.4	79 ± 2.0	82.9 ± 8.4	0.579 ± 0.027
86.8 ± 3.3	0.024 ± 0.002	11.6 ± 0.7	4.2 ± 0.3	147 ± 14	1.70 ± 0.05
71.1 ± 2.0	0.020 ± 0.003	1.3 ± 1.1	18.3 ± 0.9	115 ± 24	0.58 ± 0.07
61.8 ± 2.2	0.035 ± 0.007	< 3.7	130 ± 4	243 ± 54	1.43 ± 0.15
131 ± 3	0.088 ± 0.013	< 4.7	69 ± 4	336 ± 110	1.35 ± 0.20
70.5 ± 2.6	0.023 ± 0.007	< 3.5	59 ± 4	242 ± 61	1.28 ± 0.15
101 ± 3	0.051 ± 0.016	14.8 ± 5.9	103 ± 5	< 120	1.66 ± 0.23
72.1 ± 2.6	0.054 ± 0.010	5.9 ± 2.9	263 ± 8	363 ± 63	1.32 ± 0.14
146 ± 16	0.106 ± 0.008	4.9 ± 2.3	113 ± 5	497 ± 54	2.62 ± 0.14
87 ± 13	0.079 ± 0.018	< 7.1	116 ± 5	865 ± 164	2.45 ± 0.32
97 ± 16	0.171 ± 0.018	30.4 ± 6.6	31.3 ± 4.0	929 ± 127	3.18 ± 0.30
14.0 ± 2.0	0.014 ± 0.004	< 0.9	2.1 ± 0.4	80 ± 30	0.23 ± 0.05
3.0 ± 0.5	0.021 ± 0.008	< 2.6	3.7 ± 0.4	100 ± 60	< 0.08
3.5 ± 0.5	0.028 ± 0.006	2.8 ± 1.8	3.8 ± 0.4	130 ± 40	0.11 ± 0.04
7.2 ± 0.6	< 0.007	< 2.2	8.1 ± 0.6	160 ± 60	0.18 ± 0.09
3.4 ± 0.5	0.031 ± 0.007	3.0 ± 2.0	13.3 ± 1.0	190 ± 50	0.46 ± 0.09
14.9 ± 2.0	0.018 ± 0.010	7.8 ± 3.4	14.6 ± 1.0	230 ± 100	0.19 ± 0.14

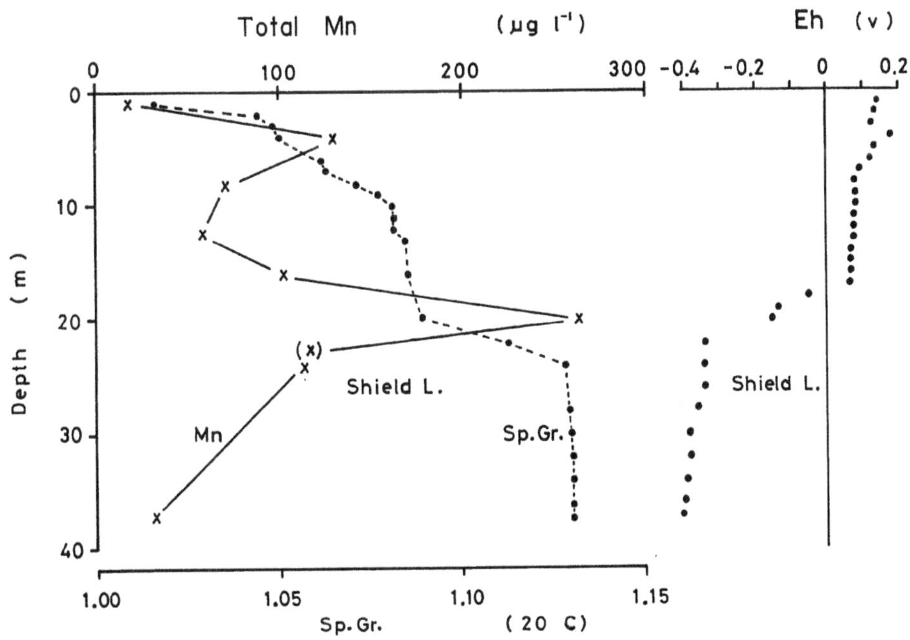

Fig. 3. Vertical distributions of total extractable manganese, specific gravity and Eh for Shield Lake.

Table 1. (Continued).

Ni (μg kg^{-1})	Cu (μg kg^{-1})	Zn (μg kg^{-1})	Se (μg kg^{-1})	Rb (μg kg^{-1})	Sr (μg kg^{-1})
ND	7.5 ± 0.8	15.0 ± 1.2	1.08 ± 0.25	27.1 ± 2.1	1480 ± 110
ND	8.5 ± 0.9	18.6 ± 2.7	7.1 ± 1.0	12.1 ± 1.6	1410 ± 120
6 ± 3	<0.5	4.9 ± 0.5	32.5 ± 2.3	28.0 ± 4.4	1890 ± 140
18 ± 9	<0.5	25.0 ± 2.8	5.8 ± 0.6	36.5 ± 4.4	2410 ± 110
ND	5.4 ± 0.7	22.1 ± 3.6	6.5 ± 0.8	45.7 ± 6.1	3500 ± 330
18 ± 12	7.6 ± 0.8	41.1 ± 7.0	47.1 ± 5.1	159 ± 20	10500 ± 1000
70 ± 40	14.4 ± 1.5	26.9 ± 7.4	35.5 ± 5.7	164 ± 22	10100 ± 900
<17	6.9 ± 0.7	14.5 ± 3.1	19.7 ± 2.5	339 ± 53	10900 ± 800
<21	10.8 ± 1.2	18.9 ± 6.3	18.8 ± 4.5	190 ± 26	13200 ± 1200
<11	4.8 ± 0.5	19.3 ± 3.6	32.2 ± 4.5	230 ± 40	6300 ± 500
<7	5.7 ± 0.6	67.2 ± 8.0	35.4 ± 3.7	250 ± 30	7100 ± 700
27 ± 25	4.7 ± 0.6	35.4 ± 10.3	45.6 ± 7.6	280 ± 40	10400 ± 1000
49 ± 34	27.5 ± 3.0	95.4 ± 15.1	48.0 ± 5.6	320 ± 40	7300 ± 700
<6	6.3 ± 0.6	8.9 ± 2.3	1.8 ± 0.7	47.1 ± 6.6	4030 ± 360
<16	1.3 ± 0.4	8.9 ± 4.0	6.5 ± 2.1	114 ± 16	9600 ± 900
ND	0.9 ± 0.3	17.7 ± 4.4	11.6 ± 1.6	87 ± 12	7300 ± 700
ND	0.9 ± 0.3	11.0 ± 3.8	6.1 ± 1.8	69 ± 10	7600 ± 700
26 ± 15	1.1 ± 0.3	56.8 ± 8.6	10.4 ± 1.5	107 ± 15	6900 ± 700
ND	1.2 ± 0.3	59.4 ± 11.2	10.5 ± 2.8	107 ± 16	11000 ± 1000

boundary was also likely to be restrained by the pycnocline. A similar profile of Mn concentration has been reported in Lake Nurume of the Syowa Oasis (Sano *et al.*, 1977) and in the Black Sea (Spencer & Brewer, 1971). The bottom concentration of total extractable Mn (31.3 μg kg^{-1} at 37 m) suggests the formation of MnS. Sulfide concentrations of 6.5 mM were measured at this depth.

The concentrations of Fe, Cr, Zn, Cu and Co, on the other hand, increased in the anoxic layer. Fe and Cr are also sensitive to oxic/anoxic conditions. However the data for this study have been obtained as total metal concentrations, including suspended solid forms. The concentrations of Fe and Cr at the bottom represent the maximum value at each of the three lakes. The results indicate that Fe and Cr may be reduced in the lake sediments and have diffused upward. This may also apply to the profiles of Zn, Cu and Co.

The origin of chemical elements

One purpose of this study was to determine the origin of salts in these antarctic saline lakes. It was thought that the dissolved chemical elements could have had three possible origins.

1) Sea-water origin: the lakes of the Vestfold Hills are believed to have been isolated from the sea by isostatic uplift following the last glacial period, possibly 4000–7000 a B.P. (Kerry *et al.*, 1977).
2) Dissolution of rock weathering origin: in 4000–7000 a, rock weathering within the catchments may also have contributed to the chemical composition of the lake water.
3) Aerosol particles origin: the contribution of aerosol particles, which may derive from sources including sea-water and crustal weathering, must also be considered.

The major chemical composition of typical antarctic saline lakes is presented in Fig. 4. Many coastal

Table 1. (Continued).

Sb ($\mu g\ kg^{-1}$)	Cs ($\mu g\ kg^{-1}$)	Ce ($\mu g\ kg^{-1}$)	Eu ($\mu g\ kg^{-1}$)	Hf ($\mu g\ kg^{-1}$)	Th ($\mu g\ kg^{-1}$)
0.09 ± 0.03	0.039 ± 0.019	ND	0.007 ± 0.003	0.046 ± 0.032	ND
0.03 ± 0.02	ND	ND	0.015 ± 0.005	< 0.017	0.082 ± 0.025
0.31 ± 0.05	0.028 ± 0.015	ND	0.012 ± 0.004	ND	ND
0.25 ± 0.05	0.10 ± 0.02	ND	0.020 ± 0.005	ND	0.28 ± 0.05
0.16 ± 0.06	< 0.014	ND	< 0.006	0.07 ± 0.06	0.14 ± 0.09
< 0.01	0.27 ± 0.08	ND	0.06 ± 0.02	< 0.14	< 0.22
0.84 ± 0.02	0.41 ± 0.16	< 4.3	0.08 ± 0.04	< 0.25	< 0.37
0.35 ± 0.16	0.10 ± 0.10	< 3.3	< 0.01	< 0.19	< 0.27
0.63 ± 0.28	< 0.19	< 5.4	0.14 ± 0.06	< 0.33	< 0.47
1.00 ± 0.23	0.32 ± 0.11	< 2.7	0.04 ± 0.02	< 0.18	< 0.26
0.32 ± 0.09	0.18 ± 0.08	31.0 ± 2.4	0.05 ± 0.02	< 0.13	< 0.17
0.70 ± 0.30	< 0.22	< 6.2	0.07 ± 0.07	< 0.40	< 0.58
1.10 ± 0.40	0.51 ± 0.21	< 3.2	0.14 ± 0.05	< 0.19	< 0.28
0.12 ± 0.06	0.11 ± 0.04	ND	< 0.008	< 0.06	< 0.09
< 0.14	0.44 ± 0.12	ND	< 0.019	< 0.16	< 0.25
< 0.13	0.17 ± 0.07	< 1.7	0.02 ± 0.01	0.1 ± 0.1	0.42 ± 0.16
0.19 ± 0.13	0.21 ± 0.09	2.0 ± 1.2	< 0.03	0.26 ± 0.14	< 0.21
0.25 ± 0.15	< 0.08	< 1.8	< 0.02	0.25 ± 0.12	< 0.15
0.47 ± 0.29	0.21 ± 0.11	10.0 ± 3.1	< 0.04	< 0.2	0.83 ± 0.31

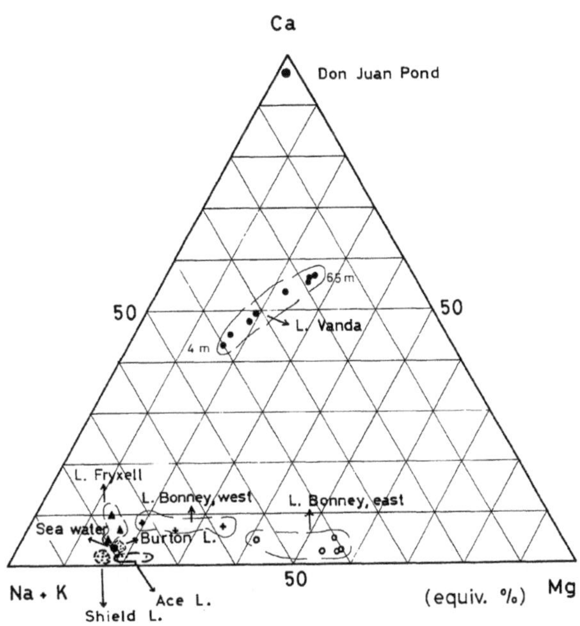

Fig. 4. Relations between Ca, Mg and Na + K in some antarctic saline lake waters and sea-water.

lakes of the antarctic oases are close in composition to that of sea-water. These include Lake Nurume of the Syowa oasis and Lake Fryxell of the Dry Valleys region.

The chemical composition of inland oases lakes, such as lakes Bonny, Vanda and Don Juan Pond have resulted from concentration under frigid conditions (Torii & Yamagata, 1981). The relationships of chemical components between two lakes and sea-water (Bruland, 1983) are presented in Fig. 5. The solid line was defined as a 1:1 line on the basis of the values for the major alkali, alkaline earth elements. The high correlations between Cl, Sb, Na, Mg, Ca, K, Sr and Rb in the lake waters and sea-water suggests that the alkali, alkaline earth elements and chlorine have been derived from sea-water or sea spray. This conclusion is also supported by the relationships expressed in Fig. 4.

The other elements are in higher concentrations in the lake waters with respect to sea-water, even if a factor of ten is allowed (Fig. 5, dotted line). These elements should lie close to the solid line if they had been derived from sea-water of the same composition as standard sea-water. Thus further consideration must be given to the origin of these trace elements.

A comparison of the chemical composition of the earths crust and two lake waters is presented in Fig. 6. The rock composition of the Vestfold Hills is not yet sufficiently known to allow direct comparison with lake trace elements. Therefore, the average crustal abundance data of Taylor (1964) was used. If Al is taken as a marker element of crustal origin, then Sc and Fe are also likely to be of crustal origin. Assuming the range of error to be a factor of ten, then Cr, Th, Eu, Hf and Cs may also have originated from rock weathering. Co and Mn fall within this range for Ace Lake. But Zn, Cu, and Se still do not appear to be attributable to a rock weathering origin or a sea-water origin.

Figure 7 shows the relationship between the chemical elements of aerosol particles at the South Pole and those of the two lake waters. Because the multi-element data of aerosol particles could not be obtained for the Vestfold Hills region, data pertaining to the South Pole (Maenhaut *et al.*, 1979) was used.

It is evident that an aerosol origin is likely for Zn, Cu and Se. However, most of the lake elements have a similar concentration ratio to that of the South Polar aerosol particles; the only clear exceptions to this are Cl, the alkaline earth elements, and alkali metals (but not Cs). These exceptions are considered to have originated from relict sea-water, as previously described. Aerosols therefore provide a reasonable pathway for many of the trace elements found in these Vestfold Hills saline lakes; this applies to those elements, such as Eu, Sc, Co, Mn, Th and Cs, for which a crustal origin is also possible, Hf and Cr, which show comparatively large errors of determi-

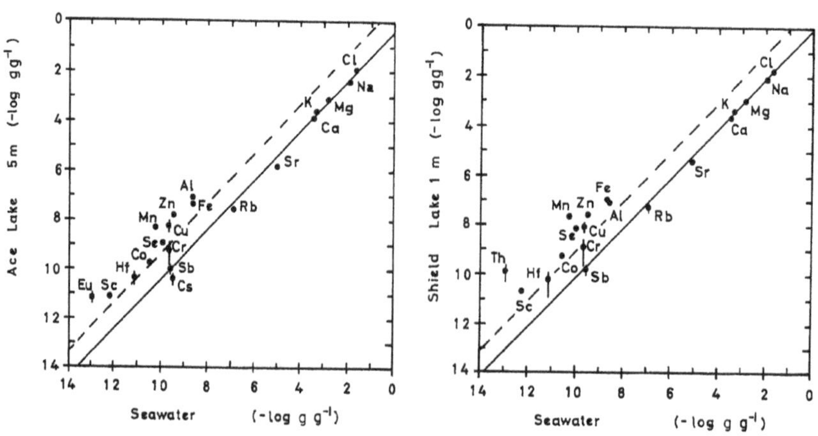

Fig. 5. The relationship between the chemical composition of sea-water and that of two Vestfold Hills saline lakes.

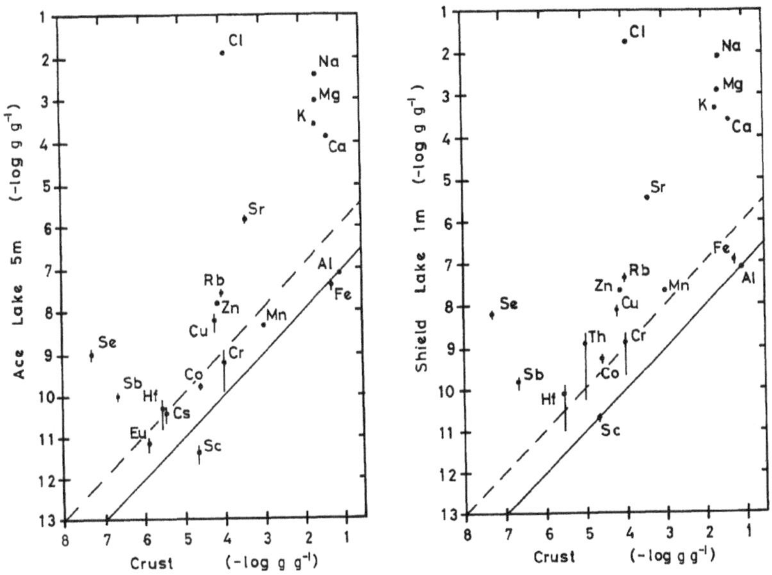

Fig. 6. The relationship between the chemical composition corresponding to average crustal abundance and that of two Vestfold Hills saline lake waters.

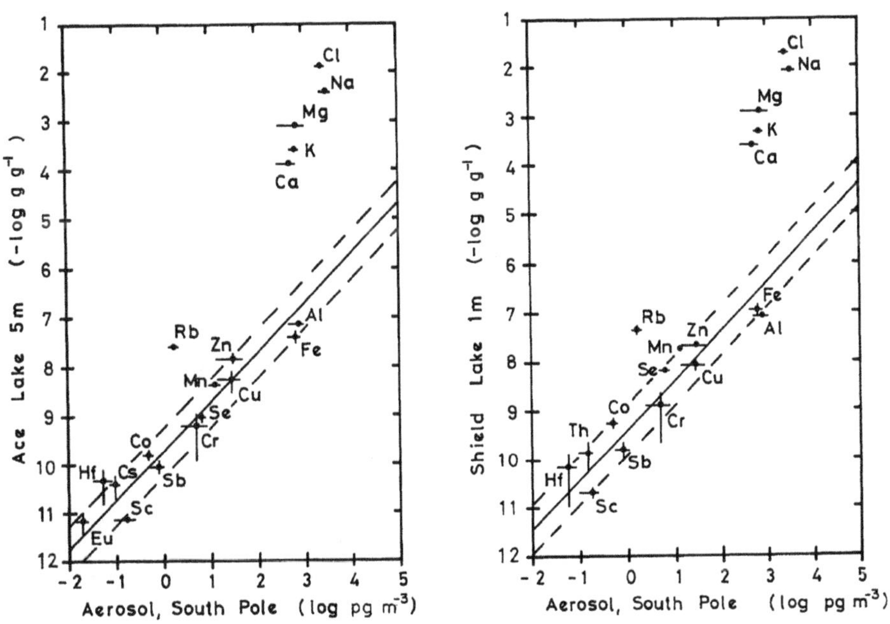

Fig. 7. The relationship between the chemical composition of aerosol particles at the South Pole and that of two Vestfold Hills saline lake waters.

nation and are attributable to all three sources, and Sb which is possibly of sea-water origin. This is explained by the fact that antarctic aerosol particles may be derived from sea-water, crustal weathering, volcanic activity and also from marine microbial production of reduced sulfur gases. This latter input is certainly occurring in the Vestfold Hills (Deprez *et al.*, 1986). However, the contributions of rock weathering and sea spray to aerosol particles, may be considered of particular significance in this dusty, coastal oasis.

To further examine the aerosol origin of trace elements, the residence time was calculated based on dry fallout. The mean residence time of any trace element in the water column for this study is the time it would theoretically take to account for the total amount of the element from annual aerosol input.

Although aerosol data from the South Pole are taken to represent the ratios between chemical elements in Antarctica generally (see Figs. 7 & 8), lakes in the dry, rocky and hence dusty Vestfold Hills exist in an environment similar to the Dry Valleys region. Therefore, aerosol data from the Dry Valleys region (Nakaya *et al.*, 1983) were used to calculate input rates in this study. Other aerosol data of antarctic coastal oases are reported from Syowa Oasis (Nakaya, 1982). The range of metal concentrations are similar to those of the Dry Valleys, but only five metals (Al, Mn, Fe, Cu, Zn) were reported. The aerosol deposition velocity used in the calculation was assumed to be 1.0 cm s^{-1} (Young & Silker, 1980). The yearly input rate (F) of an element from aerosol fallout was calculated as follows:

$$F(\mu g\ m^{-2}\ a^{-1}) = 1 \times 10^{-2}\ (m\ s^{-1}) \times 3600 \times 24 \times 365\ (s\ a^{-1}) \times \text{concentration}\ (\mu g\ m^{-3})$$

The inventory (Q) was calculated as the total mass of an element in a 1 m^2 water column at the profile site. It was the sum of all the depths that were profiled and included both oxic and anoxic waters according to the general expression:

$$Q(\mu g\ m^{-2}) = \Sigma(\text{Sp. Gr.} \times \text{depth range (m)} \times \text{concentration}\ (\mu g\ m^{-3}))$$

where the concentration ($\mu g\ m^{-3}$) was estimated as 1000 × the concentration in $\mu g\ kg^{-1}$. Mean residence times (T = Q/F) are presented in Table 2.

Na, Cl, Ca and Mg require > 3 × 10^6 a of deposition from the atmosphere to account for the total amounts of these elements presently found in the lakes. The Vestfold Hills lakes, being very close to the sea, probably have a larger contribution of aerosols derived from sea spray than is the case for the Dry Valleys. Nevertheless, the very long residence times

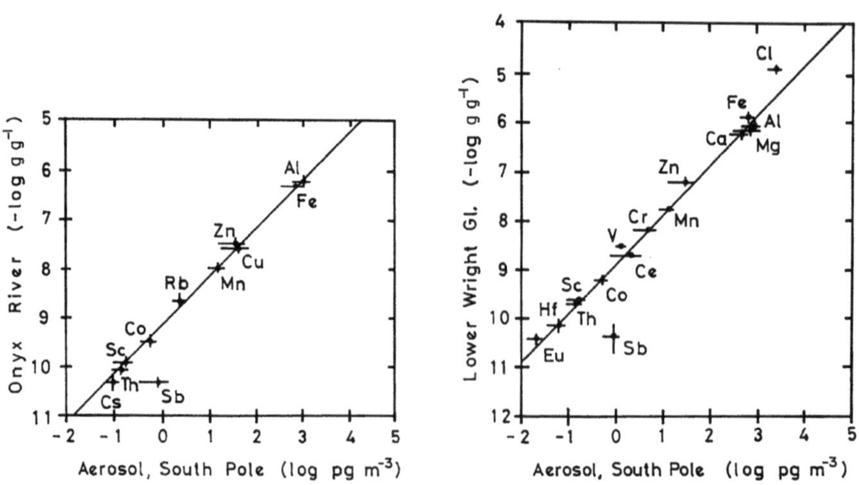

Fig. 8. The relationship between the trace element content of aerosol particles at the South Pole, and that of Onyx River water and ice from Lower Wright Glacier of the Dry Valleys, Victoria Land.

Table 2. Mean residence time in Shield Lake. Input rates were calculated from Dry Valley aerosol data and assuming a deposition velocity of 1 cm s^{-1}. In steady state; Mean residence time (T) = Inventory (Q)/Input rate (F).

	Input rate ($\mu g\ m^{-2}\ a^{-1}$)	Residence time (a)
Na	7.0×10^4	1.6×10^7
Cl	1.3×10^5	1.9×10^7
Ca	5.3×10^4	6.0×10^6
Mg	4.8×10^4	3.5×10^6
Al	1.6×10^5	58
Fe	1.8×10^5	85
Zn	1.8×10^4	79
Co	1.0×10^3	67
Mn	1.1×10^4	320

tend to confirm that the major cations and Cl were derived directly from relict sea-water (see Burton, 1981).

Al, Zn, Fe and Co need only 60–90 a of aerosol deposition to reach their greatest concentrations in Shield Lake. The mean residence times of 60–90 a are extremely small; but they may be reasonable times for coastal antarctic lakes. Mean oceanic residence times are somewhat larger, as follows: Al 620 a, Zn 510 a, Fe 54 a, and Co 340 a (Broecker & Peng, 1982). If the aerosol input to the lakes also includes catchment contribution via melt water, the residence times are even shorter. These residence time calculations indicate that aerosol deposition can readily account for the total amounts of some elements in Shield Lake, even those like Al and Fe that also have derived directly from dissolution of weathered crustal material. Taken with earlier comparisons of element ratios, it is evident that aerosol deposition provides the most general explanation for the origin of many trace elements in these Vestfold Hills saline lakes.

Figure 8 shows the relationships between the trace element composition in the Onyx River and in aerosol particles (Dry Valleys areas). The Onyx River is a main inflow into Lake Vanda and arises from meltwater leaving the Lower Wright Glacier. Figure 8 shows quite a good correlation between the Onyx River trace elements and aerosol composition. Some other lakes and ponds are also correlative in this regard (Masuda *et al.*, 1984). These results also support the conclusion that aerosol particles are important suppliers of trace elements to antarctic saline lakes.

Conclusions

1) A manganese maximum at just below the oxic/anoxic boundary layer of Shield Lake may have resulted from manganese oxide – manganous ion cycling at this boundary.
2) Our data support the view that the origin of alkali, alkaline earth elements and chlorine was relict sea-water.
3) Aerosol particles were probably an important source of trace elements for Shield and Ace Lakes. This may also be true for other antarctic saline lakes.

Acknowledgements

The authors are greatly indebted to P. Deprez and P. Franzman for assistance in obtaining the samples and to J. van den Hoff for measurements of specific gravity. The authors thank also Drs N. Yamagata, S. Tsunogai and G. I. Matsumoto for their discussions and critical readings. This work has been supported in part by the funds of Geochemical Research Foundation of Japan, Japan Polar Research Association and Common Utilization of the Research Reactor of Rikkyo University.

References

Anderson, D. M. & F. M. M. Morel, 1978. Copper sensitivity of Gonyaulax tamarensis. Limnol. Oceanogr. 23: 283–295.

Bayly, I. A. E., 1986. Ecology of the zooplankton of a meromictic antarctic lagoon with special reference to Drepanopus bispinosus (Copepoda: Calanoida). Hydrobiologia 140: 199–231.

Boswell, C. R., R. R. Brooks & A. T. Wilson, 1967a. Trace element content of antarctic lakes. Nature 213: 167–168.

Boswell, C. R., R. R. Brooks & A. T. Wilson, 1967b. Some trace elements in lakes of McMurdo Oasis, Antarctica. Geochim. Cosmochim. Acta. 31: 731–736.

Broecker, W. S. & T. H. Peng, 1982. Tracers in the sea. Eldigio Press, New York.

Bruland, K. W., 1983. Trace elements in sea-water. In J. P. Riley

& R. Chester (eds.), Chemical Oceanography, 8. Academic Press, Lond.

Burke, C. M. & H. R. Burton, this volume. The ecology of photosynthetic bacteria in Burton Lake, Vestfold Hills, Antarctica. Hydrobiologia.

Burton, H. R., 1981. Chemistry, physics and evolution of antarctic saline lakes: a review. In W. D. Williams (ed.), Salt Lakes. Junk, The Hague: 339–362.

Burton, H. R. & R. J. Barker, 1979. Sulfur chemistry and microbiological fractionation of sulfur isotopes in a saline antarctic lake. Geomicrobiol. J. 1: 329–340.

Deprez, P. P., P. D. Franzman & H. R. Burton, 1986. Determination of reduced sulfur gases in antarctic lakes and sea-water by gas chromatography after solid absorbent preconcentration. J. Chromatogr. 362: 9–21.

Franzman, P. D., G. W. Skyring, H. R. Burton & P. P. Deprez, this volume. Sulfate reduction rates and some aspects of the limnology of four lakes and a fjord in the Vestfold Hills, Antarctica. Hydrobiologia.

Horiuchi, K., T. Torii & U. Murakami, 1977. A geochemical study on the distribution of some minor elements in deposits and water samples of the antarctic oases, 1. The Ra content of DVDP 13 core and the deposits of the Vestfold Hills. Antarct. Rec. 58: 69–80.

Jackson, G. A. & J. J. Morgan, 1978. Trace metal-chelator interactions and phytoplankton growth in sea-water media: theoretical analysis and comparison with reported observations. Limnol. Oceanogr. 23: 268–282.

Kerry, K. R., D. R. Grace, R. Williams & H. R. Burton, 1977. Studies on some saline lakes of the Vestfold Hills, Antarctica. In G. A. Llano (ed.), Adaptations with Antarctic Ecosystems. Smithsonian Institution, Wash., D.C.: 839–858.

Landing, W. M. & K. W. Bruland, 1980. Manganese in the North Pacific. Earth Planet. Sci. Lett. 49: 45–56.

Maenhaut, W., W. H. Zoller, R. A. Duce & G. L. Hoffman, 1979. Concentration and size distribution of particulate trace elements in the South Polar atmosphere. J. Geophys. Res. 84: 2421–2431.

Masuda, N., S. Nakaya & T. Torii, 1984. Origin of salt in antarctic saline lake waters through trace element analysis. Mem. natn. Inst. Polar Res. Spec. Issue 33: 194–203.

Masuda, N., M. Nishimura & T. Torii, 1982a. Distribution and origin of some trace metals in Lake Vanda, Antarctica. Antarct. Rec. 75: 25–36.

Masuda, N., M. Nishimura & T. Torii, 1982b. Pathway and distribution of trace elements in Lake Vanda, Antarctica. Nature 298: 154–156.

Matsubaya, O., H. Sakai, T. Torii, H. R. Burton & K. Kerry, 1979. Antarctic saline lakes – stable isotopic ratio, chemical compositions and evolution. Geochim. Cosmochim. Acta. 43: 7–25.

Matsumoto, G., T. Torii & T. Hanya, 1983. Stenols and phytol in lake sediments from Syowa and Vestfold Oases in the Antarctic. Geochem. J. 17: 1–8.

Nakaya, S., 1982. Chemical composition of atmospheric aerosol collected at Syowa Station. Antarct. Rec. 75: 1–11.

Nakaya, S., N. Masuda, N. Torii & T. Torii, 1983. Chemical composition of aerosol particles at the Dry Valleys area, Antarctica. Abstract in Japanese of the Annual Meeting of the Geochemical Society of Japan. Tokyo, Nihon Chikyuu Kagakukai (Geochem. Soc. Jpn.): 93.

McLeod, I. R., 1964. The saline lakes of the Vestfold Hills, Princess Elizabeth Land. In R. J. Adie (ed.), Antarctic Geology. North Holland, Amsterdam: 65–72.

Ondov, J. M., W. H. Zoller, I. Olmez, N. K. Aras, G. E. Gorden, L. A. Rancitelli, K. H. Adel, R. H. Filby, K. R. Shah & R. C. Ragaini, 1975. Elemental concentrations in the National Bureau of Standards environmental coal and fly ash standard reference materials. Analyt. Chem. 47: 1102–1109.

Sano, M., N. Nakai & T. Torii, 1977. Vertical distribution of some trace metals in Lake Nurume, Antarctica. Antarct. Rec. 58: 108–115.

Skyring, G. W. & L. A. Chambers, 1976. Biological sulfate reduction in carbonate sediments of a coral reef. Aust. J. mar. Freshwat. Res. 27: 592–602.

Spencer, D. W. & P. G. Brewer, 1971. Vertical advection diffusion and redox potentials as controls on the distribution of manganese and other trace metals dissolved in waters of the Black Sea. J. Geophys. Res. 76: 5877–5892.

Strickland, J. D. H. & T. R. Parsons, 1972. A practical handbook of sea-water analysis, 2nd edn. Bull. Fish Res. Bd Can. 167, 310 pp.

Taylor, S. R., 1964. Abundance of chemical elements in the continental crust: a new table. Geochim. Cosmochim. Acta. 28: 1273–1286.

Torii, T. & N. Yamagata, 1981. Limnological studies of saline lakes in the Dry Valleys. In L. D. McGinnis (ed.), Dry Valley Drilling Project. Am. Geophys. Union. Wash., D.C. Antarct. Res. Ser. 33: 141–160.

Young, J. A. & W. B. Silker, 1980. Aerosol deposition velocities on the Pacific and Atlantic Oceans calculated from ^{7}Be measurements. Earth planet. Sci. Lett. 50: 92–104.

Hydrobiologia 165: 115–128 (1988)
J. M. Ferris, H. R. Burton, G. W. Johnstone & I. A. E. Bayly (eds) Biology of the Vestfold Hills, Antarctica
Kluwer Academic Publishers

The annual cycle of heat content and mechanical stability of hypersaline Deep Lake, Vestfold Hills, Antarctica

J. M. Ferris & H. R. Burton
Department of Science, Antarctic Division, Channel Highway, Kingston, Tasmania, 7050, Australia

Key words: heat budget, stability, saline lake, Vestfold Hills, Antarctica

Abstract

Deep Lake, a hypersaline lake of about ten times seawater concentration, rarely freezes and is characterized by a monomictic thermal cycle, Winter circulation, at c. −17 °C, lasts for two to three months. In summer, epilimnetic temperatures from 7–11 °C result in large vertical thermal gradients (21–26 °C) which combine with the enhanced rate of density change per degree Celsius, accompanying such high salt concentration, to produce a particularly stable density configuration in Deep Lake (Schmidt stability c. 8000 g-cm cm^{-2}; 0.785 J cm^{-2}). The Birgean annual heat budget (c. 24500 cal cm^{-2}; 102.7 10^3 J cm^{-2}) is comparable to that of a temperate lake with a similar mean depth, despite the comparatively high ratio of Birgean wind work to annual heat budget (0.37 g-cm cal^{-1}). Deep lake retains c. 50% of the incident solar radiation during the short summer heating period; within the range estimated for 'first class' lakes in North America. Extended daylight hours certainly contribute to the high maximum rate of heating in the lake (444 cal cm^{-2} day^{-1}; 1.86 10^3 J cm^{-2} day^{-1}). Deep Lake cools at a rate less than half its average heating rate. Partitioning the total stability into thermal and saline components shows that salinity can contribute up to c. 20% of the maximum summer Schmidt stability. In early summer, the effect of small melt-streams is to increase stability by diluting the epilimnion. In autumn, evaporative water loss can overtake this effect, creating small de-stabilizing salinity gradients. The usually short-term stabilizing influence of snowfall and drift is less predictable, but is probably more common in winter when strong winds are most frequent.

Hypersalinity has a profound effect on the physical cycle of Deep Lake, through freezing point depression and the increased rate of density change with temperature. These changes affect the lake's biota, both in relation to osmotic stress, and by effectively exposing them to a more thermally extreme environment. A comparison between Deep Lake and a smaller lake of similar salinity (Lake Hunazoko, Skarvs Nes), demonstrates that it is inappropriate to consider the biological effects of salinity in isolation. The smaller lake offers warmer epilimnetic conditions for at least part of the summer, which may explain the much greater limnetic algal production in Lake Hunazoko.

Introduction

The study of empirical parameters, such as heat budget and mechanical stability, which can be derived from thermal profiles and a knowledge of lake morphometry, is uncommon for antarctic lakes. A study of the annual cycle of heat content and stability of an antarctic saline lake is rarer still.

Deep Lake (Vestfold Hills) has a number of features that make it of particular importance in the context of such studies. The lake has similar ionic proportions to seawater but is about ten times as concentrated (Barker, 1981), it is deep (36 m) for a saline lake and very clear, having the rare dinstinction of remaining ice-free in most years. In this last fact is the suggestion that Deep Lake may provide a record of the outer limits of thermal behaviour in a climate distinguished by its extremes. Further, Deep Lake may give a more general insight into the role of temperature and dissolved salts in the annual stratification cycle of saline lakes.

Deep Lake is a biologically extreme environment and, at present, only a green alga (*Dunaliella* sp.) and two strains of a halobacterium (P. D. Franzmann, pers. commun.) have been grown in cultures using water from the lake. Lake Hunazoko (Skarvs Nes), a smaller and shallower lake which has similar chlorinity to Deep lake, is reported to support a greater diversity and standing crop of limnetic algae. Some thermal and chemical data are compared in an attempt to account for this difference between two lakes that place similarly extreme osmotic stress on their biota.

Material and methods

Thermal profiles have been intermittently recorded, for Deep Lake, since 1957 (McLeod, 1964; Kerry *et al.*, 1977). Since 1973 there have been two periods of regular sampling, October 1973 to January 1976, and January 1977 to March 1979. This latter period of relatively intensive measurement is considered in detail here. Temperature measurements were taken using a thermistor designed to give probe/battery voltage ratio, to overcome cold effects. Profiles, with a depth interval of 1 m, were taken weekly from the deepest section of the lake during 1978 and early 1979. Sampling was less frequent, both temporally and spatially in 1977. The sampling distribution is shown in Fig. 1. Temperature isopleths (Fig. 2) were plotted using SURFACE II (Sampson, 1978), implemented on a Burroughs B6800 mainframe computer. Water samples were collected with a 2.5 l Van Dorn bottle, from sufficient depths to characterize vertical salinity changes for each of the sample dates. These samples were stored in tightly sealed heavy polyethylene bottles which were returned to Australia where density (20 °C) was measured using an Anton Paar DMA55 density meter, capable of reading reproducibly to five decimal places. Salinity (g kg^{-1}) was calculated from these density measurements using a linear relationship (Salinity = 1315.7895 × density (20 °C) − 1313.4211) derived from Table 1 (Appendixes; p 505) in Whitfield & Jagner (1981). We recognize the potential weakness of using a linear regression well beyond the bounds of the data from which it was derived and have given the equation so that measured density (at 20 °C) can be recovered, if desired. The morphometry of Deep Lake was derived from echo-sounding transects made in December 1978.

The Idso (1973) modification of Schmidt stability, Birgean wind work, heat content, and volume

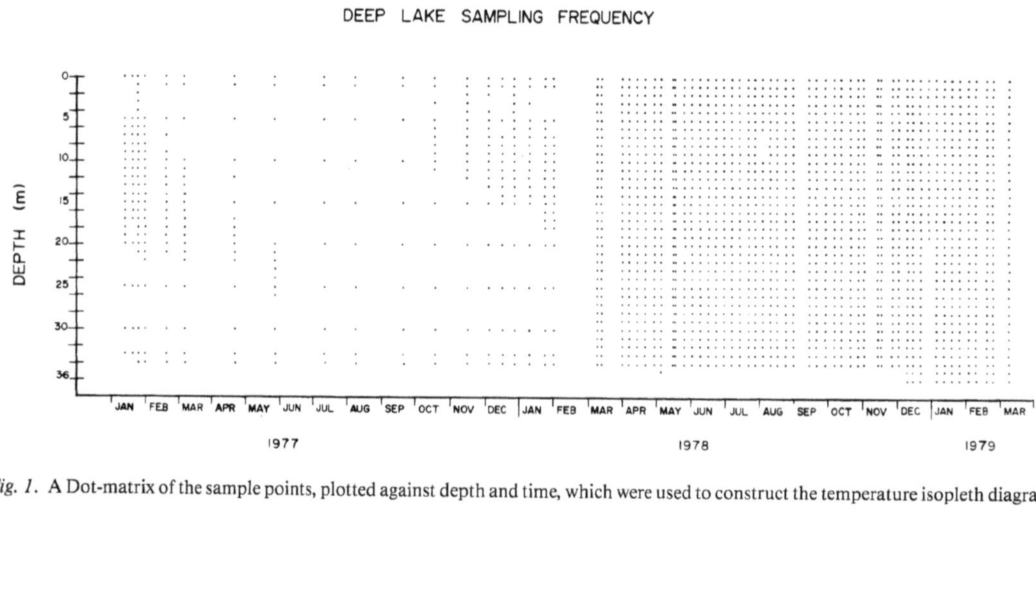

Fig. 1. A Dot-matrix of the sample points, plotted against depth and time, which were used to construct the temperature isopleth diagram.

weighted averages of temperature and salinity were calculated using a version of LIMNO (see Johnson *et al.,* 1978) modified to accept direct input of density profiles, rather than inferring density from temperature and salinity. *In situ* density was estimated from measured density (20 °C) using an experimentally determined, linear temperature/density relationship for Deep Lake, prior to use of LIMNO. Subsequently, a series of quadratic regressions was used to construct a general algorithm (DEQCLC) relating temperature and density for saline lakes in the Vestfold Hills, which range from approximately seawater salt concentration to that of Deep Lake (c. 10 × SW). Data derived from DEQCLC is presented in Fig. 3. Schmidt stability was partitioned into thermal and saline components by first calculating total stability (S_{tot}), and then declaring the water isohaline at the volume weighted mean salinity, to enable a separate calculation of thermal stability (S_t. Saline stability (S_s) equals $S_{tot} - S_t$.

Theoretical considerations

Heat budget, stability and wind work are empirical quantities derived from thermal and salinity profiles, and a knowledge of lake morphometry.

The Annual heat budget of a lake is the term used to denote the change in stored heat between the time of minimum heat content and the subsequent time of maximum heat content. It is the amount of heat stored by the lake during the annual heating period, expressed per unit surface area of the lake. The sequential nature of Birge's heat budget often means that it must be calculated as the difference between the winter minimum of one calendar year and the maximum occurring in the following year. The hypersalinity of Deep Lake introduces a potential complication in calculating the heat content, as stored heat is a function of mass, temperature and specific heat. In a freshwater system, mass and specific heat are unity; so that the heat content (calories) is calculated simply as temperature × volume. In hypersaline water, however, both mass and specific heat are significantly altered. Fortunately, it appears that increased mass (1.176 g cm^{-3}) is approximately balanced by lowered specific heat (c. 0.82 cal g-$°C^{-1}$; unpublished data), and we have assumed that temperature × volume adequately reflects heat content in Deep Lake. We believe that any error in this assumption is less than 5%. Mason (1967) adopted the same approximation for calculating the heat content of Mono Lake, California (c. 2 × SW).

Birge's work of the wind, and Schmidt's stability are complementary quantities expressing work done through the lake surface: first, to establish a given density stratification from an assumed initial condition, and then to destroy that stratification without further exchange of heat or solute. References which deal with the mathematical and conceptual framework of these quantities in more detail than is given here include Birge (1916), Hutchinson (1957), Idso (1973), Walker (1974), Johnson *et al.* (1978), and Viner (1984).

Birgean wind work is limited by its conceptual time dependence, in that a starting point is assumed several months prior to the period of maximum stratification in summer. Simplified assumptions are necessary concerning the processes which led to the final maximum departure from the initial condition. In the context of Deep lake, the likely contribution of salinity to this density stratification complicates the interpretation of Birgean wind work. In general, processes that contribute to density layering but do not operate through the surface of the lake, for example inputs of water of different temperature, turbidity or salinity, fall outside the conceptual framework of Birgean wind work.

Schmidt stability, on the other hand, is conceptually instantaneous and is therefore not limited by assumptions concerning the establishment of the existing density stratification. It does however assume that the means of destroying the stratification is turbulent energy, transferred to the lake through its surface (i.e. by the wind). Finally it should be noted that all these quantities are minima. That is, the Heat budget is the minimum quantity of heat stored in the lake during the period of heat income. Birgean wind work and Schmidt stability are minimum estimates of energy transferred to a lake through its surface, by the wind.

Results and discussion

Thermal stratification

Descriptive reports of thermal stratification in Deep Lake, based on data collected between 1973 and 1976, are contained in Kerry *et al.* (1977), Campbell (1978), Burton & Campbell (1980), and Barker (1981). Temperature isopleths for the period 1977–1979 are shown in Fig. 2. In the absence of ice-cover, Deep Lake exhibits an essentially monomictic thermal cycle. Vertical circulation is normally complete for a period of two to three months between July and September (winter), when water temperature falls to between −17 and −18 °C. Maximum surface temperatures range from 7–11.5 °C for brief periods in summer. Bottom temperatures rarely warm to −14 °C (except for a small pocket below 34 m) giving a total range of 4–5 °C annually. The maximum vertical temperature gradients in summer, from 21–26 °C, are large in relation to most other lakes. A further point is that temperatures exceeding 0 °C are only recorded in the upper 10–15 m of the water column, and for three to four months of each year.

Barker (1981) noted the existence of a deep layer of more saline water, that appeared to survive the 1974 mixing period (which extended only to between 25 m and 30 m), but was obliterated in the winter mixing of 1975. The extension of sampling to 36 m, in 1978, revealed a small pocket of more saline water which supported an inverse temperature gradient. This has been found to exist on occasional visits to the lake since that time. It is possible that a small pool, of variable vertical extent, exists for much of the time in Deep Lake. This pool may be generated by dissolution of crystalline deposits of mirabilite ($Na_2SO_4 \cdot 10H_2O$) which lie exposed on the lake bottom below 30 m.

Annual terms

Table 1 contains a list of mainly high latitude lakes for which some data concerning annual heat budget, or annual maximum of either Birgean wind work or Schmidt stability is published.

Deep Lake has an annual heat budget similar to that of Mono Lake (38°N), but large in comparison with other high latitude lakes. Exceptions to this are Great Slave Lake which is both deeper and much more extensive than Deep Lake, and Lake Chandler if the latent heat of fusion is taken into account. Goreham (1964) provides predictive regressions relating lake morphometry to the magnitude of the annual heat budget for 71 temperate lakes. In this context, Deep Lake's annual heat budget closely matches that predicted for lakes of similar mean

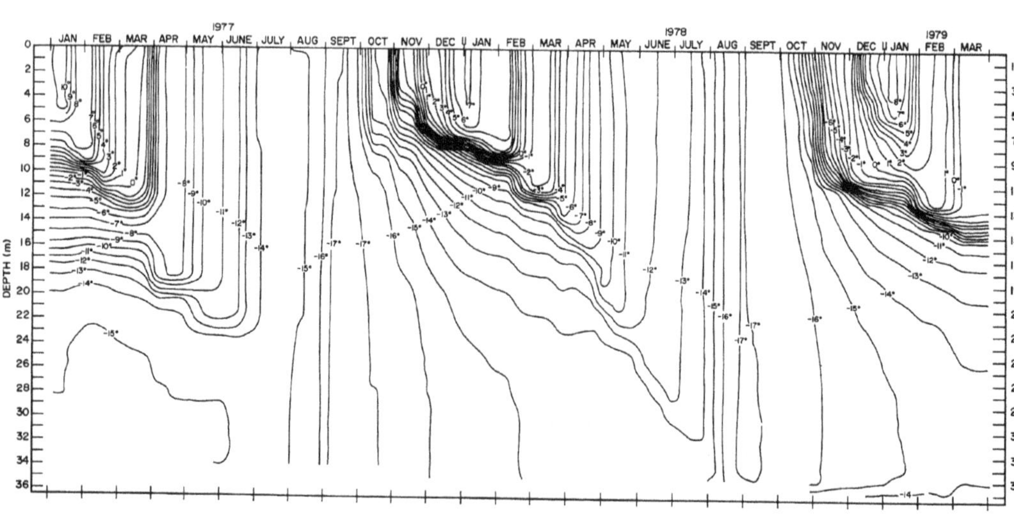

Fig. 2. Temperature isopleths (°C) for Deep Lake, from January 1977 to March 1979.

Table 1. Comparative data for annual heat budget, Birgean work, and mechanical stability, for mainly high latitude lakes. Ordered, where possible by mean depth.

Lake	Latitude	Altitude (m. ASL)	Area (km^2)	Depth Maximum (m)	Depth Mean (m)	Annual heat budget (cal cm^{-2})	Birgean wind work (gm-cm cm^{-2})	Heating efficiency[a] (gm-cm cal^{-1})	Schmidt stability (gm-cm cm^{-2})	Source
Deep	68°34′ S	−50	0.64	36	20.2	24600	c. 9000	0.37	8100	
Tennesvatn	67°53′ N	230	1.0	168	93.0	5000[b]				Hutchinson (1957)
Great Slave	61°31′ N	150	27200	614	62.0	24300+				Hutchinson (1957)
Schrader	Alaska		13.2	57	33.0	8980				Wetzel (1983)
Mono	38°00′ N	1979	199.5	51.5	19.0	22300	5750	0.26	3500	Mason (1967)
Chandler	67°40′ N	886	15.0	21	13.5	5760[b,c]	175	0.03		Hutchinson (1957)
Burton	68°34′ S	0	1.35	18.3	7.2				1520	Burton (unpublished)
Irizar	62°58′ S	7	0.046	18	4.3	1430			770	Bienati (1967)
Joyce	c. 77°30′ S								2610	Parker & Simmons (1978)
Fryxell	77°30′ S								1140	Parker & Simmons (1978)
Chad	77°39′ S								550	Parker & Simmons (1978)

[a] Defined after Hutchinson (1957) as Birgean wind work divided by summer heat income; [b] Summer heat income, less than the annual heat budget for a dimictic lake; [c] Heat budget = c. 27000 cal cm^{-2}, if the energy needed to melt the ice cover is included.

depth; but it is considerably larger (c. 9000 cal cm^{-2}; 37.7 10^3 J cm^{-2}) than would be predicted on the basis of either surface area or volume.

The difficulty of interpreting Birgean wind work for a lake where salinity can play a significant role in determining the density stratification has already been mentioned. Nevertheless, as the most important salinity effects in Deep Lake relate to dilution of the surface water, it is apparent that work is done in mixing this lighter water downwards in a way analogous to that for water warmed by the sun. Consequently, the estimate of Birgean wind work (Table 1) incorporates the influence of both temperature and salinity on the change from the previous period of minimum stratification to the subsequent time of maximum deviation from this initial condition. Calculation of Birgean wind work for polar lakes is rare, presumably because most such lakes are ice-covered for the greater part of the year. However, the Birgean wind work value for Deep Lake (9000 g-cm cm^{-2}; 0.883 J cm^{-2}) is unusually large compared to that of hypersaline Mono Lake, and also in the broader context of temperate and tropical lakes throughout the world. Large Freshwater lakes, such as Lake Riñihue (39°50′S: Campos *et al.*, 1978) and Lake Powell (37°30′N: Johnson & Merritt, 1979) yield values of Birgean wind work no greater than 8000 g-cm cm^{-2} (0.785 J cm^{-2}).

Hutchinson (1957) used the ratio of Birgean wind work to summer heat income (equivalent to heat budget for monomictic lakes) to give an efficiency term (g-cm cal^{-1}) describing the work required to account for each calorie of heat stored by the lake; a low ratio indicates efficient heating, such as for Lake Chandler (Table 1), while the higher ratio for Mono Lake denotes relatively inefficient heating. Any reference, here, to the efficiency or inefficiency of heating follows the restricted sense of Hutchinson's (1957) definition. Data for temperate and tropical lakes range from 0.04 (Lunzer Untersee; 47°57′N; Hutchinson, 1957;) to 0.56 (L. Valencia; 10°N; from data given by Lewis, 1984). Although Deep Lake (0.37) lies within this range, the efficiency ratio is second only to that of Lake Valencia, indicating particularly inefficient heating relative to most lakes for which data are available.

Deep Lake is characterized by a relatively high stability compared to lakes represented in Table 1, with an annual maximum stability of similar magnitude to that reported by Hutchinson (1957) for much larger and deeper Pyramid Lake.

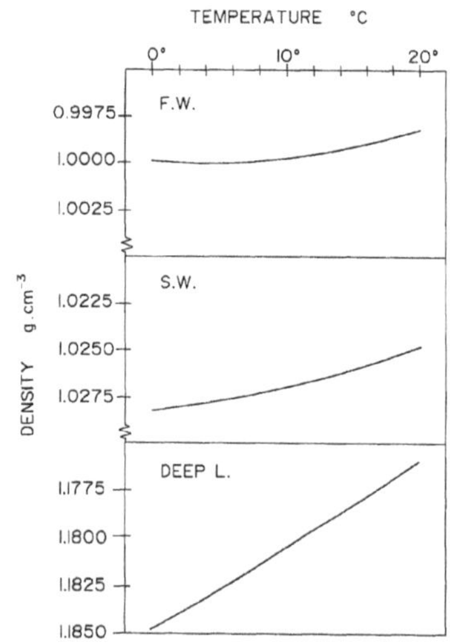

Fig. 3. A comparison of density (g cm^{-3})/temperature (°C) curves for freshwater (upper panel), seawater (middle panel), and Deep Lake water (lower panel), in the range from 0 °C to 20 °C, and plotted such that each of the density axes covers the same total range of density.

Density change temperature

The principle factor contributing to the high stability and inefficient heating of Deep Lake is the very high concentration of dissolved salts. Mason (1967) observed for Mono Lake (c. 2 × SW) '... a very rapid change of density with temperature.' Deep Lake has a concentration of approximately 10 × SW (Barker, 1981), and shows a much greater density change per degree Celsius. The rate of density change with temperature (0 – 20 °C) for freshwater, seawater and Deep Lake water, is illustrated by the gradients of the curves in Fig. 3. Direct comparison of equivalent volumes of fresh and Deep Lake water warmed from 4 °C to 9 °C (within the range of temperatures recorded for Deep Lake) shows the change in density to be approximately 11 times greater for Deep Lake water. The density change per degree Celsius, of

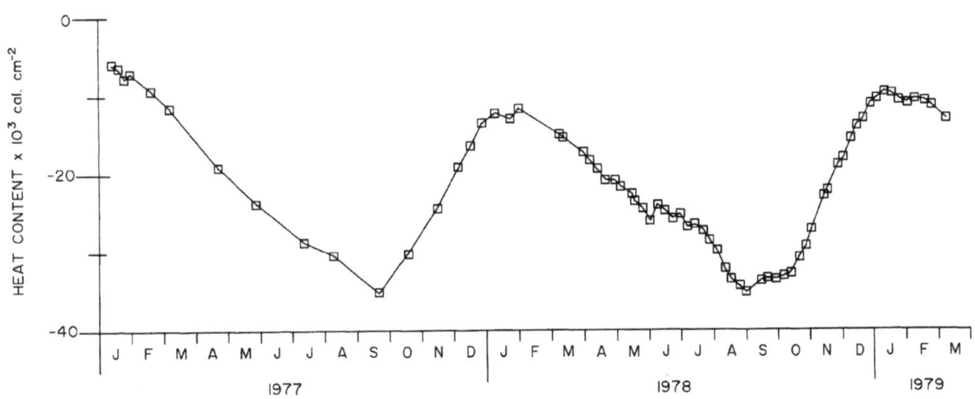

Fig. 4. The seasonal progression of whole lake heat content (cal cm^{-2}; $\times$ 4.19 = Joules cm^{-2}), from January 1977 to March 1979.

freshwater, increases with temperature above 4 °C, in the range of temperatures normally encountered in lakes (Hutchinson, 1957). This change in the density of Deep Lake water (between 4 °C and 9 °C) is comparable (c. 1.3 times) with that of freshwater warmed from 30 °C to 35 °C. Therefore, Deep Lake will tend to stratify as a result of quite small thermal gradients, in much the same way as a tropical freshwater lake. The generally low stability of tropical lakes reflects the very small thermal gradients forming under conditions of diminished seasonality. In Deep Lake, large summer thermal gradients combine with rapid density change per degree Celsius to give a quite stable density configuration explaining both high stability and inefficient heating. The thermal density gradient recorded on 14 January 1977 was approximately twice that of a freshwater lake stratified from 35 °C to 4 °C.

Annual cycles

Figures 4 & 5 show the annual progression of heat content and stability in Deep Lake. The stability has been partitioned such that total and thermal stability are shown, the space between the two lines (in Fig. 5) represents the stability resulting from saline gradients in the water column. The apparent smoothness of the 1977 curves is a result of less frequent sampling in that year.

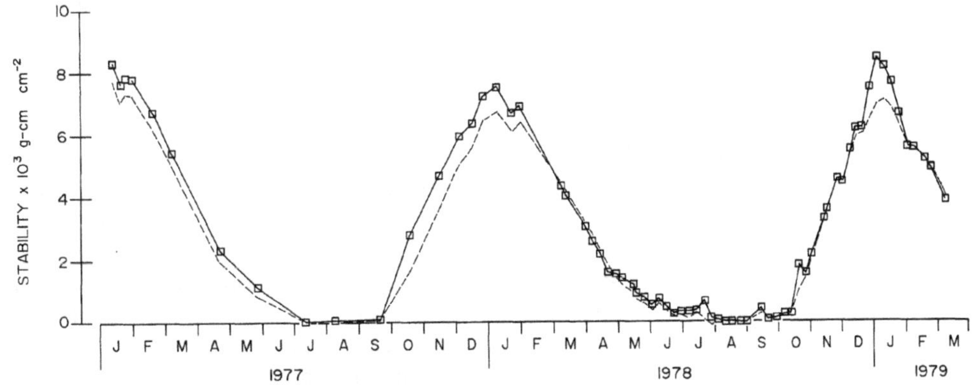

Fig. 5. The seasonal progression of total stability (S_{tot}), and thermal stability (S_t: g-cm cm^{-2}; $\times$ 9.807 10^{-5} = Joules cm^{-2}). The region between the two curves is equivalent to the saline stability (S_s).

Table 2. Heating and cooling rates for various lakes (change in heat content per unit time).

Lake	Heating period	Heating rate (cal cm^{-2} day^{-1})		Cooling period	Cooling rate (cal cm^{-2} day^{-1})		Source
		Mean	Maximum		Mean	Maximum	
Deep (Antarctica)	Sep. – Dec.	252	444	Jan. – Sep.	– 107	– 339	
Powell (USA)	Apr. – Aug.	368	439	Sep. – Jan.	– 345	– 392	Johnson & Merritt (1979)
Kamloops (Br. Col.)	June		c. 400	January		c. – 400	Carmack *et al.* (1979)
Baikal (USSR)	117 days	362					Hutchinson (1957)
Klammingen (Sweden)	May		332	October		– 201	Hutchinson (1957)
Riñihue (S. Am.)[a]	December		300	July		– 260	Campos *et al.* (1978)
Green (USA)	May – Aug.	215					Hutchinson (1957)
Mono (USA)	May – Aug.	176					Mason (1967)
Ladoga (USSR)		160					Hutchinson (1957)
Windemere (Gr. Br.)		160					Hutchinson (1957)
Titicaca (S. Am.)	Sep. – Dec.	158		May – Jul.	– 172		Richerson *et al.* (1975)
Burragorang (Aust.)	Sep. – Dec.	113	125	Apr. – Jul.	– 157	– 169	Ferris (unpublished)
Neagh (Ireland)[b]	May		106	November		– 106	Gibson & Stewart (1973)
Augher (Ireland)	Apr. – May		76				Rippey (1983)

[a] These rates apply to the surface layer only; [b] Predicted values derived from an empirical model.

Heat content, and therefore volume weighted average temperature, never exceed zero. Heat content shows an almost saw-toothed progression characterized by near linear heating and cooling phases with rapid interchanges between them. It is also obvious that heating proceeds more quickly but for a shorter time than the cooling phase. Table 2 contains comparative data on heating and cooling rates from various lakes around the world, most of which are freshwater. These rates represent the change in heat content per unit time for periods of approximately linear gain or loss of heat (cal cm^{-2} day^{-1}). Deep Lake has a relatively rapid period of heat gain from September to December, though it is less than for lakes Powell and Baikal. The maximum rate, however, is very large and this most probably reflects the very long hours of daylight in summer.

Deep Lake cools at a rate less than half of its heating rate, and is slow to cool compared with other lakes (Table 2). The maximum rate of heat loss, however, is relatively large. The percentage of incident radiation retained by temperate lakes has been estimated by Hutchinson (1957), in his consideration of lake thermal capacity. He noted that 40 – 60% retention was characteristic of 'supposedly first class lakes' in North America, though 70% or more was possible for large lakes such as Baikal and Michigan. Solar radiation measurements made at nearby (9 km) Davis station (Burton & Campbell, 1980) indicate that Deep Lake retained 50% of incident solar radiation in the period September – December 1977. This is despite the lake's inefficient heating (*sensu* Hutchinson, 1957), its small size, and the comparatively low angle of direct solar radiation. Mason (1967) estimated that Mono lake retained only 22% of incident radiation during the heating period from March – August. Climatic differences may be invoked to explain this latter comparison, as the mean wind speed at Davis (5 m sec^{-1}) is about twice that for Mono lake. Further, records of winds greater than 15 m sec^{-1} are rare for Mono lake, while winds at Davis frequently exceed this figure.

It is evident from Fig. 5 that temperature is the dominant contributor to S_{tot}, but that salinity is significant at times. In winter, when S_t is minimal, salinity gradients can be the dominant term in S_{tot}. Total stability shows a three part annual progression. A rapid, and approximately linear decline in stability (January – April) follows the maximum of S_{tot} in January. Between April and about July the decline

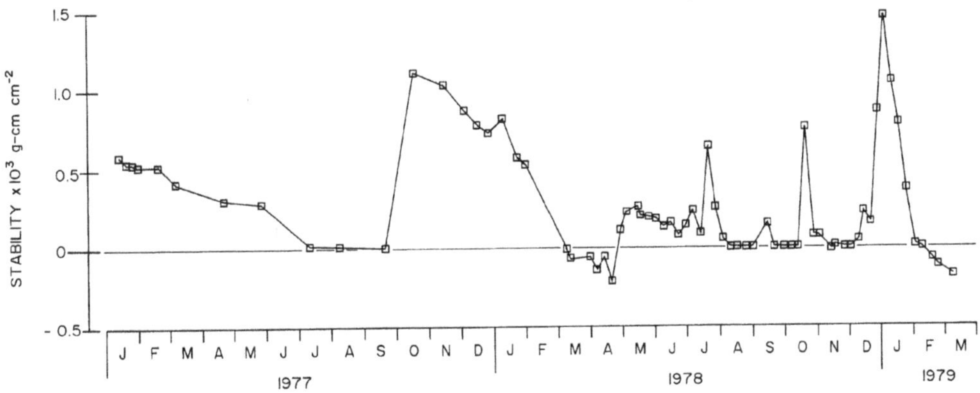

Fig. 6. Stability due to the salinity gradient (S_s: g-cm cm^{-2}), calculated as $S_{tot} - S_t$, and plotted on an expanded scale, for the period January 1977 to March 1979.

in S_{tot} is slower. July–September contains the period of minimum winter stability. The ensuing heating period (October–December) results in a near linear change (increase) in S_{tot}, at a rate greater than for any other part of the year. The asymmetry of this annual cycle of stability is in accord with that suggested by a statistical treatment of data for 11 New Zealand lakes by Viner (1984).

Regardless of different sampling frequencies, it is apparent, from Fig. 5, that the contribution of salinity to S_{tot} was qualitatively different in 1977, compared with 1978. In 1977 S_{tot} had a consistent component due to salinity, except for the period of winter mixing, whereas in 1978 the influence of salinity was irregular and episodic.

In Fig. 6 saline stability (S_s is presented on an enlarged scale. Figure 7 shows the variation in surface and 34 m salinity concentrations. The absolute magnitude of stability resulting from salinity gradients is evident from Fig. 6. The maximum effect, 1470 g-cm cm^{-2} (0.144 J cm^{-2}) occurred in January 1979 when the saline stability was approximately 20% of the maximum stability for that summer. Direct comparison of Figs. 6 & 7 gives clear indication that fluctuations in surface salinity generate the changes in S_s. These two graphs are essentially mirror images. with the sudden spikes of stability reflecting equally short-lived changes in surface water salinity. Despite this irregular progression of saline stability, there are consistent features. The data indicate that a salinity

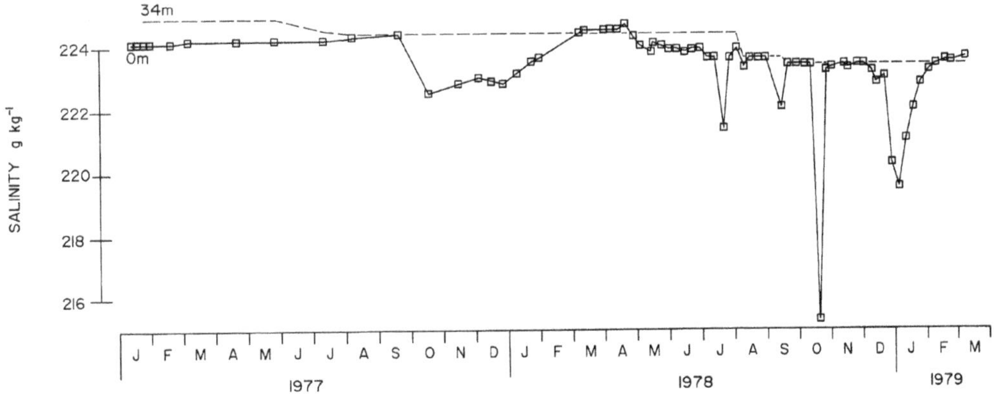

Fig. 7. The seasonal progression of salinity (g kg^{-1}), at the lake surface (0 m), and 34 m below the lake surface.

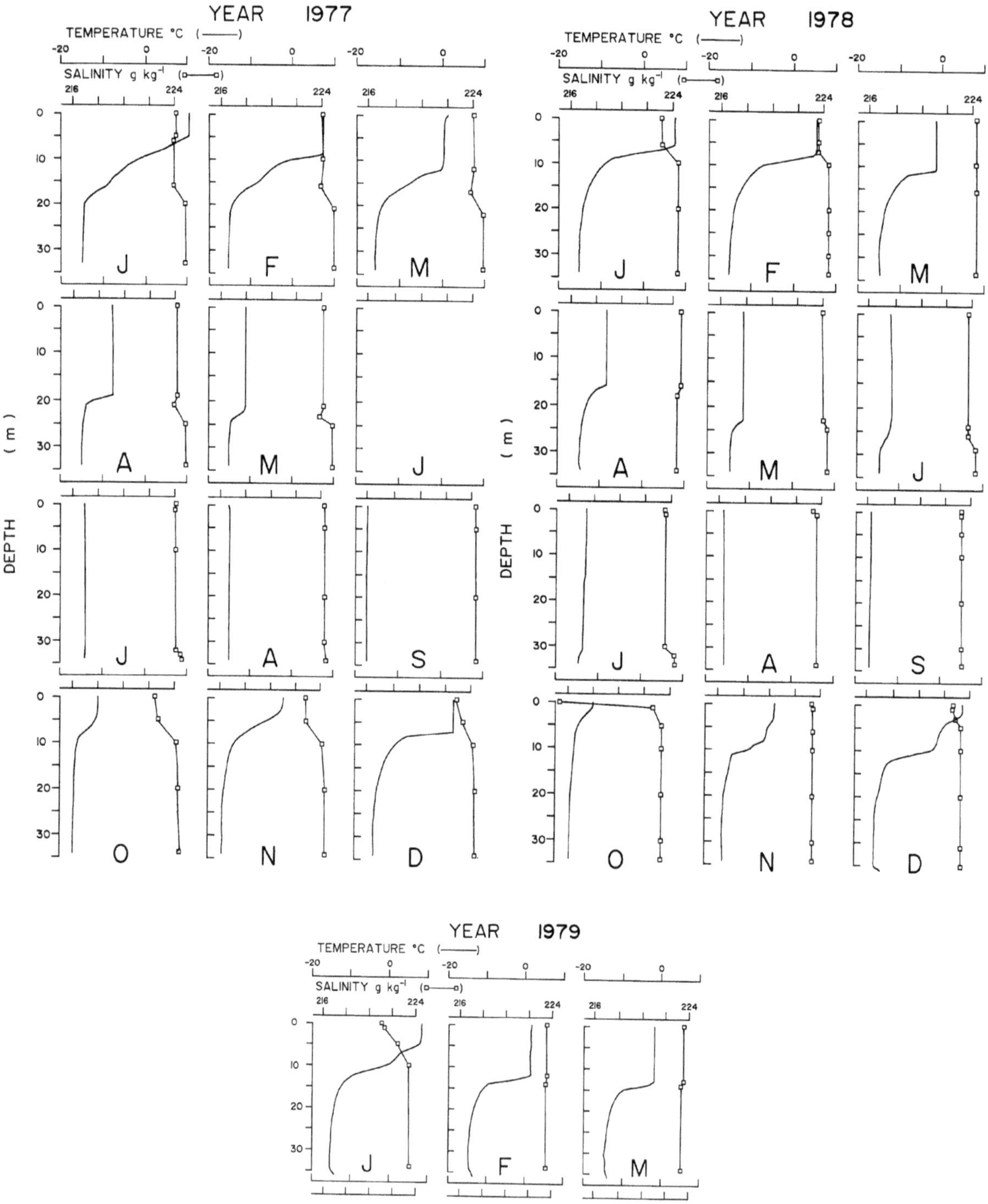

Fig. 8. Representative profiles (1 month^{-1}) of temperature (°C) and salinity (g kg^{-1}) for 1977, 1978, and early 1979.

gradient is usually found in the summer, particularly December–January. This probably results from the small melt-streams which drain into Deep Lake for one to two months in summer (Barker, 1981). This salinity gradient declines between January and April, partly as the superficial fresher layer is mixed into the lake, and partly as a result of evaporative concentration during this period when there are fewer strong winds and, therefore, less drift snow (Burton & Campbell, 1980). This latter effect causes the slightly negative S_s recorded between February and April of both 1978 and 1979. The slightly elevated surface salinity, which accompanies these destabilizing salinity gradients, is shown in Fig. 7. A maximum negative stability of -210 g-cm cm^{-2} (-0.021 J cm^{-2}) was recorded in April 1978.

Dilution of the surface water by direct snowfall, or snowdrift during strong winds, can occur at any time of the year. Burton & Campbell (1980) suggest that Deep Lake acts as a snow drift-trap, in which case the input of drift snow will be greatest in winter when strong winds are most common. This probably explains the sudden increases in saline stability in October 1977 and the 'spikes' in July, September and October of 1978. This stabilizing effect of precipitation or inflow is likely to characterize saline lakes in general. In Deep Lake, this gives rise to the paradox that periods of strong wind do not simply promote vertical mixing, as in freshwater lakes. Although this is certainly one effect of strong winds; the effect of fallen and drifted snow is to form a superficial layer of less dense water that serves to dampen wind-driven mixing.

Profiles of temperature and salinity

Figure 8 contains representative profiles of temperature and salinity for each month of the study period. These illustrate the general dependence of salinity gradients on the thermal structure in Deep Lake. The profiles clearly show that salinity changes are surface phenomena whose vertical extent usually coincides with that of the epilimnion. The early part of 1977 appears to be an exception to this pattern, as the main halocline lies well below the epilimnion. Closer examination, however, reveals that the halocline coincides with a secondary, deeper, thermocline. The formation of secondary thermoclines appears to be a common phenomenon in Deep Lake during the heating period (e.g. December 1978, Fig. 8), and presumably results from the interplay of calm and windy weather at a time when thermal stratification is being re-established (see also Burton & Campbell, 1980). The mixed depth of a lake is subject to considerable variability, especially during the early stages of the heating period (Darbyshire & Edwards, 1972; Mortimer, 1974; Wetzel, 1983). In a hypersaline lake the stabilizing effect of precipitation will certainly play some part in determining the mixed depth as well.

The effect of snow fall or drift on the lake is best illustrated by the October 1978 profile which shows a very thin layer of relatively less saline water. Figure 8 also shows the effect of evaporation on the epilimnetic salinity; slightly higher salt concentrations are evident in the surface waters during February and March of 1979.

Deep Lake and Lake Hunazoko

Deep Lake (Vestfold Hills) and Lake Hunazoko (Skarvs Nes) are lakes of similar salinity, but markedly different limnetic algal production. Selected comparative data for the two lakes are presented in Table 3.

Despite some disagreement over the species diversity in these lakes, a problem common to the investigation of hypersaline lakes throughout the world (see Nissenbaum, 1979; Heywood, 1984), it is accepted that both lakes support active populations of the halophilic green alga *Dunaliella* sp. Further, it it evident that Lake Hunazoko supports limnetic populations of *Dunaliella* reaching biomass concentrations (chlorophyll *a*) that exceed those of Deep lake by about an order of magnitude (see Akiyama, 1975; Tominaga & Fukui, 1981; Wright & Burton, 1981). The extremely low limnetic production in Deep Lake has been discussed by several authors (see Campbell, 1978; Williams, 1979; Hand, 1980; Wright & Burton, 1981), and various factors have been proposed as growth limiting in the lake. These include nutrient availability, salinity and temperature.

Table 3. Comparative data for Deep Lake and Lake Hunazoko. Thermal data for Deep Lake are from the present study, and the remainder from Barker (1981). Unless otherwise indicated, data for Lake Hunazoko are from Tominaga & Fukui (1981).

Parameter	(units)	Deep Lake	Lake Hunazoko
Surface area	km^2	0.64	0.142
Maximum depth	m	36.0	9.2
Mean depth	m	20.2	c. 4 – 5
Ice cover		none	probably none
Temperature	°C	7.5 to − 15.7 (26.1.77)	12.0 to − 13.0 (24.1.77)
Minimum temp.	°C	− 17.5	− 18.3
Depth of thermocline	m	9 – 18	3 – 4
D.O.	$ml\ l^{-1}$	2.1 – 2.6	2.6 – 2.8
Chloride	$g\ l^{-1}$	154.8 – 168.3	73.1 – 158.0[a]
pH		7.4 – 7.5	7.3 – 7.7[a]
Euphotic depth	m	>36	3
Phosphate	$\mu g\ l^{-1}$	56.7	10.9
Nitrate	$\mu g\ l^{-1}$	82.2	20.3

[a] Data also taken from Watanuki & Ohno (1975).

Concentrations of nitrogen and phosphorus tend to be higher in Deep Lake than in Lake Hunazoko (Table 3), but Campbell (1978) and Williams (1979) have suggested that, *in situ*, nutrients may be unavailable in Deep Lake. The absence of detectable ammonia, probably the preferred form of nitrogen for planktonic assimilation (Brezonik, 1972), in Deep Lake, compared to levels of 24 – 84 $\mu g\ l^{-1}$ in Lake Hunazoko (Tominaga & Fukui, 1981) may indicate that nitrogen is less readily available to the phytoplankton of Deep Lake.

Wright & Burton (1981) suggest that the existence of a lake similar salinity, but with much greater limnetic algal production (L. Hunazoko) shows that '... salinity alone is not limiting productivity in Deep Lake.' However, there are salinity differences between the two lakes, in that Lake Hunazoko has a much more variable surface salinity, and only reaches salinities within the range found in Deep Lake, in its hypolimnion. This indicates that summer inflow of meltwater constitutes a greater percentage of the total lake volume in Lake Hunazoko, which could have a variety of ecological effects in the lake.

The most obvious differences between the two lakes, however, are that Lake Hunazoko is smaller, shallower, and more turbid. Thermal profiles, taken two days apart in January 1977, show that Lake Hunazoko had a shallower and warmer epilimnion (Table 3). Watanuki & Ohno (1975) report a surface temperature some 3 °C greater than any recorded for Deep Lake in the period 1973 – 1979. How long this situation persists, in any year, cannot be determined from the available data. Lake Hunazoko might be expected to heat and cool more rapidly than Deep Lake, but during the cooling phase, when epilimnetic temperature is determined by the progressive incorporation of cold hypolimnetic water as well by surface heat exchange, it is possible that epilimnetic temperature would decline most rapidly in Deep Lake where a much greater percentage of the total lake volume remains cold throughout the year.

In general, therefore, the effects of salinity are not independent of other physical factors. The effect of increasing lake size and volume in antarctic hypersaline lakes will probably be to restrict the summer epilimnial temperature, and to shorten the period in which relatively warm conditions are available for algal growth. Parallel investigation of these lakes, for at least one annual cycle, would be valuable in defining the interaction of factors which limit algal growth in antarctic hypersaline lakes.

Conclusions

Deep Lake is the most ionically concentrated lake of the Vestfold Hills, with relative ionic proportions similar to sea water. As a direct result of its hypersalinity, the freezing point of Deep Lake water is depressed to the point where the lake virtually never freezes, despite water temparatures of −18 °C, during winter circulation. A second consequence of the lakes high salt content is an increase in the gradient of the temperature/density relationship. This factor together with particularly high summer thermal gradients results in a very stable density configuration in Deep Lake. A further effect of this enhanced rate of density change with temperature is that the lake has very inefficient heating, as indicated by the ratio of Birgean wind work to summer heat income (see Hutchinson, 1957). Nevertheless, Deep Lake has an annual heat budget close to that expected of temperate lakes with similar mean depth. Further, Deep lake captures c. 50% of the radiation incident upon it during the short heating period; this is whithin the range estimated for 'first-class' lakes in North America (see Hutchinson, 1957). It is suggested that water clarity and relatively frequent strong winds counter-balance the density and viscosity factors tending to retard downward mixing in the lake. The maximum rate of heating (444 cal cm^{-2} day^{-1}; 1.86 10^3 J cm^{-2} day^{-1})inferred from changing heat content with time is comparable to that of any lake in the world, despite the inefficiency (*sensu* Hutchinson, 1957) of heating and low solar angle. Extended summer daylight hours are probably very important in this. Deep Lake cools at a rate less than half of its average heating rate.

The annual cycle of mechanical stability is a function primarily of thermal stratification, which sets this lake apart from the many meromictic lakes of this region. Salinity gradients are capable of significant contribution to total stability, especially early in summer when the effect of small melt-streams is to increase stability by diluting epilimnial water. Evaporative water loss from the lake can reverse this dilution effect and create small de-stabilizing salinity gradients in late summer and in autumn. Salinity gradients do support inverse thermal gradients in Deep Lake. These are mostly transient features of the cooling phase, but a small pool (below 34 m) of markedly greater salinity was found in 1978, and may be persistent. Surface dilution by falling and drifting snow is a less predictable salinity effect which usually generates short-term increases in saline stability. These events may be more common in winter when strong winds, and hence drifting snow, are most frequent (Burton & Campbell, 1980). If surface dilution occurs during the period of thermocline re-formation, then relict thermoclines with associated haloclines may persist until the following cooling phase (e.g. in 1977).

It is evident that hypersalinity has a profound effect on the physical cycle of a lake, particularly a polar lake, by depressing the freezing point and increasing the rate of density change per degree Celsius compared to freshwater lakes. These changes affect the biota of the lake also, both in relation to osmotic stress and by effectively exposing them to a more thermally extreme environment. Deep Lake is characterized by a brief period when epilimnetic temperatures exceed 0 °C, in any year. A comparison between Deep Lake and a smaller lake of similar salinity (Lake Hunazoko, Skarvs Nes) demonstrates that it is inappropriate to consider the biological effects of salinity in isolation. The smaller lake offers warmer epilimnetic conditions for at least part of the summer, which may explain the much greater limnetic algal production in Lake Hunazoko. However, further investigation of both lakes is needed to determine the relative significance of thermal regime and other potentially limiting factors, such as nitrogen availability in Deep Lake.

References

Akiyama, M., 1975. Plankton and bottom deposits of Lake Funazoko-ike in Skarvs Nes, Antarctica. Shimane University Education Dept. Lett. 9: 29–42.

Barker, R. J., 1981. Physical and chemical parameters of Deep Lake, Vestfold Hills, Antarctica. ANARE Scientific Report, Publication No. 130. Aust. Govt. Publ. Service, Canberra, 73 pp.

Bienati, N. L., 1967. Estudio limnologico del lago Irizar, Isla Decepcion, Shetland del Sur. Contribucion del Instituto Antarctico Argentino No. 111: 1–36.

Birge, E. A., 1916. The work of the wind in warming a lake. Trans. Wis. Acad. Sci. Arts Lett. 18(2): 341–391.

Brezonick, P. L., 1972. Nitrogen: Sources and transformations in natural waters. In H. E. Allen & J. R. Kramer (eds) Nutrients in Natural Waters. J. Wiley & sons, N.Y.: 1–50.

Burton, H. R., 1981. Chemistry, physics and evolution of antarctic saline lakes – a review. Hydrobiologia 82: 339–362.

Burton, H. R. & P. J. Campbell, 1980. The climate of the Vestfold

Hills, Davis Station, Antarctica, with a note on its effect on the hydrology of hypersaline Deep Lake. ANARE Scientific Report, Publication No. 129. Aust. Govt. Publ. Service, Canberra., 50 pp.

Campbell, P. J., 1978. Primary productivity of a hypersaline antarctic lake. Aust. J. Mar. Freshwat. Res. 29: 717–724.

Campos, H., J. Arenas, W. Steffen & G. Agüero, 1978. Physical and chemical limnology of Lake Riñihue (Valdivia, Chile). Arch. Hydrobiol. 84: 405–429.

Carmack, E. C., C. B. J. Gray, C. H. Pharo & R. J. Daley, 1979. Importance of lake-river interactions on seasonal patterns in the general circulation of Kamloops Lake, British Columbia. Limnol. Oceanogr. 24: 634–644.

Darbyshire, J. & A. Edwards, 1972. Seasonal formation and movement of the thermocline in lakes. Pure & Applied Geophysics, 93: 141–150.

Gibson, C. E. & D. A. Stewart, 1973. The annual temperature cycle of Lough Neagh. Limnol. Oceanogr. 18: 791–793.

Goreham, E., 1964. Morphometric control of annual heat budgets in temperate lakes. Limnol. Oceanogr. 9: 525–529.

Hand, R. M., 1980. Bacterial populations of two saline antarctic lakes. In P. A. Trudinger & M. R. Walters (eds), Biogeochemistry of Ancient and Modern Environments. Proceedings of the Fourth International Symposium on Environmental Biogeochemistry (ISEB). Australian Academy of Science, Canberra: 123–129.

Heywood, R. B., 1984. Antarctic inland waters. In R. M. Laws (ed.), Antarctic Ecology, 1. Academic Press, Lond.: 279–344.

Hutchinson, G. E., 1957. A Treatise on Limnology, 1. Geography, Physics and Chemistry. J. Wiley & sons, N.Y., 1015 pp.

Idso, S. B., 1973. On the concept of lake stability. Limnol. Oceanogr. 18: 681–683.

Johnson, N. M. & D. H. Merritt, 1979. Convective and advective circulation of Lake Powell, Utah-Arizona, During 1972–1975. Wat. Resour. Res. 15: 873–884.

Johnson, N. M., J. S. Eaton & J. E. Richey, 1978. Analysis of five North American lake ecosystems II. Thermal energy and mechanical stability. Verh. Int. Ver. Limnol. 20: 562–567.

Kerry, K. R., D. R. Grace, R. Williams & H. R. Burton, 1977. Studies on some saline lakes of the Vestfold Hills, Antarctica. In G. A. Llano (ed.) Adaptations within Antarctic Ecosystem. Smithsonian Institution, Washington D.C.: 839–858.

Lewis, W. M., Jr., 1984. A five year record of temperature, mixing, and stability for a tropical lake (Lake Valencia, Venezuela). Arch. Hydrobiol. 99: 340–346.

Mason, D. T., 1967. Limnology of Mono Lake, California. University of California Press, Berkeley and Los Angeles, 110 pp.

McLeod, I. R., 1964. The saline lakes of the Vestfold Hills, Princess Elizabeth Land. In R. J. Adie (ed.) Antarctic Geology. North Holland, Amsterdam: 65–72.

Mortimer, C. H., 1974. Lake hydromechanics. Mitt. Int. Ver. Limnol. 20: 124–197.

Nissenbaum, A., 1979. Life in a dead sea – fables, allegories, and scientific search. Bioscience. 29: 153–157.

Parker, B. C. & G. M. Simmons, Jr., 1978. Ecosystem comparisons of oasis lakes and soils. Antarct. J. U.S. 13(4): 168–169.

Richerson, P. J., C. Widmer, T. Kittel, & A. Landa C., 1975. A survey of the physical and chemical limnology of Lake Titicaca. Verh. int. Ver. Limnol. 19: 1498–1503.

Rippey, B., 1983. The physical limnology of Augher Lough (Northern Ireland). Freshwat. Biol. 13: 353–362.

Sampson, R. J., 1978. Surface II graphics system (revised). Kansas Geological Survey Lawrence, 240 pp.

Tominage, H. & F. Fukui, 1981. Saline lakes at Syowa Oasis, Antarctica. Hydrobiologia 82: 375–389.

Viner, A. B., 1984. Resistance to mixing in New Zealand lakes. N.Z.J. Mar. Freshwat. Res. 18: 73–82.

Walker, K. F., 1974. The stability of meromictic lakes in central Washington. Limnol. Oceanogr. 19: 209–222.

Watanuki, T. & M. Ohno, 1975. Cultivation of antarctic microalgae (2). Isolation and culture of antarctic diatom Acnanthes brevipes var. intermedia from the bottom sand of the salt lakes at Skarvs Nes in Lützow-Holm Bay, Antarctica. Antarctic Record 54: 94–100.

Wetzel, R. G., 1983. Limnology, (2nd edition). Saunders College Publishing, N.Y., 767 pp.

Williams, R., 1979. Phytoplankton populations in an antarctic saline lake. M.Sc. thesis, University of Melbourne.

Whitfield, M. & D. Jagner (eds), 1981. Marine Electrochemistry. A Practical Introduction. J. Wiley & sons, N.Y., 529 pp.

Wright, S. W. & H. R. Burton, 1981. The biology of antarctic saline lakes. Hydrobiologia 82: 319–338.

Hydrobiologia 165: 129–139 (1988)
J. M. Ferris, H. R. Burton, G. W. Johnstone & I. A. E. Bayly (eds) Biology of the Vestfold Hills, Antarctica
Kluwer Academic Publishers

The inshore marine ecosystem off the Vestfold Hills, Antarctica

M. J. Tucker[1,2] & H. R. Burton[1]
[1]*Antarctic Division, Department of Science, Kingston, Tasmania, 7050, Australia;* [2]*Present address: Australian National Parks and Wildlife Service, G.P.O. Box 636, Canberra, 2601, Australia*

Key words: Antarctica, coastal, marine, ecology, zooplankton, epontic

Abstract

The planktonic, ice/water interface, and benthic communities at three sites off the coast of the Vestfold Hills, Antarctica, were examined over a complete year.

The planktonic flora and fauna were composed predominantly of oceanic species with diatoms and copepods the numerically dominant groups. Primary production was largely restricted to the summer months except for epontic algae which developed in spring. The zooplankton exhibited a similar seasonal cycle but lagged some months behind that of the phytoplankton.

The ice/water interface (epontic) fauna consisted of species from the plankton and benthos. Copepods were major contributors; however, two amphipod species dominated. Seasonality of the fauna in this habitat was determined by ice formation and breakout, and development of ice algae.

Each of the benthic substrates supported a characteristic macrofaunal assemblage, although infaunal amphipods and tanaids were similar at each site. Infauna exhibited a distinct seasonal cycle related to that of the primary producers whereas macrofauna showed no seasonal changes in abundance.

Species composition of each community in this coastal antarctic region was comparable with that of similar habitats in other antarctic coastal areas, supporting the circumpolarity of antarctic marine communities.

Introduction

Few studies have investigated antarctic shallow coastal areas and seasonal changes in the coastal community under ice (e.g. Bunt, 1960; Gruzov *et al.*, 1967; Gruzov, 1977). Although species composition and, to a limited extent, ecology of the biota have been determined in these studies, quantitative data, representative of coastal Antarctica, are still lacking. Unlike the relatively short and simple food chains of antarctic oceanic waters, those of coastal areas are complex. In addition to phytoplankton, other producers are ice algae, macrophytes and epiphytic diatoms. The zooplankton consists of relatively few species and individuals and the benthic habitat with its varied and abundant fauna, plays an important role.

Apart from the benthic work of Everitt *et al.* (1980), the marine coastal area off the Vestfold Hills had not been previously investigated. The present study was undertaken to determine species composition of the planktonic, ice/water interface and benthic communities and changes in the numerical abundance of the more common species of each community over a complete year. Other relevant reports of various aspects of the inshore marine ecosystem are contained in the present volume. Dhargalkar *et al.* (this volume) describe the invertebrates associated with common macrophyte algae of this area. Williams (this volume) gives an account of the fish that have been found in the region since 1978. Thomas & Jiang (1986) and Everitt & Thomas (1986) describe the epiphytic, planktonic and ice associated diatom flora, while Perrin *et al.* (1987) give a general account of seasonal changes in the phytoplankton.

Description of sites studied

Three sites on a transect from Davis Station to Gardner Island (Fig. 1) were examined monthly from January to December 1982. Site A had a water depth of 9 m below M.S.L., a flat sandy substrate and scattered rocks to which macrophytes were attached. Site B had a depth of 20 m with a flat muddy, silty substrate and no macrophyte cover. Site C had a depth of 15 m with a sloping smooth rock bottom and small scattered pockets of sand and shell fragments; for most of the year this site had approximately 20% macrophyte cover, but in the summer months large macrophytes overlaid almost the entire substrate.

Sea ice broke out from the bay in December 1981, reformed in early March 1982 and broke out again in early January 1983 after reaching a thickness of almost 2 m. During its formation in March, the sea ice was broken up several times by strong winds preventing safe access until April. Icebergs rarely enter the bay with only one encountered during the present study.

Materials and methods

The planktonic, ice/water interface (epontic), and benthic biotic associations were examined monthly at each site throughout 1982. Water samples were taken at selected depth intervals with a 2 l Kemmerer bottle for determinations of phytoplankton cell densities, chlorophyll content, salinity, nutrients and water temperature. After sea ice formation, samples were taken through a 10 cm diameter SIPRE corer

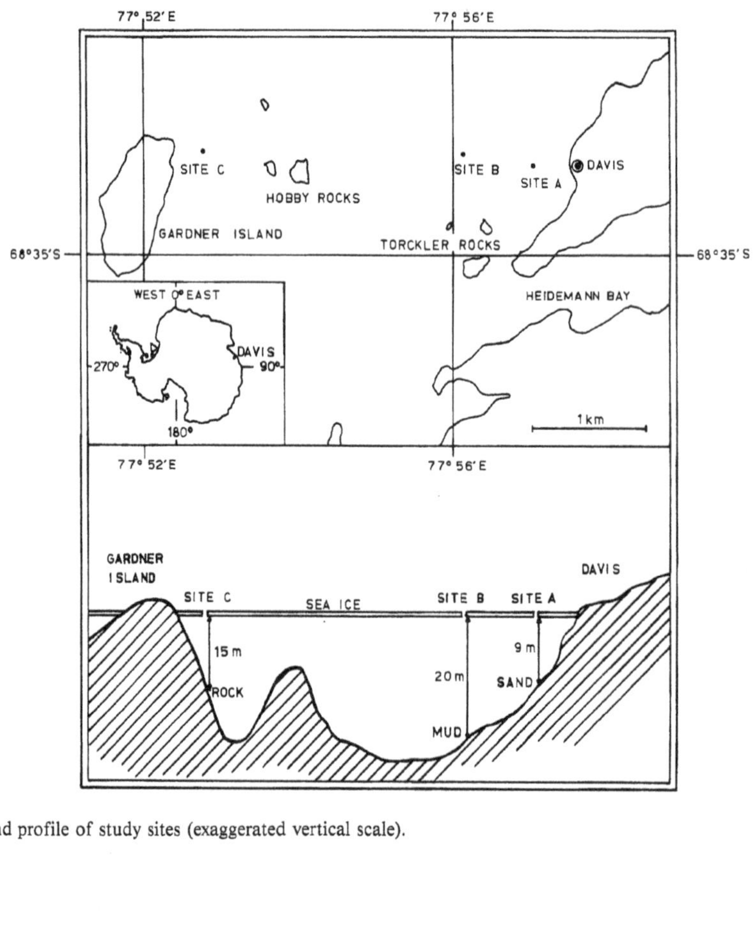

Fig. 1. Location and profile of study sites (exaggerated vertical scale).

hole (Journal of Glaciology, 1958) with the ice core retained for ice algae analysis. Details of sampling techniques and analysis are given in Perrin *et al.* (1987).

Zooplankton collection was by pump sampler (210 μm mesh net) and a collapsible net (210 μm mesh net) deployed through a SIPRE core hole during the period of ice cover. Details of the zooplankton sampling apparatus can be found in Tucker & Burton (1987).

Collections from the ice/water interface and benthos were made by divers. At the interface, a 12.5 cm diameter hand-held net (95 μm mesh) was scraped along the under surface of the ice. On several occasions collections were also made by divers scraping the friable ice layer into large glass jars. Due to the nature of these sampling techniques it was not possible to quantitatively measure the volume or surface area of the interface sampled.

Divers monitored the densities of the most numerous macroscopic benthic animals on each substrate type. Sediment samples were also taken and sifted through a 600 μm Endecott sieve with the retained portion examined for the infauna (i.e. the fauna in the mud or sand substrate). The details of the benthic sampling are given by Tucker (this volume) and details of the diving operation are given by Dick (1984).

All samples were transported to the laboratory in insulated containers to prevent freezing. Sorting was carried out under a Wild M5 microscope.

Fish were captured in a 5 cm mesh gill net, set by divers on the substrate at right angles to the prevailing water current. Fish traps, using several different meat baits, were also laid. Each net and trap was retrieved after approximately 24 h.

Results

Primary producers

Phytoplankton numbers (Fig. 2) were low for much of the year (February to November). The standing crop was much greater in the summer months (December and January) when production became so high that the waters of the bay resembled thick green

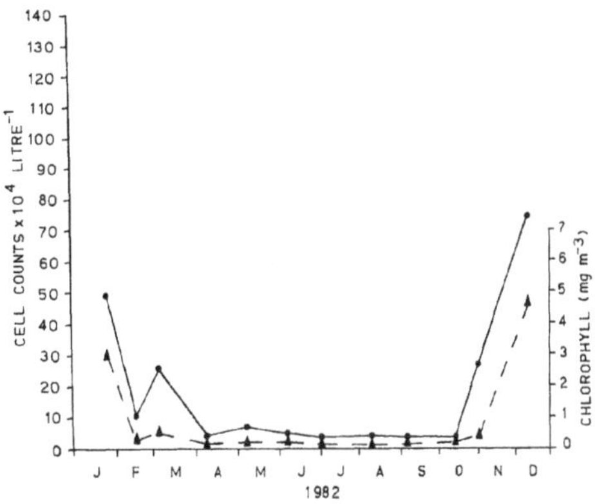

Fig. 2. Phytoplankton seasonal cycle, Site A. Plotted points represent averages for the entire water column. ● cell counts × 10^4; ▲ chlorophyll *a*.

soup. The prymnesiophyte, *Phaeocystis pouchetii* (Har.) Lagerh. dominated the early summer bloom but was replaced in mid-January by a diatom, *Nitzschia lanceolata* Wm. Smith, (Perrin *et al.*, 1987).

Diatoms, mainly *Nitzschia cylindrus* (Grun.) Hasle and *Navicula* sp., colonized the developing sea ice almost immediately in March and April (Perrin *et al.*, 1987). A distinct brown band had formed in the bottom 2–4 cm of the ice by May. This discoloured layer quickly disappeared in winter as snow cover increased and solar radiation decreased. Development of the ice algae resumed in October (Fig. 3) and was dominated by the diatoms *Am-*

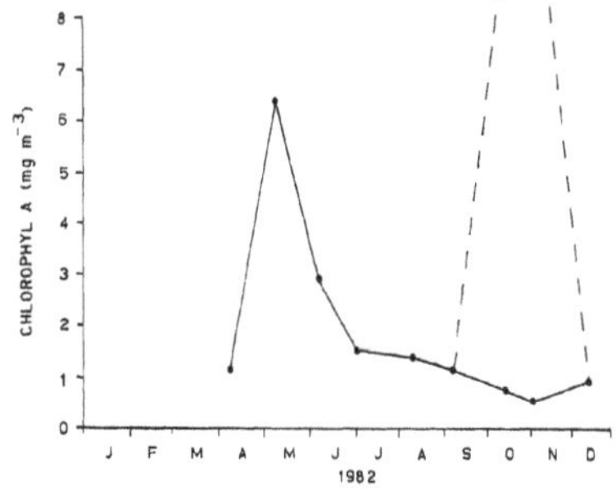

Fig. 3. Seasonal changes in chlorophyll *a* content of sea ice. ● chlorophyll *a* values from snow covered ice; ▲ chlorophyll *a* values from ice with no snow cover.

phiprora kjellmannii (= *Entomoneis kjellmannii* (Cleve) Thomas *in* Thomas & Jiang, 1986) and *Nitzschia frigida* (Grun.) in Cleve & Grunow (Perrin *et al.*, 1987). Although the development of ice algae preceded that of the phytoplankton by several months, these ice-locked algae did not contribute trophically to the plankton and benthos until ice melt began in November and December. During this spring period of maximum chlorophyll concentration, ice algal development was very patchy, there being a direct relationship observed between the absence of snow cover and ice algal growth.

Divers observed that macrophyte cover was greatest during the ice-free months of December to March. With increasing ice formation and declining sunlight hours, subsurface illumination decreased and the macrophytes deteriorated at site C, decreasing from an almost 100% bottom cover in late sum-

Table 1. Zooplankton species recorded from Davis Bay.

Taxa	Date of collection	Method of collection
Hydromedusa		
Solmudella bitentaculata Quoy & Gaimard	Apr.	Net
Halitiara spp.	Nov.	Net
siphonophore sp.	Nov.	Net
Chromatonema rubrum Fewkes	Nov.	Net
Scyphomedusa		
Desmonema gaudichaudi Lesson	Nov.	Diver
Ctenophora		
ctenophore sp.	Apr., May	Net
Polychaeta		
polychaete spp.	Mar. – May, July, Aug. Oct. – Dec.	Pump and net
polychaete larvae	Jan., Feb., July, Sept. – Dec.	Pump and net
Mollusca		
pteropod spp.	Nov., Dec.	Net
Copepoda: Calanoida		
Calanoides acutus (Giesbrecht)	Jan. – Aug.	Pump and net
Calanus propinquus Brady	Jan. – Sept., Nov.	Pump and net
Ctenocalanus vanus Giesbrecht	Jan. – Nov.	Pump and net
Metridia gerlachei Giesbrecht	Mar. – July, Sept., Oct.	Pump and net
Stephos longipes Giesbrecht	All year	Pump and net
Stephos sp.	Jan., May., Aug.	Pump and net
Paralabidocera antarctica (J. C. Thompson)	Jan., Oct. – Dec.	Pump and net
Drepanopus bispinosus Bayly	Jan., Aug., Sept.	Pump and net
Cyclopoida		
Oithona similis Claus	All year	Pump and net
Oncaea curvata Giesbrecht	All year	Pump and net
Harpacticoida		
Microstella sp.	Aug. – Nov.	Pump and net
harpacticoids	All year	Pump and net
Euphausiacea		
Euphausia crystallorophias (juvs) Holt & Tattersall	Apr.	Net
Thysanoessa macrura (juvs) G. O. Sars	Apr.	Net
Amphipoda		
Paramoera walkeri Stebbing	Apr. – Aug., Oct. – Dec.	Pump and net
Orchomene cf. *plebs* (Hurley)	July – Oct.	Pump and net
Echinodermata		
echinoderm larvae	Jan., Nov., Dec.	Pump and net
Ascidacea		
ascidian larvae	Jan., Oct., Nov.	Pump and net
Larvacea		
Oikopleura sp.	Feb., Apr., May, Nov., Dec.	Pump and net

mer to a low of approximately 20% for the winter. No new growth was observed until the following December.

Cocconeis fasciolata (Ehr.) Brown, *Trigonium arcticum* (Brightw.) Cleve and *Arachnoidiscus ehrenbergii* J. W. Bailey were common epiphytic diatoms (see Thomas & Jiang, 1986). No evaluation was made of epiphytic diatom production although it did appear to be an important factor affecting the benthic fauna.

Zooplankton

Copepods constituted almost 100% of zooplankton numbers with 11 species found (Table 1). The copepods as a group exhibited a distinct seasonal cycle (Fig. 4), increasing in number during January and February and reaching a peak between March and May. Numbers decreased almost to zero over winter and remained so until the following December. Species which contributed to this generalized pattern did not reach simultaneous maxima of abundance (Tucker, unpubl. data). *Stephos longipes* and *Calanoides acutus* attained maximum concentrations first in March. These species were followed by *Oithona similis* from March to May and *Calanus propinquus, Ctenocalanus vanus, Metridia gerlachei* and *Oncaea curvata* in May. Exceptions to this general seasonal pattern for the Copepoda were

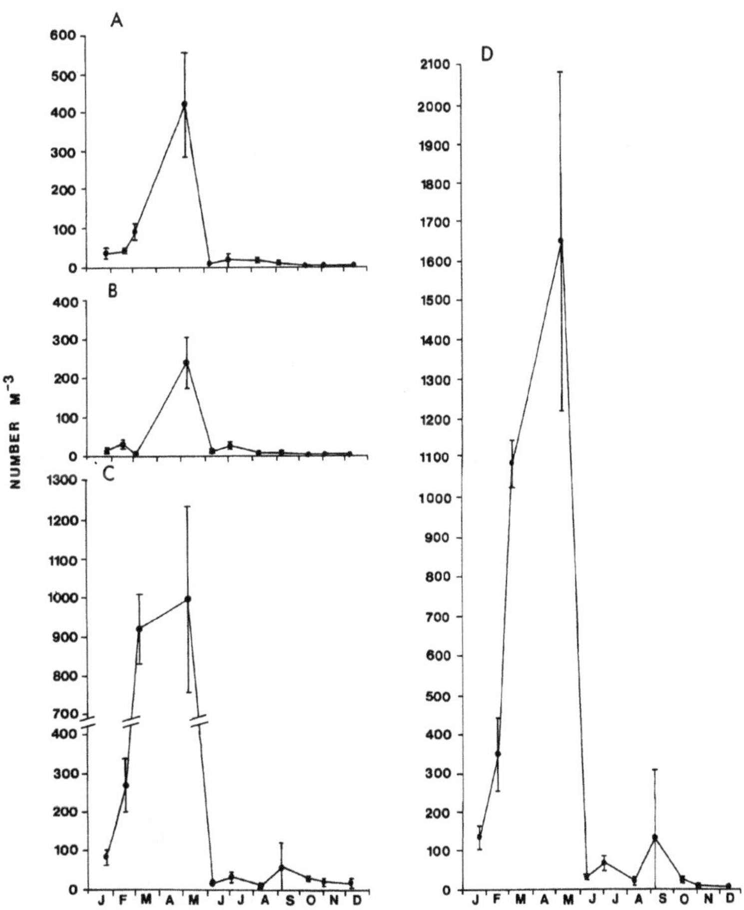

Fig. 4. Seasonal changes in copepod abundance, Site A. A females; B males; C juveniles; D total copepods. (Plotted mean values and standard deviations for entire water column.)

Paralabidocera antarctica, Drepanopus bispinosus, Stephos sp. and *Microstella* sp. Apart from *P. antarctica*, which is associated with the ice/water interface (see below), these latter species appeared in numbers too low for seasonal patterns to be established. However, the collection of *D. bispinosus* during late winter and spring was important as this species had only been reported previously from the lakes and fjords of the Vestfold Hills (Bayly, 1982).

Species other than the planktonic copepods were numerically insignificant contributors to the total zooplankton; but most also exhibited seasonal cycles. Copepod predators (medusae, ctenophores and pteropods) occurred during the period of peak copepod numbers (April and May). Larvae of benthic animals (polychaetes, echinoderms and ascidians) occurred mainly in summer, whereas adult polychaetes and harpacticoids were taken in small numbers for most of the year. Amphipod collections were related to their seasonal association and migration to the ice/water interface (see below). Only two specimens of euphausiids were taken, although a swarm of *Euphausia crystallorophias* was observed in Davis Bay in January 1983.

Ice/water interface

Species taken at the ice/water interface and their dates of capture are given in Table 2. Immediately after ice formation in April, the ice/water interface was dominated by the two cyclopoid copepod species and *S. longipes* (Table 3). From May onwards, with their migration from the benthos to the under surface of the ice, the amphipods *Paramoera walkeri* and *Orchomene* cf. *plebs* dominated. At times these amphipods were so numerous as to almost cover the under surface of the ice. *P. walkeri* was the most abundant amphipod near the shoreline, but at site C, 4 km offshore, dominance fluctuated between the two species. Each of the remaining three amphipod species (*Orchomene* sp., *Hyperia macrocephala* Dana and *Eusiris* cf. *fragilis* Birstein & Vinogradov) was represented by captures of only single specimens.

Divers noted that *P. walkeri* congregated on algal patches formed on the under surface of the ice in spring. Here they fed on filiform algae (also noted by Rakusa-Suszczewski, 1972) changing colouration from the characteristic red and white bands of winter, to brown and green. Ovigerous females, first observed in May, carried the slowly developing young through winter. First releases of juveniles were noted in October; but the majority were released in November and December. Some brooding females were still taken in late December.

Similarly, *O.* cf. *plebs* brooded its young over winter at the under surface of the ice but no newly released juveniles were observed and only two adults were captured after October at site C.

The copepod *P. antarctica* was first collected in

Table 2. Species captured at the ice/water interface.

Taxa	Date of collection
Nematoda	
nematode sp.	Nov.
Polychaeta	
polychaete larvae	Apr. – Nov.
Mollusca	
gastropod sp.	Apr.
pteropod sp.	Nov.
Copepoda: Calanoida	
Calanoides acutus	Apr. – June
Calanus propinquus	Apr. – July
Ctenocalanus vanus	Apr. – May, July – Sept., Nov.
Metridia gerlachei	Apr. – June
Stephos longipes	Apr. – Dec., i.e. all months of ice cover
Paralabidocera antarctica	Oct., Nov.
Drepanopus bispinosus	Aug.
Cyclopoida	
Oithona similis	Apr. – Dec.
Oncaea curvata	Apr. – July
Harpacticoida	
harpacticoid spp.	Apr. – Aug., Oct. – Dec.
copepod nauplii	Apr., June, July, Nov.
Amphipoda	
Paramoera walkeri	Apr. – Dec.
Orchomene cf. *plebs*	May. – Oct.
Orchomene sp.	Aug.
Hyperia macrocephala	Oct.
Eusirus cf. *fragilis*	Nov.
Larvacea	
Oikopleura sp.	Apr.
Pisces	
Pagothenia borchgrevinki	June, July

Table 3. Percentage contribution to total numbers at the ice/water interface, Site A.

	A	M	J	J	A	S	O	N	D
Copepoda									
Calanoides acutus (juvs)	11	1	2	0	0	0	0	0	0
Calanus propinquus (juvs)	8	1	4	4	0	0	0	0	0
Ctenocalanus vanus (juvs)	6	1	0	2	2[a]	0	0	0	0
Metridia gerlachei (juvs)	1	0	2	0	0	0	0	0	0
Stephos longipes	20	2	0	4	0	5	50	>10	0
Drepanopus bispinosus	0	0	0	0	3	0	0	0	0
Paralabidocera antarctica	0	0	0	0	0	0	5	>5	0
Oncaea curvata	21	4	3	0	0	1	0	0	0
Oithona similis	21	1	1	5	7	0	29	[b]	0
harpacticoids	0	1	1	1	1	0	6	1	0
copepod nauplii	1	0	1	1	0	0	0	1	0
Polychaeta									
polychaeta larvae	1	0	0	0	0	0	0	1	0
Amphipoda									
Paramoera walkeri	10	89	81	82	[b]	72	6	[b]	38
	0	12[c]	25[c]	36[c]		45[c]	6[c]		2[c]
newly released *P. walkeri*	0	0	0	0	0	0	0	[b]	72
Orchomene cf. *plebs*	0	0	4	0	[b]	22	4	0	0
other amphipods	0	0	1	1	0	0	0	0	0

[a] adults; [b] present (no quantitative data); [c] % of *P. walkeri* that were ovigerous females.

September appearing in large numbers in the scrapings from the under surface of the ice. No further collections of *P. antarctica* were made after the ice melt began in December. *S. longipes* was present at the interface for almost the entire period of ice cover, reaching large numbers in October and November. Captures of the remaining copepod species were in accordance with their appearance times in the plankton (Table 1). These species, however, were never major contributors to the interface fauna.

Fry of the nototheniid fish, *Pagothenia borchgrevinki* (Boulenger), were first observed in June among large ice pendants that resembled hollow dome-shaped stalactites, up to 1 m deep, on the under surface of the ice. In summer, in areas of open water, *P. borchgrevinki* fry were closely associated with benthic macrophytes, indicating migration from the ice/water interface (ice cover) to the benthos (no ice cover) in a manner similar to that of the amphipods *P. walkeri* and *O.* cf. *plebs*.

Benthos

Over 200 animal species were collected from the coastal water off the Vestfold Hills (Tucker & Burton, 1987) with the majority of these from the benthos. Most species that were collected were widely distributed along the sublittoral of the Vestfold Hills coastline (Tucker & Burton, 1987). To a certain extent, the three different substrate types investigated had a characteristic macrofaunal assemblage. However, the most common infaunal amphipods and tanaids were shared (Table 4).

Two general patterns of abundance emerged throughout the year, (Tucker, this volume). Firstly, a pattern of fluctuation in population numbers but with no discernable seasonal trend, e.g. most of the macrofauna. Secondly, the infaunal and smaller species showed a marked seasonal cycle with large numbers in summer (corresponding to the phytoplankton bloom), falling to a minimum in winter and spring and increasing in numbers again in December, e.g. the amphipods and tanaids. Individuals of all the infaunal amphipod and tanaid species brooded young throughout winter with release appearing to be in early December.

The fish species captured most often was *Pagothenia bernacchii* (Boulenger). The species *Pagothenia borchgrevinki, Pagothenia hansoni* (Boulenger),

Table 4. Numerically dominant benthic species at each site.

Site	Substrate	Epifauna	Infauna
A	sand	Echinoida *Abatus shackletoni* Koehler Ophiuroida *Ophiurolepsis martensi* (Studer) *Ophiura flexibilis* (Koehler)	Amphipoda *Heterophoxus videns* Barnard *Orchomene franklini* (Walker) Tanaidacea *Nototanais antarcticus* Hodgson
B	mud	Isopoda *Glyptonotus antarcticus* Eights Mollusca *Laternula elliptica* (King & Broderip)	Amphipoda *Ampelisca barnardi* Nicholls *Ampelisca* sp. *Heterophoxus videns* *Orchomene franklini* Tanaidacea *Nototanais antarcticus* *Nototanais dimorphus* (Beddard)
C	rock	Echinoida *Sterechinus neumayeri* (Meissner) Holothuria *Ekmocucumis spatha* (Cherbonnier) Ophiuroida *Ophiurolepsis martensi* Anthozoa *Glyphoperidium bursa* Roule Pycnogonida *Nymphon australe* (Hodgson)	Amphipoda *Heterophoxus videns* *Orchomene franklini* Tanaidacea *Nototanais antarcticus*

Chionodraco hamatus (Lonnberg) and *Gymnodraco acuticeps* Boulenger were next in respective abundance. Specimens of *Trematomus scotti* (Boulenger) and *Artedidraco* sp. were found in debris after a large storm in February (see also Williams, this volume). All specimens observed by divers, apart from the fry of *P. borchgrevinki* at the ice/water interface, were either in or just above the macrophyte cover. The fish appeared to rely on camouflage or concealment amongst macrophytes when disturbed. This habit enabled practiced divers to capture specimens by hand. The stomachs of *P. bernacchii* contained small benthic invertebrates (mostly amphipods) from the area.

Discussion

Primary producers (plankton, macrophytes, epiphytes) of this shallow coastal area can be classified into groups largely similar to those noted in other coastal regions of the world; but, in addition to these, epontic algae are also found. Production is markedly seasonal and is mainly concentrated in the summer months, with the exception of epontic algae which develop several months prior to the phytoplankton bloom. This seasonal cycle is largely determined by seasonal changes in solar radiation, timing of ice formation and breakout, and depth of sea ice and its snow cover (Perrin *et al.*, 1987).

Many of the inshore planktonic species are also important in the antarctic oceanic plankton communities. In fact, it appears that most of the coastal plankton species are derived from oceanic waters. Factors such as the short and steep continental slope, uniformity of coastal and oceanic regions, minimal terrestrial runoff, and absence of an intertidal community would all contribute to this lack of a distinctive coastal plankton population. It does, however, differ from the oceanic plankton in the summer occurrence of larvae from benthic animals, the addition of species associated with the ice/water interface and the occurrence of *Drepanopus bispinosus.*

Appearance of the oceanic zooplankters, *Calanoides acutus, Calanus propinquus, Ctenocalanus vanus, Metridia gerlachei,* euphausiids etc., in the waters off the Vestfold Hills is determined by the life histories of each species, their annual vertical migration patterns, and the surface water circulation of Prydz Bay. In response to the phytoplankton developing in oceanic waters in November, the zooplankton ascend to the surface waters to feed and reproduce, but descend back to the deeper waters in April when phytoplankton production decreases (Mackintosh, 1934; Ottestad, 1936; Andrews, 1966). During this period of congregation at the surface in summer, a large cyclonic gyre in Prydz Bay (Grigoryev, 1968) brings these oceanic species past the Vestfold Hills.

Although the remaining copepod species have also been recorded from oceanic waters; unlike the preceding species, sexually mature adults as well as juveniles of all stages were captured in the coastal waters at some time during the study. *Oithona similis* and *Oncaea curvata* are regarded as deep water species (Hardy & Gunther, 1935; Seno *et al.*, 1966) but in this coastal area they were the most numerous zooplankters captured (Tucker, unpubl. data). Perhaps the adaptations required to exist in deep aphotic waters (adaptations to low temperatures and no phytoplankton production) are also suitable for inhabiting the antarctic coastal area with its very short season of phyto-production. *Paralabidocera antarctica* was predominantly associated with the ice/water interface during the period of epontic algal development. *Stephos longipes* was also associated with the under surface of the ice and, although its largest numbers coincided with epontic algal growth, it was present for the entire period of ice cover. Unlike *P. antarctica,* it reverted to the plankton after ice melt began, appearing in the water column in large numbers during the ice-free months of February and March (Tucker, unpubl. data).

Temporal changes in zooplankton abundance were found to be closely related to those of the phytoplankton. However, between the peak periods of phytoplankton production (Fig. 2) and zooplankton abundance (Fig. 4) there was a lag of some months. Factors such as currents, location of breeding populations and ontogenetic changes in behaviour, such as annual vertical migration, could contribute to this delay.

Unlike other antarctic coastal habitats, the ice/water interface is much less stable and predictable from year to year. It forms and disappears on an overall seasonal basis but can quickly breakup and disperse, especially during the first months of development. Species at the interface must also possess some euryhalinity. During sea ice formation, salt is excluded and drainage channels of highly saline brine form (Lake & Lewis, 1970). Also, during the summer thaw in December, divers observed a distinct layer of less saline water (discernable by its different refractive properties) extending down to 1 m below the under surface of the ice.

Nonetheless, the ice/water interface did support a dense fauna composed of representatives from the plankton and benthos. Benthic species dominate the interface fauna with brood protecting amphipod species and fry of *Pagothenia borchgrevinki* characteristic of this habitat (Gruzov, 1977; Rakusa-Suszczewski, 1972). Most of the zooplankton species were at some time taken from the ice/water interface; but, apart from *P. antarctica* and *S. longipes*, never in large numbers. Whilst several species appeared to be able to 'visit' this habitat, *P. antarctica* and *S. longipes* appeared to be 'true' inhabitants of the interface in spring for both were poorly represented in the water column during the same period. *O. similis* may also be regarded as a 'true' inhabitant. Although it did not show a clear difference in population densities between the interface and water column, adult and juvenile stages were present at the interface in October and December when only juveniles were present in the plankton (Tucker, unpubl. data).

Seasonality of the interface biota is controlled firstly by the formation and breakup of the habitat. Due to the sampling techniques which were dependent upon a secure ice cover and were not capable of yielding abundance data on an ice area basis, seasonal changes in numerical abundance of the interface fauna could not be established quantitatively. Species from the zooplankton, apart from *P. antarctica* and *S. longipes*, occurred in the interface at times corresponding to their appearance in the plankton. *P. antarctica* and *S. longipes* appeared to have a distinct seasonal cycle with maximum population numbers in spring corresponding to those of the epontic algae, and low numbers (or complete absence for *P. antarctica*) for the remainder of the year. After migration of the amphipods to the under surface of the ice, their population numbers remained relatively stable until the ensuing ice breakup. However, there appeared to be a distinct seasonal reproductive cycle with release of advanced juveniles timed to the peak periods of primary productivity.

Substrate type appeared to determine the more common macroscopic benthic species found at each site. However, the assemblages of infaunal amphipods and tanaids were very similar where they occurred. Undoubtedly, the seasonal cycles of these infaunal species are related to those of the primary producers (Tucker, this volume). Due to the paucity of zooplankton during the bloom period, much of the phytoplankton falls through the water column to the benthos and so provides a rich, but seasonal, food supply.

The absence of seasonal cycles of abundance for the large species can be related to their feeding behaviour and life histories. The echinoids and ophiuroids are opportunistic feeders being omnivores and scavengers (Dearborn, 1977), as are the great majority of antarctic benthic animals (Arnaud, 1977). Suspension feeders, such as the holothurians and *L. elliptica*, appear to quiesce over the winter period (Gruzov, 1977). Population stability for these larger and long-lived invertebrates would also be maintained by low rates of predation and juvenile recruitment (Arnaud, 1977).

In comparison with other coastal areas of the earth, coastal Antarctica has a relatively stable temperature environment. But there are large variations in light and surface salinities. The most influential variable appears to be the very seasonal cycle of primary productivity. The fauna appears to have several responses to the seasonal extremes of primary production, e.g. migration (oceanic zooplankton), wide ranging diet (most echinoderms), quiescence (holothurians), migration between feeding areas (those species associated with the ice/water interface in spring and the plankton or benthos in summer) and brooding of young so that release of juveniles coincides with the periods of greatest primary production (the amphipods *P. walkeri* and *O.* cf. *plebs*).

The coastal ecosystem of the Vestfold Hills is by no means unique in Antarctica. None of the collected species, apart from *D. bispinosus*, are endemic to the area and relatively similar communities of almost identical species composition are found in comparable habitats of other antarctic coastal regions, e.g. Mawson (Bunt, 1960), Syowa (Fukuchi & Tanimura, 1981; Nakajima *et al.*, 1982). Molodezhnaya and Mirny (Zvereva, 1972; Andriashev, 1967; Gruzov *et al.*, 1967; Gruzov & Pushkin, 1970). This supports the theory of circumpolarity for marine communities in Antarctica.

Acknowledgements

I would like to thank Messrs R. Perrin and P. Lu for their assistance and permission for the use of results and figures. My thanks to the divers, Messrs K. Beinssen and R. Perrin. I would also like to thank Dr D. Dunn, Mr J. Kirkwood, Dr J. Lowry, Mr D. O'Sullivan, Drs W. Ponder, and G. Poore, F. Rowe, and D. Staples and Mr R. Williams for their taxonomic assistance.

References

Andrews, K. J. H., 1966. The distribution and life history of Calanoides acutus (Geisbrecht). Discovery Rep. 34: 117–162.

Andriashev, A. P., 1967. Microflora and fauna associated with the Antarctic fast ice. Zool. Zh. 46: 1585–1593.

Arnaud, P. M., 1977. Adaptations within the Antarctic marine benthic ecosystem. In G. A. Llano (ed.), Adaptations within Antarctic Ecosystems. Smithsonian Institution, Wash., D.C.: 135–158.

Bayly, I. A. E., 1982. The genus Drepanopus (Copepoda:Calanoida): a review of species in Antarctic and sub-Antarctic waters, with a description of D. bispinosus, sp. nov. Aust. J. mar. Freshwat. Res. 33: 161–172.

Bunt, J. S., 1960. Introductory studies: Hydrology and plankton, Mawson, June 1956–February 1957. Aust. nat. antarct. Res. Exped. Rep. Ser. B, III, Publ. 56. Antarct. Div., Dept. External Affairs, Melbourne, 135 pp.

Dearborn, J. H., 1977. Foods and feeding characteristics of Antarctic asteroids and ophiuroids. In G. A. Llano (ed.), Adaptations within Antarctic Ecosystems. Smithsonian Institution, Wash. D.C.: 293–326.

Dhargalkar, V. K., H. R. Burton & J. K. Kirkwood, this volume. Animal associations with the dominant species of shallow water macrophytes along the coastline of the Vestfold Hills, Antarctica. Hydrobiologia.

Dick, A. F., 1984. Thermal loss in antarctic divers. Med. J. Aust. 140: 351–354.

Everitt, D. A. & D. P. Thomas, 1986. Observations of the seasonal changes in diatoms at inshore localities near Davis Station, East Antarctica. Hydrobiologia 139: 3–12.

Everitt, D. A., G. C. B. Poore & J. Pickard, 1980. Marine benthos from Davis Station, Eastern Antarctica. Aust. J. mar. Freshwat. Res. 31: 829–836.

Fukuchi, M. & A. Tanimura, 1981. A preliminary note on the occurrence of copepods under sea ice near Syowa Station, Antarctica. Mem. natn. Inst. Polar Res. 34: 37–43.

Grigoryev, Y. A., 1968. Circulation of surface waters in Prydz Bay. Sov. antarct. Exped. Info. Bull. 7: 74–76.

Gruzov, Y. N., 1977. Seasonal alterations in coastal communities in the Davis Sea. In G. A. Llano (ed.), Adaptations within Antarctic Ecosystems. Smithsonian Institution, Wash., D.C.: 263–278.

Gruzov, Y. N. & A. F. Pushkin, 1970. Bottom communities of the upper sublittoral of Enderby Land and the south Shetland Islands. In M. W. Holdgate (ed.), Antarctic Ecology, 1. Academic Press, Lond.: 235–238.

Gruzov, Y. N., M. V. Propp & A. F. Pushkin, 1967. Biological associations of coastal areas of the Davis Sea (based on the observations of divers). Sov. antarct. Exped. Info. Bull. 6: 523–533.

Hardy, A. C. & E. R. Gunther, 1935. The plankton of the South Georgia whaling grounds and adjacent waters, 1926–1927. Discovery Rep. 11: 1–456.

Journal of Glaciology, 1958. Instruments and methods: Ice drills and corers. J. Glaciol. 3: 30.

Lake, R. A. & E. L. Lewis, 1970. Salt rejection by sea ice during growth. J. geophys. Res. 75: 583–597.

Mackintosh, N. A., 1934. Distribution of the macroplankton in the Atlantic sector of the Antarctic. Discovery Rep. 9: 65–160.

Nakajima, Y., K. Wantanabe & Y. Naito, 1982. Diving observations of the marine benthos at Syowa Station, Antarctica. In T. Hoshiai & Y. Naito (eds), Proceedings of the 5th symposium on Antarctic Biology. Mem. natn. Inst. Polar Res. Spec. Issue 23: 44–54.

Ottestad, P., 1936. On Antarctic copepods from the 'Norvegia' Expeditions 1930–1931. Sci. Results Norw. antarct. Exped. 15: 1–44.

Perrin, R. A., P. Lu & H. J. Marchant, 1987. Seasonal variation in marine phytoplankton and ice algae at a shallow antarctic coastal site. Hydrobiologia 146: 33–46.

Rakusa-Suszczewski, S., 1972. The biology of Paramoera walkeri Stebbing (Amphipoda) and the Antarctic sub-fast ice community. Pol. Arch. Hydrobiol. 19: 11–36.

Seno, J., Y. Komahi & A. Takeda, 1966. Report on the biology of the 'Umitaka–Maru' Expedition. Plankton collected in the Antarctic Ocean and adjacent waters by the closing net, with special references to copepods. J. Tokyo Univ. Fish. 52: 1–16.

Thomas, D. P. & J. Jiang, 1986. Epiphytic diatoms of the inshore marine area near Davis Station. Hydrobiologia 140: 193–198.

Tucker, M. J., this volume. Temporal distribution and brooding behaviour of selected benthic species from the shallow marine waters off the Vestfolds Hills, Antarctica. Hydrobiologia.

Tucker, M. J. & H. R. Burton, 1987. A survey of the marine fauna in shallow coastal waters of the Vestfold Hills and Rauer Islands, Antarctica. ANARE Res. Notes. No. 55. Antarct. Div., Dept. Sci., Hobart, 24 pp.

Williams, R., this volume. The inshore marine fishes of the Vestfold Hills region, Antarctica. Hydrobiologia.

Zvereva, Z. A. (ed.), 1972. Seasonal changes of Antarctic plankton in the Molodezhnaya and Mirny region. Exploration of marine fauna 12. Geographical and seasonal variability of marine plankton. Israel program for scientific translations, Jerusalem 1975: 248–262.

Hydrobiologia 165: 141–150 (1988)
J. M. Ferris, H. R. Burton, G. W. Johnstone & I. A. E. Bayly (eds) Biology of the Vestfold Hills, Antarctica
Kluwer Academic Publishers

Animal associations with the dominant species of shallow water macrophytes along the coastline of the Vestfold Hills, Antarctica

V. K. Dhargalkar[1], H. R. Burton[2] & J. M. Kirkwood[3]
[1]*National Institute of Oceanography, Dona Paula, 403 004, Goa, India;* [2]*Antarctic Division, Department of Science, Kingston, 7050, Tasmania, Australia;* [3]*Zoology Department, Monash University, Clayton, Victoria, 3168, Australia*

Key words: macrophytes, epibionts, epiphytes, thalli, fjord, Antarctica

Abstract

Benthic grab samples, taken through core holes in sea-ice 1983, returned eight species of marine macrophytes which had associated fauna on at least one sample. A total of 24 animal taxa were found on these eight macrophytes. *Phyllophora antarctica* had a significantly greater biomass of attached fauna per unit of macrophyte biomass than any of the other seven species. It also had the greatest variety of epiphytic fauna (21 taxa). *Phyllophora antarctica* was the only macrophyte collected from comparatively protected Ellis Fjord, apart from filamentous algae, including *Cladophora subsimplex* and *Geminocarpus austrogeorgiae*, which were found in the most landward basin. It therefore played an important role in the ecology of the fjord.

Palmaria decipiens was commonly collected at more exposed coastal sites and was dominant in Long Fjord, characterized here as intermediate in terms of water movement. Sixteen animal taxa were collected from this macrophyte with the greatest numbers being found in July and August. Only four animal taxa were collected from *Iridaea cordata*. The other five macrophyte species from which epiphytic animals were collected (*Porphyra endiviifolium, Plocamium cartilagineum, Desmarestia menziesii, Himantothallus grandifolius* and *Cladophora subsimplex* had animals associated with them on less than half the occasions when they were found.

Introduction

The importance of marine macrophytes in coastal ecosystems as major primary producers, and therefore to certain species of grazing animals, has been long recognized. The coastal belt of marine macrophytes harbours diverse animal communities (Mukai, 1971; Sheppard *et al.*, 1980). However, the relationships between macrophytes and epiphytic dependent/adherent animal species are still not well understood (Gore *et al.*, 1981).

It has been reported that the distribution of epiphytic fauna is primarily influenced by algal shape, water depth, wave exposure and water movement (Sarma & Ganapati, 1972; Hicks, 1976; Dahl, 1948; Norton, 1971). However the abundance of the fauna is largely dependent upon seasonal factors (Mukai, 1971; Hagerman, 1966).

Tropical and temperate macrophytes have been investigated for epiphytic animal composition, abundance and species diversity (Sarma & Ganapati, 1972; Norton, 1971). The most numerous taxa found on these macrophytes were amphipods, molluscs, isopods and polychaetes. The species composition of these animals and their abundance are seasonal and depend largely on the availability of the food resource of the algal thallus (Hagerman, 1966; Jansson, 1967; Fauchald & Jumars, 1979).

Many areas of rocky antarctic coastline have extensive marine algal belts (Delepine *et al.*, 1966;

Zaneveld, 1966) inhabited by various invertebrate species; but there is even less known about the relationships between macrophytes and epiphytic animal species from these waters then elsewhere on the globe. In a general account of research into antarctic marine benthos, White (1984) noted that the fauna and flora of shallow marine sites were 'greatly undersampled' and also that a requirement existed for 'basic (but now unfashionable) descriptive systematic and ecological studies'. The present study helps to address this need by describing the taxonomic composition, seasonal abundance and distribution of the fauna associated with eight macrophyte species collected from the nearshore area of a coastal oasis in East Antarctica.

Materials and methods

Collections were made through sea-ice (up to 1.9 m thick) by lowering a three pronged grab to the bottom through 11 cm holes cut by a SIPRE corer (Journal of Glaciology, 1958). A similar grab and its operation are described by Heath & Singleton (this volume). Collected macrophytes were immediately separated into different species with minimal disturbance to attached fauna and were transported to the laboratory in separate polythene bags.

In the laboratory, thalli were placed in a tray and carefully picked over to remove adhering animals. These animals were then preserved in 4% formaldehyde solution pending further examination, while the washed macrophyte thalli were blotted dry, weighed and measured. The preserved animals were separated into taxa, counted and then blotted dry and weighed.

Regular sampling was concentrated in Ellis Fjord and Burton Lake (Fig. 1) for the period May 1983–January 1984. Other sites were sampled, intensively, only in summer.

Results

Eight species of marine macrophyte, collected from the nearshore waters and marine fjords of the Vestfold Hills, Antarctica, had associated fauna on at

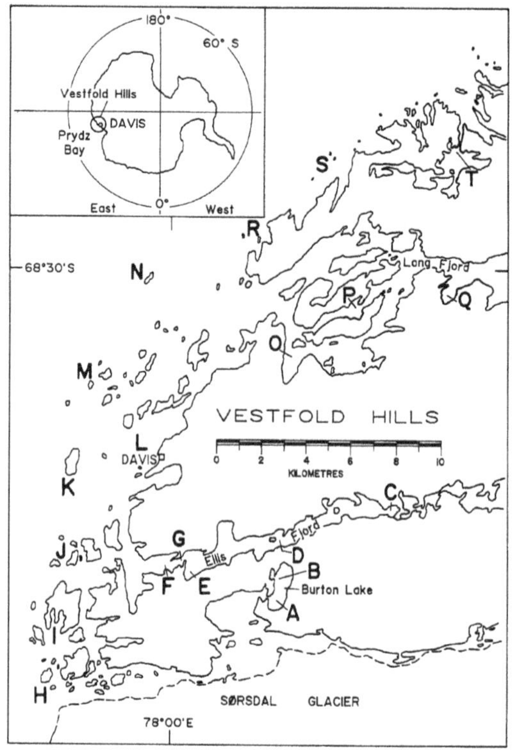

Fig. 1. Locations of the twenty sampling sites along the coastline of the Vestfold Hills, Antarctica.

least one sample (Table 1). These were *Palmaria decipiens* (Reinsch) Ricker comb. nov., *Phyllophora antarctica* A & E.S. Gepp, *Iridaea cordata* (Turner) Bary, *Porphyra endiviifolium* (A & E.S. Gepp) Chamberlain, *Plocamium cartilagineum* (Linnaeus) Dixon, *Desmarestia menziesii* J. Agardh, *Himantothallus grandifolius* (A & E.S. Gepp) Zinova and *Cladophora subsimplex* Kutzing.

The twenty sites sampled during this study fell aproximately into three categories on the basis of exposure to water movement, subjectively assessed from our experience of the region. Sites G, H, I, J, K, L, M, N, R, S & T were directly exposed to the turbulence associated with wave action or strong tidal flow in summer. Sites A, B, C, D & E, on the other hand, were comparatively protected from these sources of mechanical disturbance; sites A, B & C were almost lake-like in their isolation from the sea, being tidal only in the summer months (Burke &

Table 1. Animal biomass (mg) (left figure) and numbers (right figure) g^{-1} of plant on eight macrophyte species. An empty space indicates that the macrophyte was not collected on that sampling date. A dash indicates no animals were present in that macrophyte sample.

Collection Date	Rhodophyta					Chlorophyta	Phaeophyta	
	Palmaria decipiens	*Phyllophora antarctica*	*Iridaea cordata*	*Porphyra endiviifolium*	*Plocamium cartilagineum*	*Cladophora subsimplex*	*Desmarestia menziesii*	*Himantothallus grandifolius*
6/ 5/83	-	1176.27/1.21		-				
8/ 5/83	88.78/2.14				-		1.04/0.075	-
15/ 5/83	59.84/0.47							
27/ 5/83	11.44/0.50			-	-	-		
23/ 6/83	0.51/0.29			0.79/0.26			-	-
29/ 6/83	6.02/0.31		12.61/2.10		6.36/14.55		-	-
23/ 7/83	15.90/4.20				-	29.33/18.67		-
10/ 8/83		31.76/2.66						
22/ 8/83	12.69/3.54					-		
12/ 9/83		1109.92/1.98						
16/ 9/83		31.74/2.93				-		
3/10/83	-	235.75/1.30						
25/10/83	49.50/6.50	-	-					
29/10/83	1.73/0.66							
30/10/83	1.98/0.37		2.41/0.46					
1/11/83		887.95/2.24						
14/11/83	3.19/0.53		7.81/1.02		-			
18/11/83	1.43/0.30	13.21/1.32	2.59/0.77		-			0.19/0.045
26/11/83	9.25/1.28		0.83/0.38		-			
6/12/83		1.58/0.18						
8/12/83	3.42/0.49	112.17/8.83			-	-	10.65/1.70	5.54/0.031
19/12/83	2.20/0.46		2.85/0.82			-		
29/12/83	11.23/0.98						-	-
30/12/83	0.30/0.32		-	-				-
31/12/83	0.54/0.055		-	-	-	-	-	-
8/ 1/84	0.65/0.20		0.64/0.12					
10/ 1/84	1.20/0.40						16.21/0.75	
12/ 1/84	0.04/0.010		0.04/0.012	0.04/0.006	3.82/1.99	-	0.57/0.029	0.56/0.19
Mean	11.74/1.00	327.30/2.06	2.71/0.52	0.14/0.04	1.02/1.65	3.67/2.33	3.56/0.32	0.63/0.03

Burton, this volume; J. B. Gallagher, pers. commun.). Sites F, O, P & Q were regarded as intermediate.

The distribution and abundance of some macrophytes was related to this gradient. The thalloid Phaeophyta *Himantothallus grandifolius* and *Desmarestia menziesii* were found almost exclusively at the most exposed sites; they also occurred at one intermediate site (site F). These algae dominated the macrophyte biomass at more exposed sites in summer and probably also in winter although they could not be collected through the small diameter SIPRE ice-corer holes. The rhodophyte *Iridaea cordata* was also confined to exposed sites.

In contrast, *Phyllophora antarctica* (Rhodophyta) was only abundant in the comparatively protected waters of Ellis Fjord, although it occurred at more exposed sites. In a subsequent study, using SCUBA in 1985, this alga was the only apparent macrophyte in the main body of Ellis Fjord (Kirkwood & Burton, in press). Another rhodophyte, *Palmaria decipiens*, was the dominant contributor to macrophyte biomass in Long Fjord, an area of intermediate exposure in the above classification, and was commonly collected at all exposed and intermediate sites. Filamentous algae, including *Cladophora subsim-*

plex (Chlorophyta) and *Geminocarpus austrogeorgiae* (Phaeophyta), were the only macrophytes collected from the most protected sites (sites A, B & C).

Sea-ice was also an important factor affecting the distribution of macrophytes. Its effect was to preclude macrophytic growth in the intertidal zone and down to about 2 m below low tide level for 10–12 months of the year. In summer, the sea-ice usually breaks out along the coast and may break out of the fjords, though usually only partially. Bare rock, newly exposed when the sea-ice disappeared, was colonized by the rhodophyte *Porphyra endiviifolium* with chlorophyte and phaeophyte filamentous algae growing in rock crevices: *Enteromorpha* spp., *Urospora penicilliformis* (Roth) Areschoug, *Ulothrix* sp., *Cladophora* spp. and *Geminocarpus austrogeorgiae*.

In all, 24 animal taxa were collected from the marine macrophytes of the inshore waters of the Vestfold Hills (Table 2). To enable comparison of epiphytic fauna on macrophyte samples of very different weights (ranging from 0.55 to 2186 g wet weight), faunal composition is expressed in terms of number and wet weight of each species g^{-1} (wet weight) of macrophyte.

The number of individual animals from each taxon per unit biomass of *Palmaria decipiens*, *Phyllophora antarctica* and *Iridaea cordata* are contained in Tables 3, 4 & 5. These three Rhodophyta, considered in detail here, were the only species to have animals associated with them on more than 50% of the occasions on which they were collected (Table 1) and virtually all taxa of associated animals collected from macrophytes in this study were represented on these algae.

Table 2. List of animal taxa collected from marine macrophytes in this study.

Taxon number	
1.	Actiniaria sp. (Anthozoa)
2.	Nematoda spp
3.	*Lagisca floculosa* (Polychaeta, Polynoidae)
4.	*Serpula narconensis* Baird (Polychaeta, Serpulidae)
5.	Polychaeta spp
6.	*Lepeta coppingeri* E. A. Smith (Gastropoda, Lepetidae)
7.	*Laevilacunaria* sp. (Gastropoda, Littorinidae)
8.	*Submargarita crebrilirulata* (E. A. Smith) (Gastropoda, Trochidae)
9.	*Laternula elliptica* (King & Broderip) (Pelecypoda, Laternulidae)
10.	*Philobrya sublaevis* (Pelseneer) (Pelecypoda, Philobrydiae)
11.	Pelecypoda sp.
12.	Pycnogonida spp
13.	*Paramoera walkeri* Stebbing (Amphipoda, Lysianassidae)
14.	Gammaridea spp (Amphipoda)
15.	*Glyptonotus acutus* Eights (Isopoda, Iodoteidae)
16.	*Cymodocella tubicauda* Pfeffer (Isopoda, Sphaeromatidae)
17.	*Antarcturus* sp. (Isopoda, Arcturidae)
18.	Isopoda sp.
19.	*Chorismus antarcticus* (Pfeffer) (Decapoda, Hippolytidae)
20.	*Sterechinus neumayeri* (Meissner) (Echinoidea, Echinidae)
21.	Asteroidea sp.
22.	Ophiuroidea spp
23.	Holothuroidea sp.
24.	Ascidiacea sp.

Animal diversity

Overall, the greatest number of animal taxa were found on *Phyllophora antarctica* (21), followed by *Palmaria decipiens* (16). Only four taxa were found associated with *Iridaea cordata*. In any one month, the greatest diversity of animals was found on *Phyllophora antarctica* in November (14 taxa; Shannon diversity = 0.933), but 13 animal taxa were found on *Palmaria decipiens* in May (Table 3). Only a few of the recorded taxa were found consistently on each of these algae. The number of taxa found in more than 50% of months for which animals were recorded, was five on *Palmaria decipiens*, four on *Phyllophora antarctica* and three on *Iridaea cordata* (Tables 3, 4 & 5).

Numbers of animals

An analysis of variance found a significant difference in the number of animals g^{-1} of macrophyte

Table 3. Population density of animal taxa collected from *Palmaria decipiens* (number of individuals g^{-1}). A dash indicates an absence of that taxon in samples from that month.

Taxon	Month							
	May	June	July	Aug	Oct	Nov	Dec	Jan
2 Nematoda spp	0.003	0.073	0.280	0.187	-	0.001	0.001	0.005
3 *L. floculosa*	0.011	0.004	-	-	-	-	0.001	-
4 *S. narconensis*	0.003	-	-	-	0.001	-	-	-
5 Polychaeta spp	0.088	0.048	-	-	0.020	0.037	0.013	0.023
7 *Laevilacunaria* sp.	0.071	0.026	3.642	3.172	0.401	0.433	0.164	0.028
8 *S. crebrilirulata*	0.011	0.010	-	-	-	-	-	-
10 *P. sublaevis*	0.003	-	-	-	-	0.001	0.001	-
11 Pelecypoda sp.	-	-	-	-	-	0.001	-	-
12 Pycnogonida spp	-	-	-	-	0.001	-	-	0.002
13 *P. walkeri*	0.201	0.046	0.140	0.187	0.005	0.014	0.118	0.002
14 Gammaridea sp.	0.017	0.046	-	-	0.003	-	-	-
15 *G. acutus*	0.003	-	-	-	-	-	-	-
16 *C. tubicauda*	0.093	-	-	-	0.001	0.025	0.017	0.012
17 *Antarcturus* sp.	-	-	-	-	-	-	0.006	-
18 Isopoda sp.	0.006	-	-	-	-	-	-	-
22 Ophiuroidea spp	0.003	-	-	-	-	-	-	-
Weight of *P. decipiens* (g)	354.15	504.92	7.14	5.36	1976.2	1401.7	139.1	427.66

Table 4. Population density of animal taxa collected from *Phyllophora antarctica* (number of individuals g^{-1}) in those months when animals were present. A dash indicates an absence of that taxon in samples from that month.

Taxon	Month					
	May	Aug	Sept	Oct	Nov	Dec
1 Actiniaria sp.	-	-	-	-	0.134	-
2 Nematoda spp	-	-	-	-	0.067	-
3 *L. floculosa*	-	0.121	-	-	-	-
4 *S. narconensis*	0.115	-	-	-	0.067	0.037
5 Polychaeta spp	0.115	0.242	0.345	-	0.670	0.074
6 *Lepeta coppingeri*	0.634	1.932	1.476	1.296	0.402	0.889
7 *Laevilacunaria* sp.	-	-	-	-	0.067	-
8 *S. crebrilirulata*	-	0.121	0.049	-	-	0.111
9 *L. elliptica*	-	-	-	-	0.034	-
10 *P. sublaevis*	-	-	-	-	0.067	-
12 Pycnogonida spp	-	-	0.098	-	-	-
13 *P. walkeri*	0.230	0.121	-	-	0.034	0.074
14 Gammaridea spp	-	0.121	0.098	-	-	0.037
16 *C. tubicauda*	-	-	-	-	0.034	-
17 *Antarcturus* sp.	-	-	0.049	-	-	-
18 Isopoda sp.	-	-	-	-	0.134	-
19 *C. antarcticus*	-	-	-	-	0.034	-
20 *S. neumayeri*	0.115	-	0.148	-	0.201	0.259
21 Asteroidea sp.	-	-	-	-	-	0.037
22 Ophiuroidea spp	-	-	0.197	-	-	-
24 Ascidiacea sp.	-	-	-	-	0.134	-
Weight of *P. antarctica* (g)	17.36	8.28	20.32	3.86	29.86	26.99

Table 5. Population density of animal taxa collected from *Iridaea cordata* (number of individuals g^{-1}) in those months when animals were present.

Taxon	Month				
	June	Oct	Nov	Dec	Jan
2 Nematoda spp	0.840	-	-	-	-
5 Polychaeta spp	-	-	0.115	0.407	0.010
7 *Laevilacunaria* sp.	0.840	0.434	0.659	0.713	0.020
13 *P. walkeri*	0.420	0.024	0.029	-	-
Weight of *I. cordata* (g)	2.38	41.45	34.93	9.82	202.94

for the eight species of algae ($F = 2.15$, $P < 0.05$).

The total number of animals on *Palmaria decipiens* varied seasonally from 0.1 to 4.2 individuals g^{-1} (Fig. 2) and showed a definite peak in winter (July/August), followed by a decline to < 1 individual g^{-1} over spring and summer; this macrophyte was collected in every month of the study period except September. A similar pattern may have applied to *Phyllophora antarctica* but this macrophyte was not collected in June or July. In those months when *Phyllophora antarctica* was collected, it had from 1.2 to 2.7 individuals g^{-1}, a greater number than was found on *Palmaria decipiens* during spring and early summer (Fig. 2). In the five months for which specimens of *Iridaea cordata* were collected, from 0.03 to 2.1 individual animals g^{-1} were found associated with it (Fig. 2).

Animal biomass

The weight of animals g^{-1} of macrophyte varied significantly between the eight species of macrophyte on which animals were found ($F = 4.24$, $P < 0.001$). Further anlysis, using the Tukey multiple comparison test (Zar, 1984), showed that *Phyllophora antarctica* supported a significantly ($P < 0.001$) greater animal biomass g^{-1} than the other seven macrophyte species (Table 6). In fact, the biomass of associated animals occasionally approached or exceeded that of *Phyllophora antarctica* (Table 1) when the sea-urchin *Sterechinus neumayeri* was found.

Distribution of animal taxa

Five taxa were found on *Palmaria decipiens* in more than 50% of months for which animals were recorded: *Laevilacunaria* sp., *Paramoera walkeri*, Nematoda spp., Polychaeta spp., and *Cymodocella tubicau-*

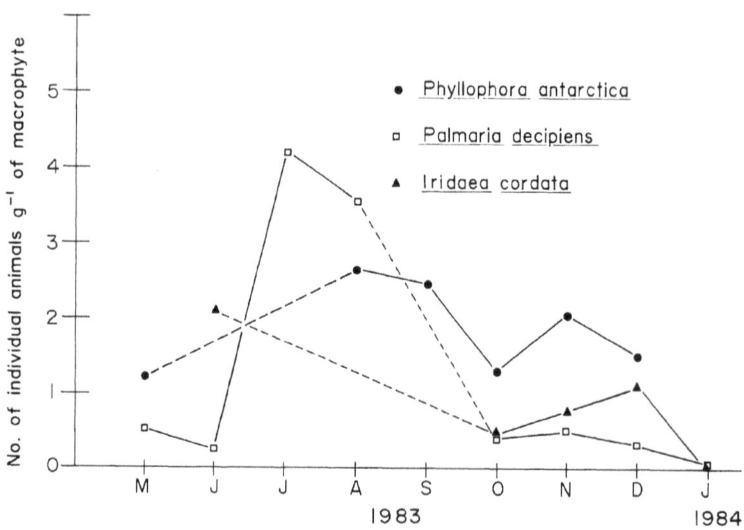

Fig. 2. Seasonal variations in the total number of epiphytic animals g^{-1} of macrophyte for the three macrophyte species supporting the greatest epiphytic animal populations.

Table 6. Comparison of mean animal biomass g^{-1} of plant tissue on eight macrophyte species.

a: Macrophytes ranked by mean animal biomass (mg g^{-1} wet weight).

Rank	Macrophyte Species	Animal Biomass (mg g^{-1})	n
1	*P. antarctica*	327.30	11
2	*P. decipiens*	11.74	24
3	*C. subsimplex*	3.67	8
4	*D. menziesii*	3.50	8
5	*I. cordata*	2.71	11
6	*P. cartilagineum*	1.02	10
7	*H. grandifolius*	0.63	10
8	*P. endiviifolium*	0.14	6

b: Tukey multiple comparison test.

Comparison (A vs B)	n_A	n_B	Difference ($X_A - X_B$)	SE	q
1 vs 2	11	24	315.56	43.74	7.21
2 vs 3	24	8	8.07	49.04	0.165
3 vs 4	8	8	0.15	60.06	0.002
4 vs 5	8	11	0.79	55.82	0.014
5 vs 6	11	10	1.69	52.48	0.032
6 vs 7	10	10	0.59	53.72	0.011
7 vs 8	10	6	0.39	62.03	0.006

$q_{(0.001)\ 60,8} = 6.268$ (Zar, 1984).

da (Tables 2 & 3). *Paramoera walkeri* and Nematoda spp. were numerically dominant in May and June respectively, and the gastropod *Laevilacunaria* sp. was numerically dominant from July onwards.

Four taxa were found on *Phyllophora antarctica* in more than 50% of months for which animals were recorded: *Lepeta coppingeri*, Polychaeta spp., *Paramoera walkeri* and the sea-urchin *Sterechinus neumayeri*. The limpet *Lepeta coppingeri* was numerically dominant in all months except November, when Polychaeta spp. dominated (Table 4).

Three of the four taxa found on *Iridaea cordata* were present in more than 50% of months for which collections were made: *Laevilacunaria* sp., Polychaeta spp. and *Paramoera walkeri*. *Laevilacunaria* sp. was numerically dominant in all months, although

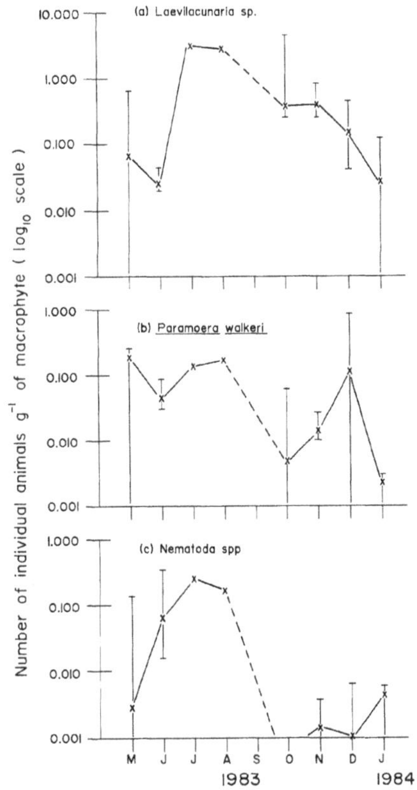

Fig. 3. Seasonal variations in the populations of the three epiphytic animal taxa occurring most consistently on *Palmaria decipiens*: a. *Laevilacunaria* sp.; b. *Paramoera walkeri*; c. Nematoda spp.

Nematoda spp. were equally frequent in June (Table 5).

It is evident that while *Paramoera walkeri* and Polychaeta spp. were frequently associated with all three of the macrophyte red algae considered here, the distribution of other taxa was macrophyte specific and/or determined by water movement. *Cymodocella tubicauda* and Nematoda spp. were a frequent component of the fauna on *Palmaria decipiens* but were rarely associated with either *Phyllophora antarctica* or *Iridaea cordata*. The gastropod *Laevilacunaria* sp. was usually numerically dominant on *Palmaria decipiens* and also on *Iridaea cordata*, macrophytes that were found at exposed or intermediate sites. *Laevilacunaria* sp. was not found in the more protected Ellis Fjord, where *Phyllophora antarctica* was abundant, instead, the limpet *Lepeta coppingeri* was dominant on this macrophyte and was not found on either *Palmaria decipiens* or *Iridaea cordata*. *Phyllophora antarctica* was the only macrophyte with which the sea-urchin *Sterechinus neumayeri* was found to be associated.

Seasonal change of animal numbers on Palmaria decipiens

The seasonal change in numbers (g^{-1} of macrophyte) of *Laevilacunaria* sp., *Paramoera walkeri* and Nematoda spp. on *Palmaria decipiens* are shown in Fig. 3; error bars indicate that the variability of animal numbers was quite large in some months. No data are available for September. Both *Laevilacunaria* sp. and Nematoda spp. showed winter maxima. *Laevilacunaria* sp. subsequently declined gradually over spring and summer, while Nematoda spp. showed a sudden decline in spring and remained in low numbers until January when a slight increase was evident (Fig. 3). *Paramoera walkeri* maintained its numbers over the period May to August, declined suddenly in spring and then increased again, peaking briefly in early summer.

Discussion

The list of 24 animal taxa which were collected from the marine macrophytes of the inshore waters of the Vestfold Hills (Table 2) is certainly incomplete, but does indicate the more common epiphytic species. Active, quickly moving animals (e.g. Amphipoda) are likely to be discriminated against by grab sampling.

Little is known of the general growth patterns of perennial antarctic macroalgae (Heywood & Whitaker, 1984). During the present study, macrophytes were found in crumpled condition during winter, indicating an absence of growth. Heywood & Whitaker (1984), reviewing earlier studies, reported that growth of the endemic brown alga *Phyllogigas grandifolius* (now *Himantothallus grandifolius*; see Skottsberg & Neushal, 1960), at Signy Island (60° 43′ S), was not measurable from June to August. The Vestfold Hills receive little solar input in winter as they are situated at 68° 30′ S and thus photosynthesis is likely to be inhibited in this period. The ragged condition of the macrophyte thalli, in winter, may also indicate active consumption by animals. The greater population density of animals, particularly the gastropod *Laevilacunaria* sp., on *Palmaria decipiens* in July and August (Fig. 2) may possibly be consistent with active grazing in these months. Seasonal changes in biomass of the macrophytes collected in this study can be inferred from the wet weight of material collected in each month and such changes are obviously important to the animals that associate with them. *Palmaria decipiens* was collected in small quantities in July and August and was not found in September; similarly, *Phyllophora antarctica* and *Iridaea cordata* were not found or were collected in relatively small quantities in winter (Tables 3, 4 & 5). Grab sampling is not sufficiently intensive a technique for us to assert that the absence of a particular macrophyte in a monthly sample denotes its absence from the environment. However, our observations seem to indicate that these macrophytes were far less abundant in winter than in summer, and that this seasonality may have reflected a balance between macrophyte growth, on one hand, and their consumption by animals, on the other. SCUBA divers observed a winter decline of macrophytes at a near-shore marine site off Davis in 1982. They estimated a decline of macrophyte cover from nearly 100% to about 20% on a rocky substrate at a depth of 15 m (Tucker & Burton, this volume).

The lower total animal numbers on *Palmaria decipiens* in spring (Fig. 2) were primarily the result of the fall in numbers of *Laevilacunaria* sp. from its winter peak, but both *Paramoera walkeri* and Nematoda spp. were also present in lower numbers than in the preceding winter (Fig. 3). The steady decline in numbers of *Laevilacunaria* sp. from October to January may represent a 'dilution effect' (Norton, 1971) with the rapid growth of the macrophyte thallus causing the epiphytic animal populations to become less dense. *Paramoera walkeri* had decreased in numbers, by October, and this may be explained by the reported tendency of this amphipod to depart benthic macrophytes in the spring to feed on dense under-ice algal growths (Rakusa-Suszczewski, 1972; Tucker & Burton, this volume). The reasons for the great increase in abundance of Nematoda on *Palmaria decipiens* in the winter and the sharp decline in spring are not known, but it may be that these Nematoda exploit different food sources at different times of the year.

Seasonal variations aside, the number of *Laevilacunaria* sp. on *Palmaria decipiens* was relatively high throughout the study period. This macrophyte usually supported a thin covering of diatoms which may possibly have provided a food source for *Laevilacunaria* sp. A large number of individuals of this gastropod were juveniles. Fretter & Manly (1977) have also reported large numbers of juveniles of the gastropods *Lacuna vincta* and *Cerithiopris tubercularis* on *Sargassum muticum*.

Polychaetes and nematodes were usually found on the lower part of the macrophyte, while gastropods, limpets and amphipods did not appear to show a preference for any particular part of the thallus. Withers *et al.* (1975) discussed the distribution of animal species over the thallus of *Sargassum muticum*, and concluded that distributions similar to those found in the present study were related to the animals' feeding habits. Other factors which may influence the distribution of a particular animal species over the macrophyte thallus are the amount of sediment accumulated on the plant surface and the turbidity of the water column (Dahl, 1948; Hagerman, 1966; Moore, 1972; Boaden *et al.*, 1975; Hicks, 1980).

The significantly higher weight of associated animals g^{-1} of *Phyllophora antarctica* may be due to the characteristic bushy growth of this species, which would be expected to afford better shelter for the animals than would other macrophyte species. This macrophyte also had a larger number of associated animal taxa (21) and higher species diversity of associated animals than other macrophytes collected in this study. The fact that in subsequent investigations *Phyllophora antarctica* was the only macrophyte recorded along all of Ellis Fjord apart from the meromictic landward basin (Kirkwood & Burton, in press) and that it harboured such a rich and diverse fauna indicate its importance in the ecology of this fjord. The present study also indicates the probable significance of *Palmaria decipiens* in nearby Long Fjord (3 sites sampled; Fig. 1), as this macrophyte was a major contributor to the benthic algal biomass there and supported a comparatively large number of individual animals g^{-1}, in winter.

Acknowledgements

We thank Mr. Ron Sherwood for designing and constructing the claw grab used in this study. We also thank all members of the 1983 wintering expedition at Davis Station for their support and their assistance with field sampling. We are grateful to Dr. R. Ricker for identifying the macrophytes and Dr. P. Hutchins for identifying the polychaetes.

References

Boaden, P. J. S., R. J. O. O'Connor & R. Seed, 1975. The composition and zonation of a Fucus serratus community in Co. Down. J. exp. mar. Biol. Ecol. 17: 111–136.

Burke, C. M. & H. R. Burton, this volume. Photosynthetic bacteria in meromictic lakes and stratified fjords of the Vestfold Hills, Antarctica. Hydrobiologia.

Dahl, E., 1948. On the smaller Arthropoda of marine algae, especially in the polyhaline waters of the Swedish West Coast. Lunds Univ. Arsskrijt. N.F. Adv. 2 Undersök. Över Oresund. 35: 1–193.

Delépine, R., I. M. Lamb & M. H. Zimmerman, 1966. Preliminary report on the marine vegetation of the Antarctic Peninsula. Proc. V. Intern. Seaweed. Symp. Halifax: 107–116.

Fauchald, K. & P. A. Jumars, 1979. The diet of worms: a study of polychaete feeding guilds. Oceanogr. Mar. Biol. Annu. Rev. 17: 193–284.

Fretter, V. & R. Manly, 1977. Algal associations of Tricolia pullus, Lacuna vincta and Cerithiopsis tubercularis (Gastropoda) with special reference to the settlement of their larvae. J. mar. biol. Ass. U.K. 57: 999–1017.

Gore, R. H., E. E. Gallaher, L. E. Scotto & K. A. Wilson, 1981. Studies on decapod crustaceans from the Indiana River region of Florida. XI Community composition, structure, biomass and species areal relationships of seagrass and drift algae – associated macrocrustaceans. Estuarine Coastal Shelf. Sci. 12: 485–508.

Hagerman, L., 1966. The macro and microfauna associated with Fucus serratus L. with some ecological remarks. Ophelia 3: 1–43.

Heath, C. W. & W. J. Singleton, this volume. A device for remote sampling of benthic algae under ice. Hydrobiologia.

Heywood, R. B. & T. M. Whitaker, 1984. The antarctic marine

flora. In R.M. Laws (ed.), Antarctic Ecology, 2. Academic Press, Lond.: 373–419.

Hicks, G. R. F., 1976. Species associations and seasonal population densities of marine phytal harpacticoid copepods from Cook Strait. N.Z. J. Mar. Freshwater Res. 11: 621–643.

Hicks, G. R. F., 1980. Structure of phytal harpacticoid copepod assemblages and the influence of habitat complexity and turbidity. J. exp. mar. Biol. Ecol. 44: 157–192.

Jansson, A. M., 1967. The food web of the Cladophora belt fauna. Helgolander Wiss. Meeresunters. 15: 574–588.

Journal of Glaciology, 1958. Instruments and methods: Ice drills and corers. J. Glaciol. 3: 30.

Kirkwood, J. M. & H. R. Burton, in press. Macrobenthic species assemblages in Ellis Fjord, Vestfold Hills, Antarctica. Mar. Biol.

Moore, P. G., 1972. Particulate matter in the sublittoral zone of an exposed coast and its ecological significance, with special reference to the fauna inhabiting kelp holdfasts. J. exp. mar. Biol. Ecol. 10: 59–80.

Mukai, H., 1971. The phytal animals on the thalli of Sargassum serratifolium in the Sargassum region, with reference to their seasonal fluctuations. Mar. Biol. 8: 170–182.

Norton, T. A., 1971. An ecological study of the fauna inhabiting the sublittoral marine alga Saccorhiza polyschides (Lightf.) Batt. Hydrobiologia. 37: 215–231.

Rakusa-Suszczewski, S., 1972. The biology of Paramoera walkeri Stebbing (Amphipoda) and the Antarctic sub-fast ice community. Polskie Archiwum Hydrobilogii 19: 11–36.

Roland, W., 1980. Epiphytism and endophytism of Macrocystis integrifolia and Nereocystis luetkeana: seasonality, succession and tactics on temporary living substrate. M.Sc. Thesis, Simon Fraser University.

Sarma, A. L. N. & P. N. Ganapati, 1972. Faunal associations of algae in the intertidal region of Visakhapatnam. Proc. Ind. Nat. Sci. Acad. 38: 380–396.

Sheppard, C. R. C., D. J. Bellamy & A. L. S. Sheppard, 1980. Study of the fauna inhabiting the holdfast of Laminaria hyperborea (Gunn.) Fosl. along some environmental and geographical gradients. Mar. Environ. Res. 4: 25–51.

Skottsberg, C. & M. Neushal, 1960. Phyllogigas and Himantothallus. Antarctic Phaeophyceae. Bot. mar. 2: 164–173.

Trotter, D., 1981. Distribution and feeding preferences of free-living nematodes associated with the kelp, Macrocystis integrifolia (Laminariales). M.Sc. Thesis, Simon Fraser University.

Tucker, M. J. & H. R. Burton, this volume. The inshore marine ecosystem off the Vestfold Hills, Antarctica. Hydrobiologia.

Withers, R. G., W. F. Farnham, S. Lewey, N. A. Jephson, J. M. Haythorn & P. W. G. Gray, 1975. The epibionts of Sargassum muticum in British waters. Mar. Biol. 31: 79–86.

White, M. G., 1984. Marine benthos. In R. M. Laws (ed.), Antarctic Ecology, 2. Academic Press, Lond.: 421–461.

Zaneveld, J. S., 1966. The occurrence of benthic marine algae under shore-fast ice in the Western Ross Sea, Antarctica. Proc. V. Intern. Seaweed Symp. Halifax: 217–231.

Zar, J. H., 1984. Biostatistical analysis. Second edition. Prentice Hall, New Jersey, 718 pp.

Hydrobiologia 165: 151–159 (1988)
J. M. Ferris, H. R. Burton, G. W. Johnstone & I. A. E. Bayly (eds) Biology of the Vestfold Hills, Antarctica
Kluwer Academic Publishers

Temporal distribution and brooding behaviour of selected benthic species from the shallow marine waters off the Vestfold Hills, Antarctica

M.J. Tucker[1]
Antarctic Division, Department of Science, Kingston, Tasmania, 7050, Australia; [1]Present address: Australian National Parks and Wildlife Service, GPO Box 636, Canberra, 2601, Australia

Key words: antarctic, benthos, seasonality, brooding, reproductive strategy

Abstract

Selected epifaunal and infaunal species from three sites of differing substrate were examined over a 12 month period. The characteristic species comprising the epifauna at each site were found to be dictated by substrate type whereas the infaunal amphipods and tanaids were relatively similar between sites. No distinct seasonality of abundance was noted for the macroscopic species; however, the infaunal amphipods and tanaids exhibited distinct seasonal cycles of abundance. These temporal changes in population size can be correlated with the seasonal cycle of primary producers. Females, of each of the infaunal species that were monitored, exhibited brood protection throughout the winter period and released juveniles at times that coincided with the period of high primary productivity. All species exhibited properties characteristic of K-strategists; a result of the very stable and predictable nature of the antarctic benthic environment.

Introduction

Investigations of the benthos of inshore antarctic regions (Dayton *et al.*, 1974; Gruzov *et al.*, 1967; Gruzov, 1977), have revealed a surprisingly rich and diverse biota. However, most studies have been limited to the short summer period or have been primarily concerned with vertical zonation.

This investigation concerns the shallow bay off Davis Station (68°35′S, 77°58′E), the floor of which consists partly of wind-derived sediment of several types (Everitt *et al.*, 1980). Muddy areas are known to consist primarily of autochthonous, diatomaceous material typified by a water content of about 65% (by weight) (Lin Jian-ping, unpubl. data). Extensive collection of the bottom fauna were made previously over a spring period with a remote sampling device (Everitt *et al.*, 1980); this allowed description of substrate types and their characteristic species. Apart from this, knowledge of the benthos of the area was limited to material collected from beach debris and from chance captures (Johnstone *et al.*, 1973).

This paper presents results from a complete year's diving under the sea ice to examine, monthly, three sites of differing substrate (sand, mud and rock). Temporal changes in numerical abundance of the dominant and characteristic species of the epifauna and infauna of each site were noted with particular reference to the life histories of the species and seasonal changes in primary production. The brooding behaviours of several species and the apparent ecological significance of these activities were also examined. The life histories of selected species are interpreted in terms of reproductive strategies.

Description of sites

From preliminary examination of Davis Bay in sum-

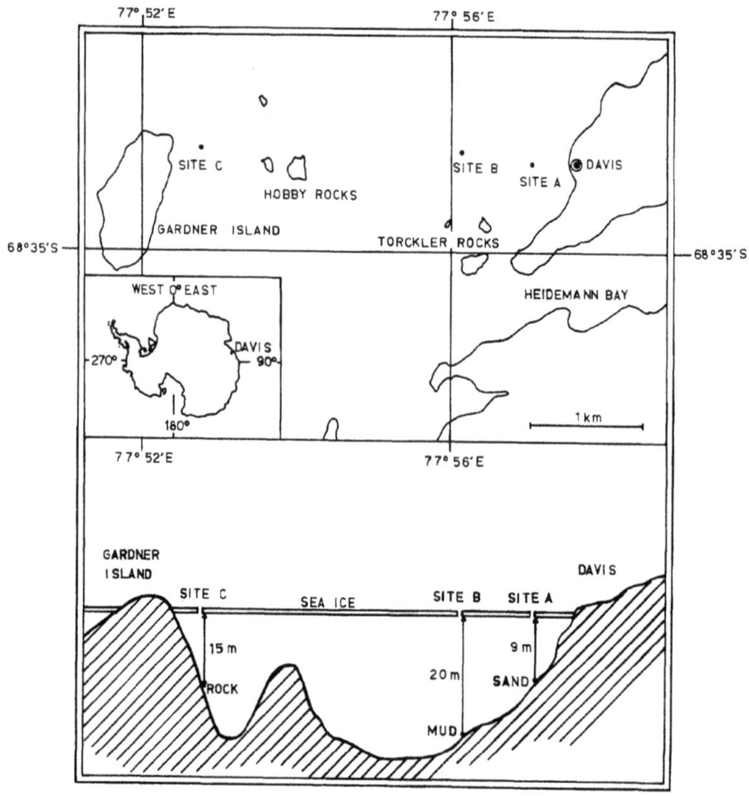

Fig. 1. Location and profile of study sites (exaggerated vertical scale).

mer 1981/1982 (Tucker & Burton, 1987), three benthic study sites were established (Fig. 1). Sites were selected on the criteria of depth, substrate type (from Everitt *et al.*, 1980) and proximity to Davis Station. Site A was at a water depth of 9 m below M.S.L. It had a flat sandy substrate and a few scattered rocks to which macrophytes were attached. Site B was at a depth of 20 m with a flat mud bottom and no macrophyte cover. Site C was at a depth of 15 m with a sloping, rocky bottom and scattered small pockets of sand and shell fragments; macrophytes covered approximately 100% of this substrate in late summer but for most of the year an incomplete cover existed. This reached a minimum value of approximately 20% in winter.

Fast-ice broke out from the bay in December 1981 which then remained clear of ice until early March 1982. During its formation in March, the newly developing sea ice was broken up several times by strong winds and swells, preventing access until April. After reaching a depth of almost 2 m in December, fast-ice was again cleared from the bay in January 1983. Icebergs rarely enter this area, but one small berg entered during the study period.

Materials and methods

Sampling of the benthos was carried out by divers using SCUBA in the ice-free months and surface supplied air during the months of ice cover. A heated shelter, transported between sites over the fast-ice, was provided for diver safety and comfort (see Dick (1984) for further details on diving equipment and procedure).

To assess seasonal changes in abundance for the numerically dominant benthic infaunal and epifaunal species, each site was examined monthly from

January to December 1982. On several occasions at the beginning of the programme, sampling was incomplete or prevented by problems with equipment and the newly developing sea ice. However, for most species, data were sufficient for general seasonal patterns to be evaluated.

Individuals of macroscopic benthic species, within a 1 m^2 frame, were counted by divers on each sampling occasion. Five or more replicates were taken by random placement of the frame within the sampling region. All counts were converted to number of individuals m^{-2} of substrate surface.

To monitor the infauna, five replicate sediment samples of approximately 100 cm^2 of substrate were taken from each site. Sediment samples were transported to the laboratory in insulated containers and preserved in 4% formalin. Each sample was sifted through a 600 μm mesh Endecott sieve and the retained portion examined under a Wild M5 microscope.

One-way analyses of variance were carried out to investigate any differences between the monthly counts of each species.

Results

Epifauna

Preliminary investigation at Site A (sandy substrate) showed the most commonly occurring macroscopic species to be the burrowing heart urchin *Abatus shackletoni* Koehler, and the ophiuroids *Ophiurolepsis martensi* (Studer) and *Ophiura flexibilis* (Koehler). Distribution of *A. shackletoni* was confined to this site but captures of *O. martensi* and *O. flexibilis* were also made at Site C. A significant difference was established between monthly counts of *A. shackletoni* (Fig. 2 & Table 1), with winter and spring months having fewer individuals than those at other times of the year. No significant seasonal variation was found for *O. martensi* (Fig. 2 & Table 1). Numbers of *O. flexibilis* were too low for quantitative comparison.

The isopod *Glyptonotus antarcticus* Eights and the bivalve *Laternula elliptica* (King & Broderip) dominated the macrofauna of the mud substrate at Site B. Distribution of *L. elliptica* was limited to this

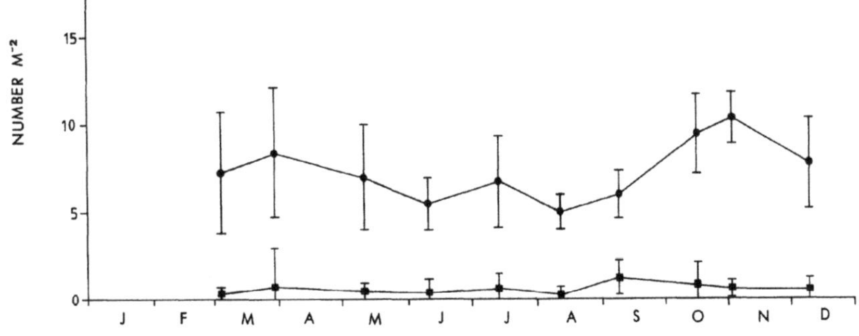

Fig. 2. Monthly changes in population numbers, Site A (sand substrate). • *Abatus shackletoni*; ▪ *Ophiurolepsis martensi* (means with standard deviations represented as error bars).

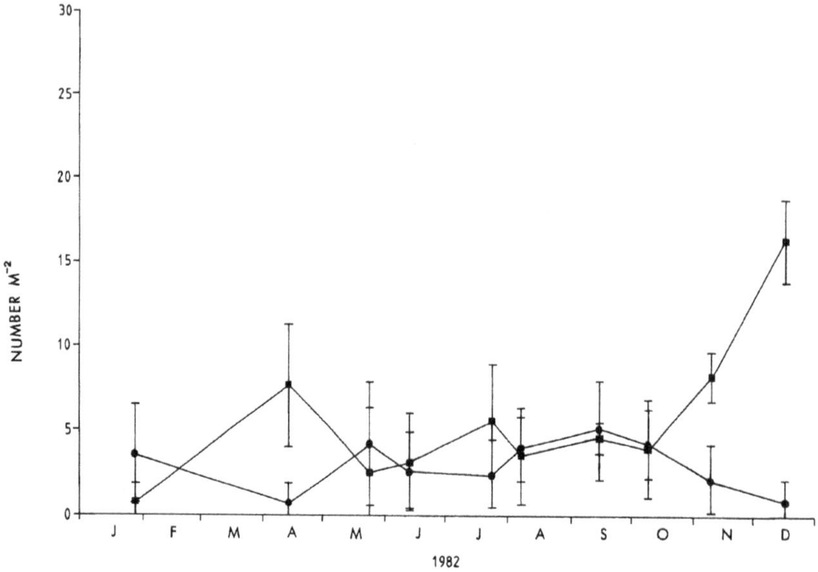

Fig. 3. Monthly changes in population numbers, Site B (mud substrate). ● *Glyptonotus antarcticus*; ■ *Laternula elliptica* (means with standard deviations represented as error bars).

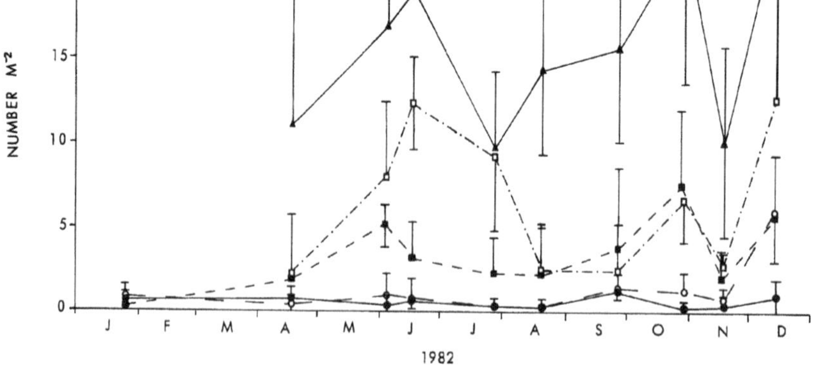

Fig. 4. Monthly changes in population numbers, Site C (rock substrate). ● *Glyphoperidium bursa*; ■ *Sterechinus neumayeri*; ▲ *Ophiurolepsis martensi*; □ *Nymphon australe*; ○ *Ekmocucumis spatha* (means with standard deviations represented as error bars).

Table 1. Probability levels from one-way analyses of variance between monthly means for each macroscopic species.

Site	Species	Variance ratio F	Probability of obtaining null hypothesis
A	*Abatus shackletoni*	2.2	0.03
A	*Ophiurolepsis martensi*	0.44	0.9
B	*Laternula elliptica*	12.19	$<10^{-8}$
B	*Glyptonotus antarcticus*	1.97	0.07
C	*Glyphoperidium bursa*	1.25	0.29
C	*Ekmocucumis spatha*	6.96	$<10^{-4}$
C	*Nymphon australe*	5.58	$<10^{-5}$
C	*Ophiurolepsis martensi*	2.55	0.03
C	*Sterechinus neumayeri*	3.12	0.01

substrate but occasional collections of *G. antarcticus* were made at Site C. *L. elliptica* exhibited a highly significant difference between monthly counts (Fig. 3 & Table 1). More individuals were observed in December than in other months. No significant difference was established among the monthly counts of *G. antarcticus* (Fig. 3 & Table 1).

The most commonly occurring and conspicuous macroscopic species on the rock substrate at Site C were the echinoid *Sterechinus neumayeri* (Meissner), the anemone *Glyphoperidium bursa* Roule, the holothurian *Ekmocucumis spatha* (Cherbonnier), the pycnogonid *Nymphon australe* (Hodgson), and the ophiuroid *Ophiurolepsis martensi*. Only *G. bursa* and *E. spatha* appeared to be restricted to this rock substrate. Although the remaining species were observed on the sand substrate (Site A), sightings were rare. Significant monthly differences were established for all species apart from *G. bursa* (Fig. 4 & Table 1). December counts of *E. spatha* were higher than those of the other months, which did not differ significantly. A distinct spring low was found for *N. australe*. Although significant differences were found for *S. neumayeri* and *O. martensi*, no consistent seasonal pattern was apparent.

Infauna

Two amphipods, *Heterophoxus videns* Barnard, and *Orchomene franklini* (Walker), and a tanaid, *Nototanais antarcticus* Hodgson, were the numerically dominant species within the sandy substrate (Site A). Each was more numerous in summer than in the remainder of the year (Fig. 5 & Table 2).

These species also inhabited the mud substrate with the amphipods *Ampelisca barnardi* Nicholls and *Ampelisca* sp., and the tanaid *Nototanais dimorphus* (Beddard). The two *Ampelisca* species were restricted to this substrate where the mud particles could be bound together to form a flexible envelope housing the amphipod. The amphipods and tanaids inhabiting this mud substrate (Site B) generally exhibited higher numbers in the summer, autumn, early winter and September compared with the remaining months (Fig. 6 & Table 2). Neither *O. franklini* nor *N. dimorphus* conformed to this pat-

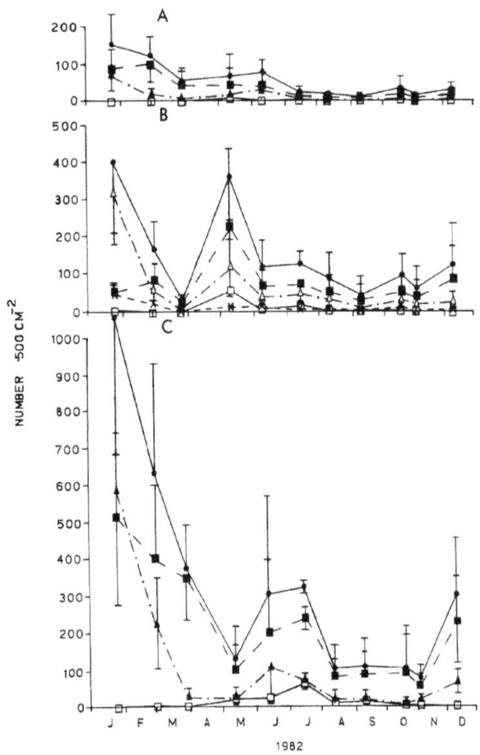

Fig. 5. Monthly changes in population numbers for infaunal species, Site A (sand substrate) A, *Heterophoxus videns*; B, *Nototanais antarcticus*; C, *Orchomene franklini*. ● total animals; ■ juveniles; ▲ adults; △ females; X males; □ ovigerous females (means with standard deviations represented as error bars. In some instances error bars are drawn only on one side of the mean).

Table 2. Probability levels from one-way analyses of variance between monthly means of infaunal amphipods and tanaids.

Site	Species	Variance ratio F	Probability of obtaining null hypothesis
A	*Heterophoxus videns*	3.93	$<10^{-4}$
A	*Orchomene franklini*	7.58	$<10^{-5}$
A	*Nototanais antarcticus*	2.57	0.02
B	*Ampelisca barnardi*	3.78	0.002
B	*Heterophoxus videns*	4.75	$<10^{-4}$
B	*Orchomene franklini*	3.85	0.002
B	*Ampelisca* sp.	3.58	0.003
B	*Nototanais antarcticus*	3.31	0.005
B	*Nototanais dimorphus*	3.13	0.007
C	*Heterophoxus videns*	2.24	0.04
C	*Orchomene franklini*	10.48	$<10^{-7}$
C	*Nototanais antarcticus*	2.70	0.02

tern however; their highest concentrations were recorded in the first half of the year.

The densities of individuals of *H. videns, O. franklini* and *N. antarcticus* were also monitored in the pockets of sand and shell fragments on the rock substrate (Site C) and showed significantly more individuals in January than in other sampling months (Fig. 7 & Table 2). Numbers were consistently low after January.

Brooding females of each species of amphipod and tanaid were taken at some stage throughout the study. For most species, brooding females first appeared in April – May and gradual development of eggs took place throughout the winter and spring until juveniles were released in November – December (Figs. 5, 6 & 7). However, a small number of brooding females of *H. videns, A. barnardi,* and *N. antarcticus* were also taken in January (Figs. 5 & 6). Depending on the species and site, brooding females constituted between 5% and 20% of the population of each species from May to November.

Discussion

The benthic faunal assemblages found within Davis Bay can be classified by presence/absence of macrophytes, water depth and substrate type (Everitt *et al.*, 1980). For the communities investigated in the present study, substrate type appeared to be the most important criterion in determining species composition. Although distribution of the most numerous macroscopic species appeared to be substrate related, several less common but mobile species were found to be widespread, e.g., the nemertean *Lineus*

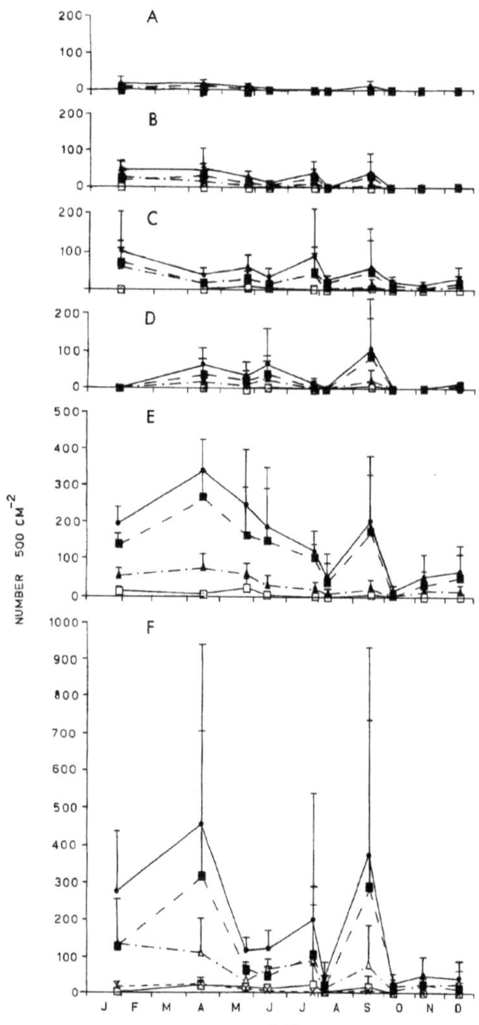

Fig. 6. Monthly changes in population numbers for infaunal species, Site B (mud substrate) A, *Nototanais dimorphus*; B, *Heterophoxus videns*; C, *Orchomene franklini*; D, *Ampelisca* sp.; E, *Ampelisca barnardi*; F, *Nototanais antarcticus*. ● total animals; ■ juveniles; ▲ adults; △ females; X males; □ ovigerous females (means with standard deviations represented as error bars. In some instances error bars are drawn only on one side of the mean).

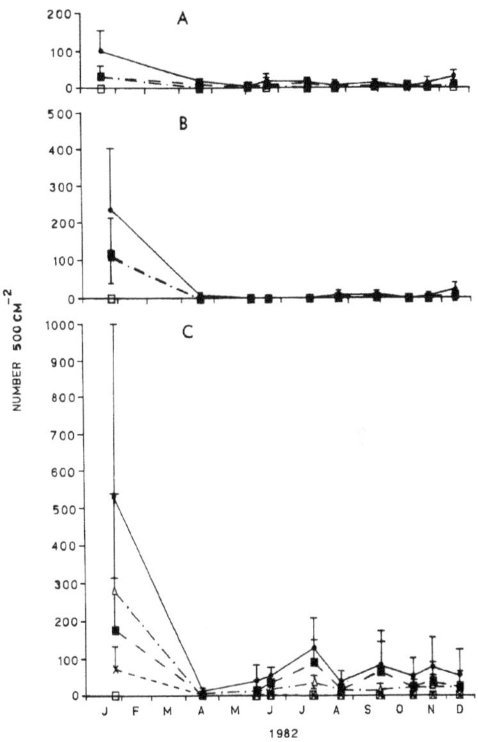

Fig. 7. Monthly changes in population numbers for infaunal species, Site C (rock substrate) A, *Heterophoxus videns*; B, *Orchomene franklini*; C, *Nototanais antarcticus*. ● total animals; ■ juveniles; ▲ adults; △ females; X males; □ ovigerous females (means with standard deviations represented as error bars. In some instances error bars are drawn only on one side of the mean).

corrugatus and the asteroid *Odontaster validus* (Tucker & Burton, 1987).

Such a fauna/substrate relationship did not appear to be the case for infaunal amphipods and tanaids, apart from the two *Ampelisca* species which were restricted to the mud substrate. The species assemblages of the infaunal amphipods and tanaids were relatively similar at all three sites, despite differences in substrate type.

The faunal assemblages described here are similar to those of comparable habitats in other antarctic coastal areas, e.g. Molodezhnaya and Mirny (Gruzov *et al.*, 1967; Gruzov & Pushkin, 1970), McMurdo (Dayton *et al.*, 1974), Syowa (Nakajima *et al.*, 1982), and Arthur Harbour (Lowry, 1975). This finding supports the concept of circumpolarity of antarctic marine communities.

Because antarctic macrobenthic species are characterized by relatively long lives, slow growth rates (Pearse, 1969a) and the production of few offspring (Hedgepeth, 1971), their population numbers were expected to be relatively stable throughout the study. However, the population size of several macroscopic species (*A. shackletoni, L. elliptica, E. spatha* and *N. australe*) did appear to vary seasonally.

This apparent seasonality may be an artifact of the method of data collection and of the organisms' behaviour rather than seasonal variations in mortality and recruitment.

For example, the counting method employed, together with a seasonal variation in feeding behaviour of *L. elliptica*, could explain the apparent increase in numbers for that species in December. Divers counted individuals by the exposed pair of siphons of each animal. For this plankton feeder (Dayton & Oliver, 1977), feeding would be at its most intense in summer when plankton concentrations are at their maximum (Perrin *et al.*, 1987). Little plankton or sedimented sea ice algae are available in winter and therefore it is quite probable that individuals become quiescent, as do suspension feeding holothurians (Gruzov, 1977), and retract their siphons at these times of food limitation. However, as some siphons were always observed in all months including winter, it appears that quiescence is not total. Perhaps the siphons are extended to vent wastes or collect some of the limited food available. Consequently, the recorded density in December is more likely to be indicative of the actual population density, than those of other months.

A. shackletoni was usually observed in the sand substrate with only part of its aboral surface exposed. The low numbers recorded for this species in winter could be due, perhaps, to several individuals burrowing deeper in the substrate, either in search of detrital material or in search of refuge during this period of zero primary production. *E. spatha* was present in higher numbers in December. It is likely that the behavioural change in December for this suspension feeder (Gruzov, 1977) from quiescence (lying flat on the substrate with mouthparts withdrawn) to standing erect with mouthparts splayed, brought it more to the attention of divers. However, it is also probable that some individuals had sheltered under rocks or in crevices during quiescence, emerging in summer to feed. The November low for *N. australe* cannot be satisfactorily explained. It is unusually low considering the densities recorded in October and December.

In general, therefore, the macroscopic benthic species lacked distinct seasonal cycles in abundance, coping with the marked seasonal cycle in food availability (Tucker & Burton, this volume) through their feeding biologies and life histories (Arnaud, 1970). Those species that rely directly and solely on the plankton as a food source have developed a period of inactivity to overcome the long duration of scarce food supplies (Gruzov, 1977). Most species, however, are omnivores and scavengers (Dearborn, 1977; Arnaud, 1977), and change their diets to accommodate whatever food is available. They do not rely solely on the phytoplankton, although most species feed on it when it is available (Arnaud, 1970).

Although the macroscopic species did not exhibit seasonally related changes in population density, the infaunal amphipods and tanaids did, showing a winter minimum and summer maximum. The seasonal cycles of abundance for these deposit and suspension feeding species (Lowry, 1975) are also related to the availability of the phytoplankton. Such seasonal cycles were not so pronounced for the infauna of the mud substrate where numbers remained consistently low only after September. It is possible that the mud had a higher organic content than the other substrates and thus could maintain a resident infaunal population over a longer period.

The release of juveniles by infaunal brooding females coincided with the beginning of phytoplankton development (Tucker & Burton, this volume). This habit of synchronizing release of juveniles with the period of maximum food availability is common amongst antarctic benthic animals (Dell, 1972). Such a brooding strategy is to an extent forced upon antarctic benthic animals in that low temperatures lengthen the development time of pelagic larvae (Thorson, 1950) making it difficult, with the short period of phytoplankton abundance, for a pelagic larva to complete development (Picken, 1980). Indeed, non-pelagic larval development appears to be the rule for benthic species of high latitudes (Thorson, 1950; Mileikovsky, 1971).

Several species of the macrofauna have been previously documented as brooding ovoviviparously (Dell, 1972), although *S. neumayeri* and *O. validus* have demersal larval stages (Pearse & Giese, 1966; Pearse, 1969b) with some, for example the predator/scavenger *G. antarcticus*, releasing young at any time throughout the year (Dearborn, 1967; White, 1970). This is in accordance with the limited dependence of this species on primary producers and so the release of juveniles is not totally dictated by the seasonal primary productivity cycle.

Some antarctic benthic animals are K-strategists (Clarke, 1979). In the present study, the epifaunal and infaunal species had several attributes characteristic of such a strategy. Several species have been reported as long-lived (up to 100 a for the large echinoderms; Pearse, 1969a), as having slow growth rates and as growing to a large size (e.g. amphipods; De Broyer, 1977). Parental care of the relatively few young, until they had reached an advanced stage (Thorson, 1950), was observed for the amphipods and tanaids. Development of such strategies by antarctic macrofaunal species is the likely result of the stable and predictable nature of the benthic environment in coastal Antarctica.

There are relatively small and predictable changes in salinity and temperature and physical disturbance is limited by the long duration of sea ice cover. Although icebergs and anchor-ice can denude areas in coastal Antarctica (Dayton *et al.*, 1969), no anchor-ice was observed at any time during this study and icebergs rarely enter the area.

Acknowledgements

I am extremely grateful to the divers, Messrs. K. Beinssen and R. Perrin, and to Mr. P. Lu for his assistance in the field. I would also like to thank Mr. D. Ratkowsky for the statistical analyses and Mr. H. Burton for his continuous support. I also thank Drs. D. Dunn, J. Lowry, W. Ponder, G. Poore, F. Rowe and Mr. D. Staples for their taxonomic assistance, and Dr. J. Greenwood for his comments on the manuscript.

References

Arnaud, P. M., 1970. Frequency and ecological significance of necrophagy among the benthic species of antarctic coastal waters. In M. W. Holdgate (ed.), Antarctic Ecology, 1. Academic Press, Lond.: 259–267.

Arnaud, P. M., 1977. Adaptations within the antarctic marine benthic ecosystem. In G.A. Llano (ed.), Adaptations within Antarctic Ecosystems. Smithsonian Institution, Wash., D.C.: 135–158.

Clarke, A., 1979. On living in cold water: K-strategies in antarctic benthos. Mar. Biol. 55: 111–119.

Dayton, P. K. & J. S. Oliver, 1977. Antarctic soft-bottom benthos in oligotrophic and eutrophic environments. Science 197: 55–58.

Dayton, P. K., G. A. Robilliard & A. L. De Vries, 1969. Anchor ice formation in McMurdo Sound, Antarctica, and its biological effects. Science 163: 273–274.

Dayton, P. K., G. A. Robilliard, R. T. Paine & L. B. Dayton, 1974. Biological accommodation in the benthic community at McMurdo Sound, Antarctica. Ecol. Monogr. 44: 105–128.

Dearborn, J. H., 1967. Food and reproduction of Glyptonotus antarcticus (Crustacea: Isopoda) at McMurdo Sound, Antarctica. Trans. r. Soc. N.Z. 8: 163–168.

Dearborn, J. H., 1977. Foods and feeding characteristics of antarctic asteroids and ophiuroids. In G. A. Llano (ed.), Adaptations within Antarctic Ecosystems. Smithsonian Institution, Wash., D.C.: 293–326.

De Broyer, C., 1977. Analysis of the gigantism and dwarfness of antarctic and subantarctic gammaridean Amphipoda. In G. A. Llano (ed.), Adaptations within Antarctic Ecosystems. Smithsonian Institution, Wash., D.C.: 327–334.

Dell, R. K., 1972. Antarctic benthos. Adv. mar. Biol. 10: 1–216.

Dick, A. F., 1984. Thermal loss in antarctic divers. Med. J. Aust. 140: 351–354.

Everitt, D. A., G. C. B. Poore & J. Pickard, 1980. Marine benthos from Davis Station, Eastern Antarctica. Aust. J. mar. Freshwat. Res. 31: 829–836.

Gruzov, Y. N., 1977. Seasonal alterations in coastal communities in the Davis Sea. In G. A. Llano (ed.), Adaptations within Antarctic Ecosystems. Smithsonian Institution, Wash., D.C.: 263–278.

Gruzov, Y. N. & A. F. Pushkin, 1970. Bottom communities of the upper sublittoral of Enderby Land and the Shetland Islands. In M. W. Holdgate (ed.), Antarctic Ecology, 1. Academic Press, Lond.: 235–238.

Gruzov, Y. N., M. V. Propp & A. F. Pushkin, 1967. Biological associations of coastal areas of the Davis Sea (based on the observations of divers). Sov. antarct. Exped. Info. Bull. 6: 523–533.

Hedgepeth, J. W., 1971. Perspectives of benthic ecology in Antarctica. In L.O. Quam (ed.), Research in the Antarctic. Am. Ass, Advan. Sci., Wash., D.C.: 93–136.

Johnstone, G. W., D. J. Lugg & D. A. Brown, 1973. The biology of the Vestfold Hills, Antarctica. ANARE Sci. Rep. Publ. 123. Antarct. Div., Dept. Sci., Melbourne, 62 pp.

Lowry, J. K., 1975. Soft bottom macrobenthic community of Arthur Harbor, Antarctica. In D. L. Pawson (ed.), Antarctic Research Series, 23. Biology of the Antarctic Seas, 5. Am. geophys. Union, Wash., D.C.: 1–19.

Mileikovsky, S. A., 1971. Types of larval development in marine bottom invertebrates, their distribution and ecological significance: a re-evaluation. Mar. Biol. 10: 193–213.

Nakajima, Y., K. Wantanabe & Y. Naito, 1982. Diving observations of the marine benthos at Syowa Station, Antarctica. In T. Hoshiai & Y. Naito (eds), Proceedings of the 5th Symposium on Antarctic Biology. Natn Inst. Polar Res., Tokyo. Spec. Issue 23: 44–54.

Pearse, J. S., 1969a. Antarctic sea star. Aust. nat. Hist. 16: 234–238.

Pearse, J. S., 1969b. Slow developing demersal embryos and larvae of the antarctic sea star Odontaster validus. Mar. Biol. 3: 110–116.

Pearse, J. S. & A. C. Giese, 1966. Food, reproduction and organic constitution of the common antarctic echinoid Sterechinus neumayeri (Meissner), Bull. mar. biol. Lab. Woods Hole 130: 387–401.

Perrin, R. A., P. Lu & H. J. Marchant, 1987. Seasonal variation in marine phytoplankton and ice-algae at a shallow antarctic coastal site. Hydrobiologia 146: 33–46.

Picken, G. B., 1980. Reproductive adaptations of antarctic benthic invertebrates. Biol. J. Linn. Soc. 14: 67–75.

Thorson, G., 1950. Reproductive and larval ecology of marine bottom invertebrates. Biol. Rev. 25: 1–45.

Tucker, M. J. & H. R. Burton, this volume. The inshore marine ecosystem off the Vestfold Hills, Antarctica. Hydrobiologia.

Tucker, M. J. & H. R. Burton, 1987. A survey of the marine fauna in shallow coastal waters of the Vestfold Hills and Rauer Islands, Antarctica. ANARE Res. Notes. No. 55. Antarct. Div., Dept. Sci., Hobart, 24 pp.

White, M. G., 1970. Aspects of the breeding biology of Glyptonotus antarcticus (Eights) (Crustacea: Isopoda) at Signy Island, South Orkney Islands. In M. W. Holdgate (ed.), Antarctic Ecology, 1. Academic Press, Lond.: 279–285.

Hydrobiologia 165: 161–167 (1988)
J. M. Ferris, H. R. Burton, G. W. Johnstone & I. A. E. Bayly (eds) Biology of the Vestfold Hills, Antarctica
Kluwer Academic Publishers

The inshore marine fishes of the Vestfold Hills region, Antarctica

R. Williams
Antarctic Division, Department of Science, Kingston, Tasmania, 7150, Australia

Keywords: fish, Antarctica, distribution, zoogeography

Abstract

Thirteen species of fish have so far been caught in the inshore waters around the Vestfold Hills, including the Rauer Islands, in depths down to approximately 100 m. Species caught depend markedly on the type of fishing gear used, but three species are clearly dominant numerically.

Pagothenia bernacchii is most abundant in the shallower (<20 m deep) weedy and rocky habitats, while *Chionodraco hamatus* is dominant in the deeper (>20 m deep) nearshore troughs and further offshore. *Pagothenia borchgrevinki* occupies the specialized habitat associated with sea ice and close-inshore areas, including fjords and Burton Lake.

The species list from the Vestfold Hills area is similar to lists from comparable locations in East Antarctica except for the major difference that *C. hamatus* has not yet been recorded from such shallow waters at the other locations, while *P. bernacchii* and *P. hansoni* are much more abundant in water deeper than 20 m at those sites than at Davis.

Introduction

Although the general distribution of antarctic fishes is quite well known (DeWitt, 1971), detailed studies of the fish fauna at individual inshore sites are relatively uncommon. Some sites in the South Atlantic area, e.g. Signy Island (South Orkney Islands) and South Georgia, have been quite thoroughly surveyed (Everson, 1969; Burchett, 1983; Burchett *et al.*, 1983), but less is known of the inshore marine fish fauna of East Antarctica. Some surveys have been done on the coast of Terre Adelie (Hureau *et al.*, 1965) and near Syowa Station (Naito & Iwami, 1982) and a small amount of work reported from the McMurdo area by various authors has concerned the shallow water fishes.

Samples of inshore fishes have been collected from the Vestfold Hills region over a number of years and there now exists a collection of over 500 specimens. This paper reports on this collection, in particular on the distribution of fish with respect to habitat and zoogeography. Later papers will deal with the biology of the commoner species.

Fish are an important part of the Vestfold Hills inshore marine ecosystem, either as potential predators or as prey of other groups. Thus recent attempts to understand such components of the shallow marine ecosystem as zooplankton and benthos (Tucker, this volume) or vertebrates, especially birds (Puddicombe & Johnstone, this volume), require a knowledge of fish to place them in context. Some inshore fish species also occur as adults or juveniles in offshore shelf waters (Williams *et al.*, 1983), highlighting the need to recognize links between the two habitats, especially where commercial exploitation of some species is a possibility.

Materials and methods

Collections have been made since 1978 and although sampling sites are widespread in the Vestfold Hills area, they come mainly from the immediate vicinity of Davis Station and from near Filla Island in the Rauer Group, overwintering site of the Oceanic Research Foundation's (ORF) Frozen Sea Expedition from February 1983 to January 1984 (Fig. 1).

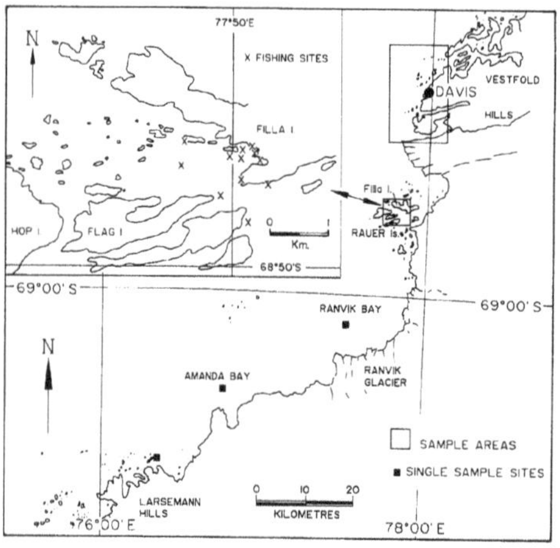

Fig. 1. Map of the Vestfold Hills and southern areas showing fish sampling locations.

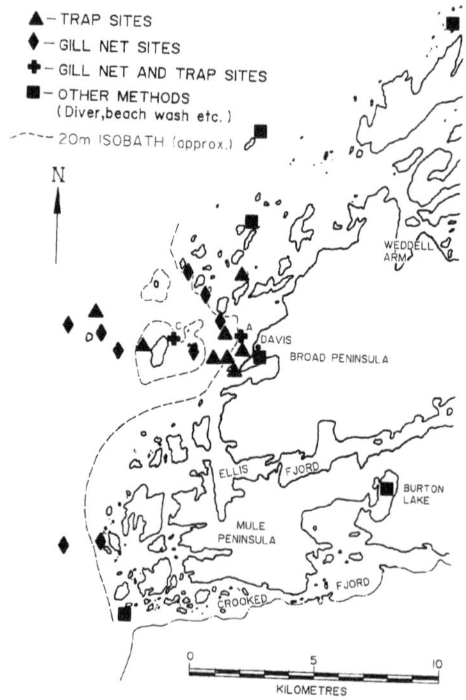

Fig. 2. Detailed map of the Vestfold Hills area showing fishing sites and equipment used.

Various methods of capture have been used. Angling, using meat or fish as bait, was used exclusively by ORF for its sampling in the Rauer Islands–Larsemann Hills area in depths down to 12 m, and has also been used occasionally in the Davis area. Traps baited with meat or fish have been used extensively near Davis. These traps were cylindrical, 1 m long by 50 cm diameter, with a funnel entrance (10 cm aperture) at one end, and constructed from 2.5 cm mesh wire netting on a steel frame. They were set for c. 24 h before being emptied and the bait replenished. Gill nets were used in two sampling programmes. Nets of 5 cm and 10 cm stretched mesh were set at sites A and C (Fig. 2) at intervals throughout 1982, when necessary with the aid of divers through holes in the sea ice. During February 1984 an intensive survey of the Davis vicinity was made using nets of 2.5, 5.0, 7.5, 10.0 and 12.5 cm stretched mesh. In all cases, nets were set on the bottom for c. 24 h at a time. A few specimens were also captured by hand by divers, or were found stranded on beaches or on the sea ice.

Many specimens were taken in summer, as is the case with most antarctic collections, but this collection also contains a significant number of specimens, at least of the commoner species, from all months of the year. Also included are some specimens from the regurgitated stomach contents of Adelie penguins *Pogoscelis adeliae* (Hombron & Jacquinot) sampled on Magnetic Island, 4 km from Davis (see Puddicombe & Johnstone, this volume), in cases where that is the only record of a species in the area. Sampling sites in the Davis area are shown in Fig. 2.

Results

Table 1 summarizes the composition of the fish collection from the Vestfold Hills region. It shows that the inshore fauna is dominated, both in number of specimens and in number of occasions sampled, by three species, *Pagothenia bernacchii* (Boulenger), *Pagothenia borchgrevinki* (Boulenger) and *Chionodraco hamatus* (Lonnberg). All other species are far less common, many being represented by only one or two specimens. Table 1 also shows that the type of fishing gear used can affect the type of fish caught. Thus while some species, such as *Pagothenia hansoni* (Boulenger) and *P. bernacchii* are caught by a wide variety of gears, 80% of the *P. borchgrevinki* were taken by angling and all but one specimen of *C. hamatus* were taken by gill net.

Table 1. Fishes collected in the Vestfold Hills region.

Species (no. of specimens)	Occasions sampled	% of specimens caught by:					
		Trap	Gill net	Rod & line	Diver	Beach,* ice	Stomach*
Trematomus newnesi (5)	3	80	20				
T. pennellii (1)	1			100			
T. centronotus (2)	2		50			50	
T. scotti (4)	1					100	
Pagothenia borchgrevinki (128)	33	1	1	80	7	11	
P. bernacchii (279)	93	31	31	24	7	8	
P. hansoni (7)	5	43	43	14			
Dissostichus eleginoides (1)	1		100				
Pleuragramma antarcticum (27)	6					100	
Artedidraco sp. (1)	1					100	
Gymnodraco acuticeps (5)	5		80	20			
Pagetopsis macropterus (4)	3						100
Chionodraco hamatus (96)	17		99		1		

* See text for explanation.

The distributions of the commoner fish species in the Davis area are shown in Figs. 3a–d. Distributions among the Rauer Islands and other southern sites have not been mapped because they were all shallow (<12 m deep) nearshore sites effectively sampled only by angling, and all had virtually identical species lists. Thus most of the *P. borchgrevinki* caught came from this area (105 specimens from 128 total) along with 74 of the 279 specimens of *P. bernacchii.*

The distributions in Fig. 3 show that some species such as *P. bernacchii* and *P. borchgrevinki,* tend to be caught near shore, while others such as *C. hamatus* and *P. hansoni* are usually found in more offshore locations. The depth distributions of the common species show that the boundary between these two types of distribution falls at about the 20 m isobath. Taking this depth as the division between nearshore and offshore, the percentage occurrence of specimens of each species within each depth category shows a clearer distinction between the shallow water species and those from deeper water than do the distribution maps (Table 2).

Table 2. Depth distributions of fish (excluding beach and ice strandings).

Species (no. of specimens)	% of specimens from depth		Depth range (m)
	<20 m	>20 m	
Trematomus newnesi (5)	100		6–8
T. pennellii (1)			5
T. centronotus (1)			90
Pagothenia borchgrevinki (114)	100		0–20
P. bernacchii (255)	93	7	0–90
P. hansoni (7)	29	71	5–58
Dissostichus eleginoides (1)			84
Gymnodraco acuticeps (5)	80	20	5–36
Chinodraco hamatus (96)	8	92	4–96

Discussion

The current species list from Davis is very similar to those from the two other well studied sites in East Antarctica, Dumont d'Urville (66°39′S, 139°55′E) and Syowa (69°01′S, 39°35′E) (Table 3). Most species occur at all three named sites; many blanks are as likely to be the result of inadequate sampling as of genuine absences. The inshore fish fauna of East Antarctica therefore appears to be quite homogeneous.

The apparent distributions of some species do deserve some comment, however. *Notothenia coriiceps neglecta* Nybelin has its centre of distribution in the Antarctic Peninsula and South Georgia region, and appears to extend westwards into East Antarctica only as far as Terre Adelie, since it has only been reported from Cape Adare (172°E) and Dumont d'Urville (DeWitt, 1971). Although it is one of the commonest taxa in shallow water (<20 m) at Terre Adelie (Hureau, 1970), despite

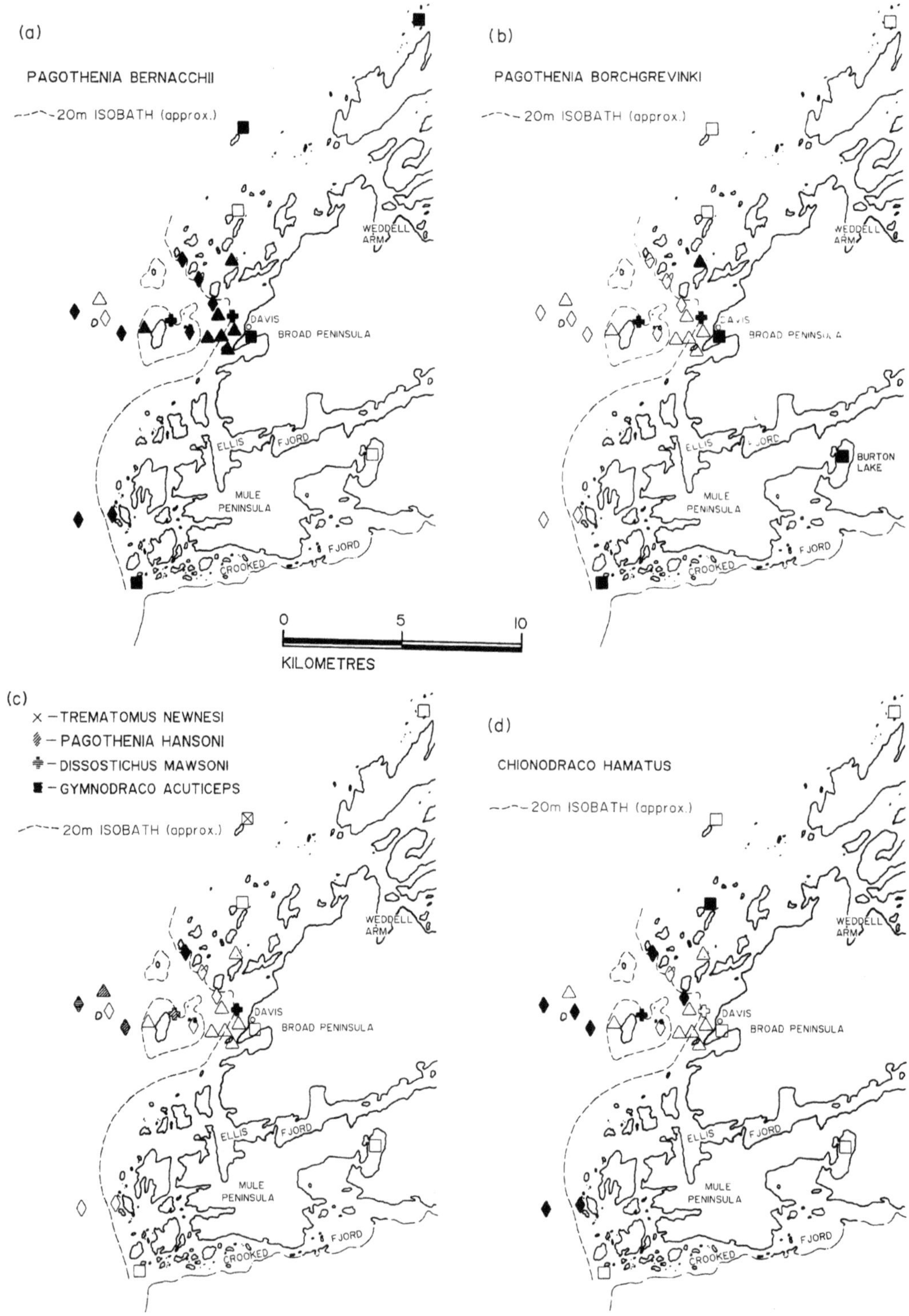

Fig. 3. Distributions of fish species at sample sites. Solid symbol = species present; empty symbol = species absent.

Table 3. Distribution of fish recorded from three sites in East Antarctica.

Species	Davis	Dumont[a] d'Urville	Syowa[b]	Known distribution & depth range[c]	
Family Nototheniidae					
Notothenia coriiceps neglecta		+		West Antarctica, Cape Adare	0 – 20
Trematomus newnesi	+	+	+	circum-antarctic & S. Orkneys	0 – 110
T. nicolai		+	+	East Antarctica	0 – 25
*T. loennbergii**	?	+	+	circum-antarctic	65 – 832
T. pennellii	+			East Antarctica	0 – 110
T. centronotus	+	+	+	East Antarctica	3 – 732
T. scotti	+		+	circum-antarctic	70 – 655
Pagothenia borchgrevinki	+	+	+	circum-antarctic & S. Orkneys	0 – ?
P. bernacchii	+	+	+	circum-antarctic	3 – 732
P. hansoni	+	+	+	circum-antarctic & S. Georgia	5 – 549
Dissostichus mawsoni Norman		+	+	circum antarctic	20 – 219
D. eleginoides	+	+	+	circum-subantarctic	72 – 830
Pleuragramma antarcticum	+			circum-antarctic	pelagic
Family Harpagiferidae					
Artedidraco sp.	+	+			
Family Bathydraconidae					
Gymnodraco acuticeps	+	+	+	East Antarctica	0 – 15
Family Channichthyidae					
Pagetopsis macropterus	+	+		East Antarctica	20 – 655
Chionodraco hamatus	+			East Antarctica	0 – 595

a From Hureau *et al.* (1965).
b From Naito & Iwami (1982).
c After De Witt (1971).
* Possibly already found in Adelie penguin stomachs at Davis.

considerable sampling effort at many other sites it had not been found further west in East Antarctica until it was recently caught at Ardery Island near Casey (Antarctic Division, unpubl. data), which extends its known distribution westward to 110°E. Some of the species are more usually found in deeper waters, and at least some of their occurrences at the coastal sites have resulted from unusual circumstances. At Davis, species in this category are *Trematomus scotti* (Boulenger) and *Artedidraco* sp., both of which were found only once, washed onto the beach after a storm; similarly *Pagetopsis macropterus* (Boulenger) has been encountered only in the stomachs of Adelie penguins and *Trematomus loennbergii* Regan may also have been found in this way, although I have not yet encountered a good enough specimen for unequivocal identification. These species, where they occur at other sites, are also rare and taken only in deeper water, probably near the upper limit of their depth ranges. Thus at Syowa, where Naito & Iwami (1982) sampled with traps and by angling to depths of 840 m, *T. scotti* and *T. loennbergii* were caught only at depths greater than 200 m. At Dumont d'Urville, Hureau *et al.* (1965) caught *Artedidraco loennbergii* Roule and *A. skottsbergi* Lonnberg only at 70–100 m depth, and the single record of *Pagetopsis macropterus* from the vicinity was recorded by Waite (1916) from 280 m depth.

By using a wide range of sampling techniques, more species have been recorded at Davis than from the other relatively well known sites in East Antarctica, so that the species list may be considered fairly complete. Apart from *Trematomus nicolai* (Boulenger), which is common elsewhere in East Antarctica but inexplicably absent from Davis, there are some other generally deeper water species which may occur infrequently in the inshore waters and which remain to be recorded. These include *Trematomus eulepidotus* Regan, a common circum-antarctic species with a depth range of 70–440 m (DeWitt, 1971), and *Prionodraco evansii* Regan, *Cygnodraco mawsoni* Waite and *Cryodraco antarcticus* Dollo which have been found as pelagic larvae or benthic adults at depths below 120 m in Prydz Bay, offshore from Davis (Williams *et al.*,

1983 and unpubl. data), and are known to occur in depths of 100 m or less elsewhere.

As noted above, the type of sampling gear used can influence the species captured. At Davis, the outstanding example is *C. hamatus,* of which 99% were caught by gill nets. Although one of the commonest inshore fish at Davis, this species has not been recorded at Syowa or Dumont d'Urville where gill nets have not been used, although Hureau *et al.* (1965) used a dredge ('dragage') at Dumont d'Urville. This fish is widespread and common in East Antarctica (DeWitt, 1971) but hitherto has been reported mostly from trawls in deeper water, for example from Leopold Bay (70°19′S, 24°15′E) in 200–300 m (Gosse, 1968) and from the Ross Sea in 415–549 m (Reseck, 1961) and 550–600 m (Iwami & Abe, 1981). This highlights the need for thorough surveys to be carried out with a variety of gear.

The distributions of the commoner fish within the Vestfold Hills area are generally similar to those observed elsewhere, but differ in some details. Thus *P. bernacchii* dominates the shallow water fauna and is most abundant on rocky substrates with macrophyte cover; 93% of specimens were taken from depths of 20 m or less (Table 1), with a few down to at least 90 m. *P. hansoni* is much less common than *P. bernacchii,* and is found mostly in the deeper water where *C. hamatus* dominates. *P. borchgrevinki* and *T. newnesi* Boulenger are cryopelagic species associated with the underside of the sea ice and very close-inshore habitats. As has been noted elsewhere (e.g. Gruzov *et al.,* 1967) the juveniles are often associated with stalactite-like formations on the underside of the sea ice, rarely straying far from their shelter (M. J. Tucker, pers. commun.). *P. borchgrevinki* was the main species (with *P. bernacchii*) caught at the inshore sites in the Rauer Islands, generally by angling through holes in the sea ice. Eighty per cent of these specimens were caught under the sea ice, especially where bergy bits were frozen into it, and were uniformly pale. The other 20%, caught near the bottom in up to 5 m of water, were a dark phase with a dark olive-brown spotted or finely vermiculated pattern above and silvery-white below (Schou, 1984). This species was even found in Burton Lake (Fig. 3b), a meromictic lake connected to Crooked Fjord by a tidal channel c. 20 m wide and up to 2 m deep. Several thousand fish were found dead on the lake shores after a blizzard in February 1982 had overturned the lake and brought anoxic water to the surface. *P. borchgrevinki* does not appear to inhabit the lake continuously, as intensive sampling in February 1984 with gill nets and traps in the lake and fyke nets in the channel failed to catch a single fish; however, these observations do show that this species will occupy the shallowest and closest-inshore habitats.

The distributions of fish at Davis are generally similar to those observed by Naito & Iwami (1982) at Syowa and Hureau (1970) at Dumont d'Urville. However, at both these locations *P. bernacchii* was common over a much wider depth range, down to 100 m and 50 m respectively, and *P. hansoni* was much more common, being the dominant species below 100 m at Syowa and below 30 m at Dumont d'Urville. As both species are easily caught by a variety of methods, these variations do not appear to result from different sampling techniques. As well as such factors as bottom topography and substrate type, the abundance of *C. hamatus* in waters deeper than 20 m at Davis may account for the relative scarcity of the *Pagothenia* species at these depths. An as yet superficial analysis of stomach contents shows that *C. hamatus* preys upon juvenile and adult *Trematomus* and *Pagothenia* spp, but to what extent cannot yet be assessed. If *C. hamatus* were as common at the other inshore sites as it is at Davis, a few occurrences at least should have been recorded, even though appropriate sampling techniques have not yet been used. Thus it may be that *C. hamatus* really is much less common, and consequently the *Pagothenia* spp are much more abundant and widespread at the other sites.

Acknowledgements

I thank all the expeditioners who collected fish for this study, in particular Alan Ward and Mark Tucker of ANARE and Jannik Schou of ORF. The army LARC detachment during summer 1983/84 enthusiastically provided essential water transport and other assistance for the gill net survey.

References

Burchett, M. S., 1983. Abundance of the nearshore fish population at South Georgia (Antarctica) sampled by trammel net. Br. antarct. Surv. Bull. 61: 39–43.

Burchett, M. S., P. J. Sayers, A. W. North & M. G. White, 1983. Some biological aspects of the nearshore fish populations at South Georgia. Br. antarct. Surv. Bull. 59: 63–74.

DeWitt, H. H., 1971. Coastal and deep-water benthic fishes of the Antarctic. In V. C. Bushnell (ed.), Antarct. Map Folio Ser., 15. Am. geogrl Soc., NY, 10 pp.

Everson, I., 1969. Inshore fishes from the South Orkney and South Shetland Islands, the Antarctic Peninsula and South Georgia. Br. antarct. Surv. Bull. 19: 89–96.

Gosse, J-P., 1968. Expéditions Antarctiques Belges. Poissons récoltés par l'expédition d'été 1967. Bull. Inst. r. Sci. nat. Belg. 44: 1–11.

Gruzov, Y. N., M. V. Propp & A. F. Pushkin, 1967. Biological associations of coastal areas of the Davis Sea (based on the observations of divers). Sov. antarct. Exped. Inf. Bull. 6: 523–533.

Hureau, J-C., 1970. Biologie comparée de quelques poissons Antarctiques (Nototheniidae). Bull. Inst. oceanogr. Monaco 68: 1–244.

Hureau, J-C., P. Arnaud & C. Dousset, 1965. A propos de deux nouvelles collections de poissons recoltées en Terre Adelie (Antarctique) en 1964 et 1965. Bull. Mus. natn. Hist. nat., Paris 37: 941–950.

Iwami, T. & T. Abe, 1981. The collection of the fishes trawled in the Ross Sea. Antarct. Rec. 71: 130–141.

Naito, Y. & T. Iwami, 1982. Fish fauna in the northeastern parts of Lutzow-Holm Bay with some notes on the stomach contents. Mem. natn. Inst. polar Res. Tokyo Spec. Issue 23: 64–72.

Puddicombe, R. & G. W. Johnstone, this volume. The diet of Adelie penguins in breeding colonies near Davis, Antarctica. Hydrobiologia.

Schou, J., 1984. Fishing programme. In D. Lewis & M. George (eds.), The initial scientific reports of the Mawson Anniversary and Frozen Sea Expeditions. Oceanic Res. Fdn occ. Publ. 1: 55–59.

Tucker, M. J., this volume. The inshore marine ecosystem off the Vestfold Hills, Antarctica. Hydrobiologia.

Waite, E. R., 1916. Fishes. Australas. antarct. Exped. sci. Rep. Ser. C 3: 1–100.

Williams, R., J. M. Kirkwood & D. O'Sullivan, 1983. FIBEX cruise zooplankton data. Aust. natn. antarct. Res. Exped. Res. Notes 7. Antarct. Div., Dept. Science, Hobart, 163 pp.

Hydrobiologia 165: 169–171 (1988)
J. M. Ferris, H. R. Burton, G. W. Johnstone & I. A. E. Bayly (eds) Biology of the Vestfold Hills, Antarctica
Kluwer Academic Publishers

A device for remote sampling of benthic algae under ice

C. W. Heath[1,2] & W. J. Singleton[1]
[1]*Antarctic Division, Department of Science, Kingston, Tasmania, 7050, Australia;* [2]*Present address: Metropolitan Water Sewerage & Drainage Board, Scientific Laboratories, P.O. Box 73, West Ryde, Sydney, 2114, Australia*

Key words: benthic, algae, sampling, ice, Antarctica

Abstract

A benthic algae sampling device designed for use through surface ice, uses the principle of pivoting arms. It is designed to be activated below ice sheets and is capable of passing easily through a hole of 11 cm diameter bored by a SIPRE ice auger.

Introduction

A diverse range of devices exist to facilitate the remote sampling of benthic material. These generally fall into the categories of grabs, corers and dredges (see reviews of Holme & McIntyre, 1971 and Menzies *et al.*, 1973). Usually employed under open water conditions however, compactness is not a major concern of their design and construction. The use of such corers and grabs in under-ice conditions therefore requires large holes to be bored in the ice, while the need to tow dredges precludes their use under ice.

Holes in ice can be produced by several means including steam jets (Science News, 1979), augers mounted on tracked vehicles, hand held motor driven augers, by explosive charges or by manually digging holes. These methods, with the exception of the latter which is very time consuming, are either bulky, require some form of logistic support or are restricted in their range. In conditions where the integrity of the ice sheet is important, such as primary productivity studies, where underwater light regimes should closely resemble ambient, it is an advantage to restrict the size of the hole produced, and to reduce the extent to which the ice sheet adjacent to the hole is damaged. The SIPRE ice auger (Journal of Glaciology, 1958) overcomes most of the deficiencies encountered with other techniques. It is compact, easily transportable and amenable to back packing into areas with poor accessability. It cores a small hole (11.09 cm diameter) doing so at a rate of around 1 m in 10–15 min, with minimal disruption to the surrounding ice sheet.

Description of the grab and its use

Described here is a device for sampling benthic algal mats, which was designed to pass through an ice hole bored by a SIPRE auger. Its advantages are its small size, capacity to be 'set' below the ice sheet and sampling technique which avoids the collection of large amounts of sediment and detritus. It was used successfully throughout the 1981 antarctic winter, collecting samples of filamentous benthic algae which were subsequently used in primary productivity studies (see Heath, this volume). Physiological perturbation of the algae, a major concern in productivity studies, was kept to a minimum by the ease with which the samples were removed from the grab in a bucket of water adjacent to the ice hole. Wide working tolerances enabled the grab to be used in water temperatures of 0.7 °C and air temperatures of −25 °C without freezeup problems. The lack of de-

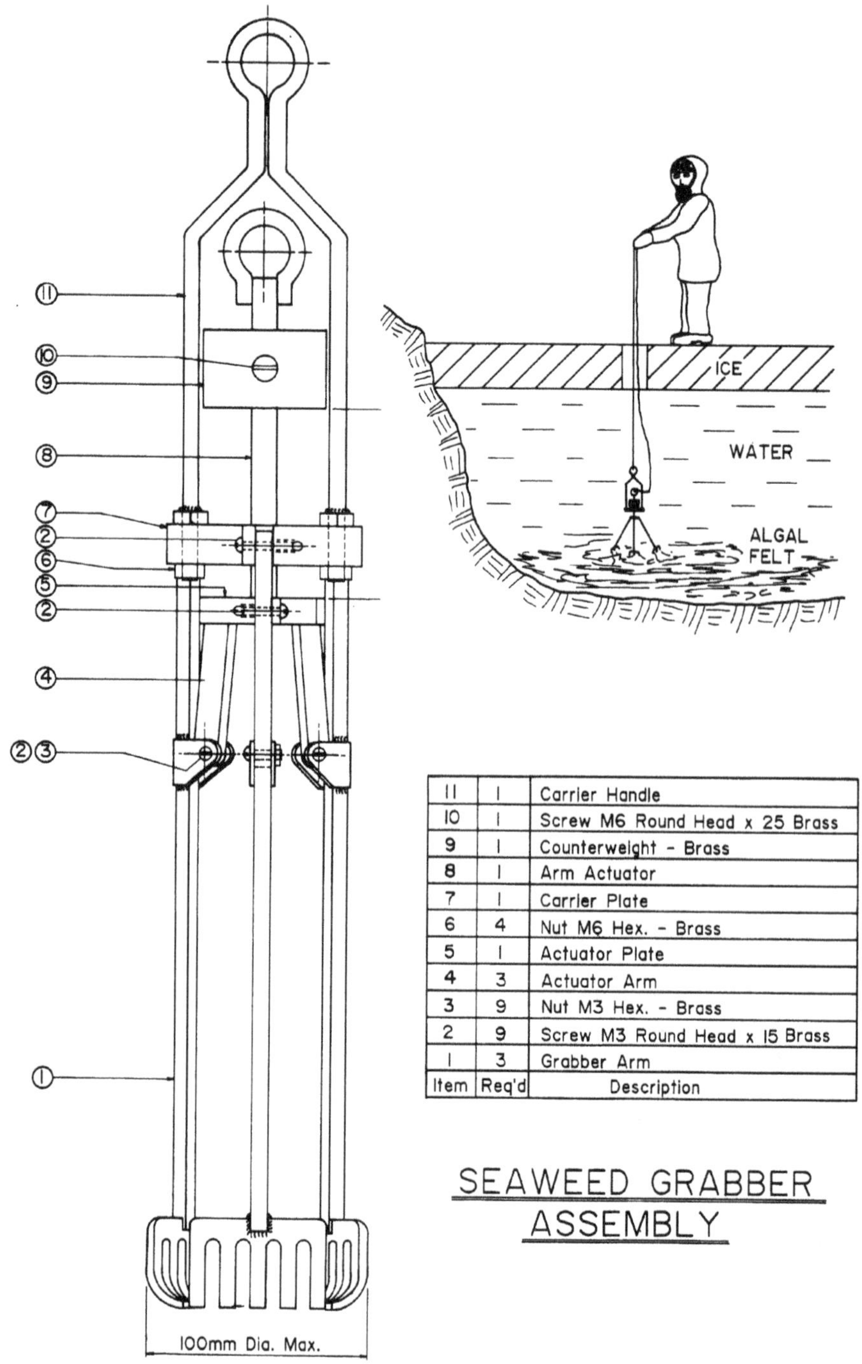

Item	Req'd	Description
11	1	Carrier Handle
10	1	Screw M6 Round Head x 25 Brass
9	1	Counterweight - Brass
8	1	Arm Actuator
7	1	Carrier Plate
6	4	Nut M6 Hex. - Brass
5	1	Actuator Plate
4	3	Actuator Arm
3	9	Nut M3 Hex. - Brass
2	9	Screw M3 Round Head x 15 Brass
1	3	Grabber Arm

Fig. 1. Technical diagram of the benthic grab.

pendance on a 'messenger' to activate the grab enabled its operation in 0.5 m of water under 2 m of ice, while it was also successfully operated at depths of 12 m.

The brass grab (Fig. 1), is operated by two lines, the carrier line and the actuator (counterweight) line. The weight of the frame is taken by the carrier line and the actuator line is attached to the actuator arm. It is lowered through the ice in the 'closed' position (Fig. 2b) by taking the weight on the actuator line. Below the ice, the weight is transferred to the carrier line. The counterweight on the actuator arm (Fig. 1) causes it to slide through the carrier plate. As the actuator arm moves down, the attached actuator plate drops also. This causes the pivot arms to push the grab arms open. The grab arms are pivoted at their top in the carrier plate. In this open position (Fig. 2a), the grab is lowered into the benthic algal mat where the weight is again taken on the actuator line, drawing the arms of the grab closed (Fig. 2b). Maintaining weight on the actuator line keeps the grab closed, preventing the sampled benthic mat from falling free during ascent to the surface. The extent to which the grab arms open, although limited by the length of the pivot arms, can be varied by altering the position of the counterweight.

References

Heath, C. W., this volume. Annual primary productivity of an antarctic continental lake: Phytoplankton and benthic algal mat production strategies. Hydrobiologia.

Holme, N. A. & A. D. McIntyre, 1971. Methods for the study of marine benthos. IBP Handbook 16. Blackwell.

Journal of Glaciology, 1958. Instruments and methods: Ice drills and corers. J. Glaciol. 3(21): 30.

Menzies, R. J., R. Y. George & G. T. Rowe, 1973. Abyssal environment and ecology of the world oceans. Wiley.

Science News, 1979. Antarctic ice keeps algae in the dark. Sci. News 116(4): 71.

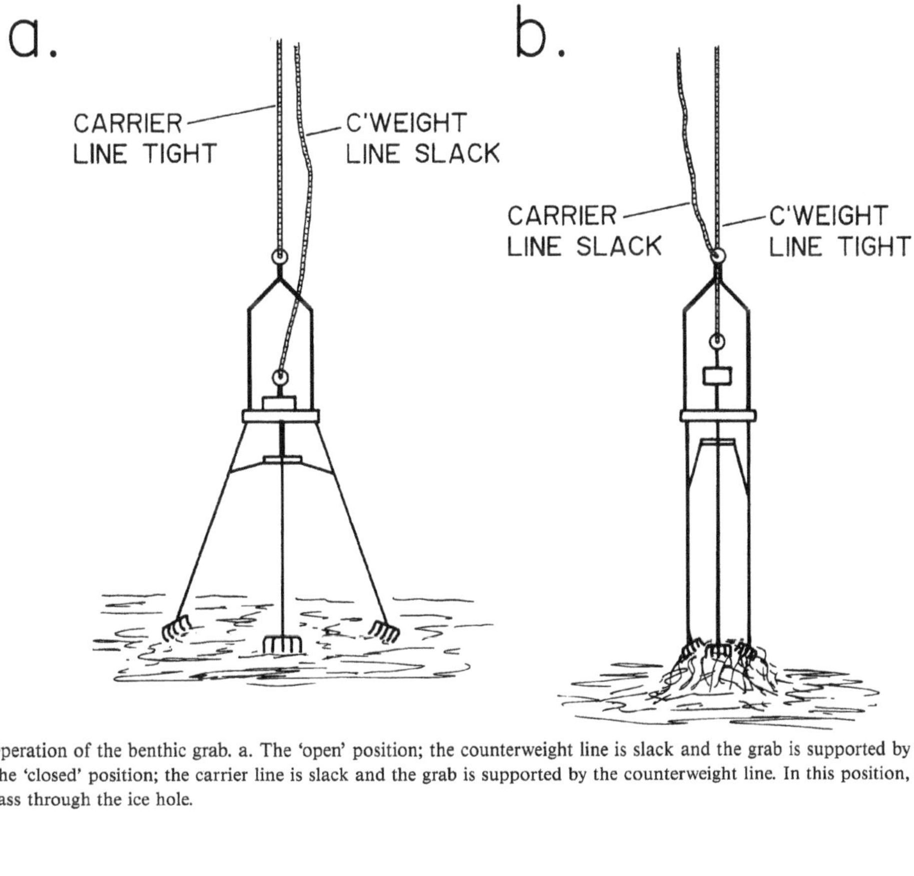

Fig. 2. Operation of the benthic grab. a. The 'open' position; the counterweight line is slack and the grab is supported by the carrier line. b. The 'closed' position; the carrier line is slack and the grab is supported by the counterweight line. In this position, the grab is able to pass through the ice hole.

Hydrobiologia 165: 173–175 (1988)
J. M. Ferris, H. R. Burton, G. W. Johnstone & I. A. E. Bayly (eds) Biology of the Vestfold Hills, Antarctica
Kluwer Academic Publishers

A device for reaming holes in ice

C. W. Heath[1,2], R. Sidebottom[1] & W. J. Singleton[1]
[1]*Antarctic Division, Department of Science, Kingston, Tasmania, 7050, Australia;* [2]*Present address: Metropolitan Water Sewerage & Drainage Board, Scientific Laboratories, P.O. Box 73, West Ryde, Sydney, 2114, Australia*

Key words: ice reaming device, ice

Abstract

A device for reaming holes cored through ice sheets is described. Its advantage being that its operation is possible while lines to equipment below the ice, such as used in *in situ* productivity studies in lakes, remain in place, without damage. This device was designed to maintain the dimensions of a SIPRE ice auger hole, similar devices of differing dimensions would serve to maintain other hole diameters.

Introduction

The ^{14}C technique for measuring primary production has been widely employed since it was first introduced by Steeman Nielson (1952). Virtually all aquatic systems from ephemeral pools and hot springs to permanently ice covered lakes and the open ocean have been studied using this technique. For such *in situ* production studies, water is sampled, containerized, treated according to the specific experimental requirements, placed in some form of retaining device and then replaced at the site of sampling.

Conducting such studies in ice covered lakes presents the problem of gaining access to the water through the ice. There are a number of methods employed to achieve this, including steam jets (Science News, 1979), motor driven augers, either hand held or mounted on tracked vehicles, hand operated augers, explosives or by manually digging holes. The appropriateness of these techniques depends to a large extent upon the location of the study site and the methods employed to deploy the samples used in the productivity studies.

With *in situ* studies, approximation to ambient environmental conditions is of major concern. Thus, large holes in the ice will disrupt the underwater light regime in the vicinity of the hole. This is not a problem if SCUBA is used, because deployment of the samples for production estimation, can be effected at some distance from the hole. Where SCUBA is not used and samples are obtained and deployed remotely through the ice, holes of minimum dimension are an advantage. In this context, the SIPRE ice auger (Journal of Glaciology, 1958), is an ideal device, as it cores a hole of 11 cm diameter with little or no disruption to the surrounding ice sheet. It is manually operated and easily transported, thus enabling studies to be conducted at remote sites. The hole cored is large enough to enable 300 ml pyrex reagent bottles, commonly used in productivity studies, contained within specially constructed cages, to pass through the hole, for deployment for the period of incubation (see Heath, this volume). A drawback of such a small diameter hole however, is that refreezing, a problem in antarctic winter conditions, rapidly reduces the diameter of the hole, rendering retrieval of equipment from below the ice impossible unless the hole can be returned to its original diameter.

Description of the reamer and its use

Described here is a device constructed to ream a

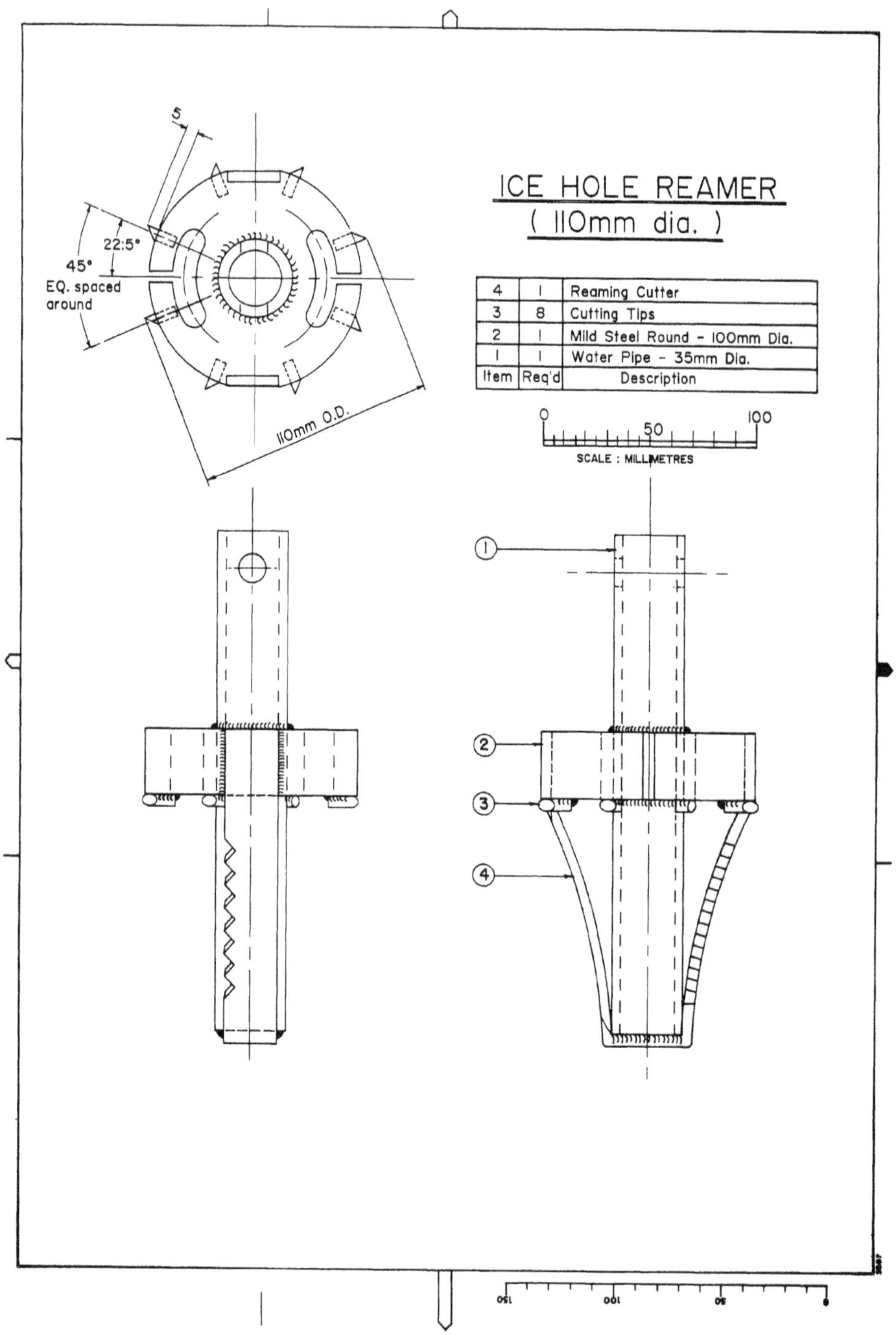

Fig. 1. Technical diagram of the ice hole reamer.

SIPRE auger hole to its original dimensions after undergoing refreezing. For convenience it is designed to be compatible with a SIPRE extension tube and handle, being attached by a pin in the same fashion as the SIPRE auger itself.

The reamer consists of a central tube to fit the SIPRE extension (Fig. 1), a sizing ring with eight protruding teeth and a strengthening bar with serated leading edges. Its operation is effective either on a back and forth or clockwise twisting motion. The serations on the angled strengthening bar act to reduce the thickness of the refrozen ice at the air/water interface. As refreezing at this interface is greater than further down the hole, removing this extra thickness of ice facilitates more rapid reaming. The teeth on the sizing ring rapidly remove any remaining ice, working down the entire ice column, returning the hole to its original dimensions. Slots provided in the sizing ring enable lines suspending any equipment underwater, to be taken inside the ring. Thus preventing them from being cut by the teeth on the outside of the ring.

Holes in ice up to 2 m thick with approximately 20% refreeze have been reamed to their original dimensions in 5–10 minutes. With some effort, the reamer was also used to recore a hole 24 hours old which had 10–12 cm of solid ice at the surface. Although initially designed to facilitate retrieval of primary productivity racks deployed below ice for 4 h periods, the reamer was regularly used throughout the 1981 antarctic winter to ream SIPRE holes used for other purposes. The reamer proved superior to the SIPRE auger under these conditions as water in the hole caused the cutting blades of the SIPRE auger to bind with ice, greatly reducing its efficiency. Such difficulties were not encountered with the reamer described here.

References

Heath, C. W., this volume. Annual primary productivity of an antarctic continental lake: Phytoplankton and benthic algal mat production strategies. Hydrobiologia.

Journal of Glaciology, 1958. Instruments and methods: Ice drills and corers. J. Glaciol. 3: 30.

Science News, 1979. Antarctic ice keeps algae in the dark. Sci. News 116(4): 71.

Steeman Nielsen E., 1952. The use of radioactive carbon (C-14) for measuring organic production in the sea. J. Cons. int. Explore. 18: 117–140.

Hydrobiologia 165: 177–184 (1988)
J. M. Ferris, H. R. Burton, G. W. Johnstone & I. A. E. Bayly (eds) Biology of the Vestfold Hills, Antarctica
Kluwer Academic Publishers

Antarctic terrestrial ecosystems: The Vestfold Hills in context

R. D. Seppelt[1] & P. A. Broady[2]
[1]*Antarctic Division, Department of Science, Kingston, Tasmania, 7050, Australia;* [2]*Botany Department, University of Canterbury, Christchurch, New Zealand*

Key words: Antarctica, terrestrial ecosystem, mosses, lichens, algae

Abstract

Antarctic terrestrial ecosystems are briefly described, with emphasis on Signy Island in the maritime antarctic region, and the McMurdo oasis, southern Victoria Land, and Vestfold Hills in the continental antarctic region.

As the largest and best known coastal ice-free oasis, the Vestfold Hills contain excellent examples of terrestrial sublithic, epilithic, chasmoendolithic, epiphytic and terricolous algal communities, as well as epilithic, endolithic, and epiphytic lichen communities, and moss communities. Many of the numerous lakes support dense communities of aquatic algae.

Introduction

Little more than 1% of the 14 million km^2 of the antarctic continent is ice-free. Terrestrial plant life is restricted to the ice-free areas of coastal outcrops and offshore islands, inland nunataks, mountain ranges and 'oases'. The flora is dominated by lichens, bryophytes and algae; only two species of vascular plants occur on the continent, both restricted to the north-west parts of the Antarctic Peninsula.

The distribution of antarctic terrestrial plant life is determined by three principal factors – climatic (temperature, incident radiation, availability of moisture, wind, exposure, etc.), edaphic (substrate characteristics) and biotic (animals and other plants).

Schemes of phytogeographic subdivision of Antarctica and adjacent ocean areas have been reviewed by Pickard & Seppelt (1984a) and Smith (1984a). The southern polar regions have been broadly subdivided into a subantarctic zone, generally between the Antarctic Convergence and Divergence, and two antarctic zones (maritime and continental) generally south of the Antarctic Divergence.

There appears to be general agreement that at least the north-west coast of the Antarctic Peninsula belongs to the maritime antarctic zone, which includes Bouvetøya, the South Sandwich, South Orkney and South Shetland Islands (Holdgate, 1964; Korotkevich, 1966; Longton, 1967; Markov *et al.*, 1970; Lindsay, 1977; Pickard & Seppelt, 1984a; Smith, 1984a) The two vascular plants occur throughout this region (except Bouvetøya) as far south as Marguerite Bay at the south of the Antarctic Peninsula. The eastern coast of the Antarctic Peninsula and the remainder of the Continent together with Peter I Øy and the Balleny Islands form the continental antarctic zone.

A biogeographic zoning of Greater or East Antarctica was proposed by Gollerbakh & Syroechkovskii (1958 – see Markov *et al.*, 1970: 322). They distinguished four major 'life arenas' the continental ice sheet, young antarctic oases, mature antarctic oases, and offshore islands and sea ice. Bugaev (1960 – see Markov *et al.*, 1970: 180) subdivided continental Antarctica into three climatic subregions: the central ice plateau above an altitude of about 2800 m, the ice slopes from this elevation to the sea, and the coastal subregion. Weyant (1966) independently proposed identical subregions which were later adopted by Holdgate (1970, 1977) and modified by Smith (1984a).

The continental ice sheet appears devoid of life. The ice slope subregion included isolated nunataks and mountain ranges. Scattered local populations of lichens, bryophytes and algae with associated invertebrates occur on these outcrops and snow algae may be found in surficial layers of ice and snow. The coastal subregion includes ice-free oases, coastal outcrops and islands, as well as ice shelves. The distribution of plant life in this subregion may be influenced significantly by wind-borne sea spray or the activities of birds and seals, in regions where no ice shelf or permanent dense pack-ice persists.

The climate of these subregions is described in detail by Markov *et al.* (1970: 180) and by Walton (1984). In the coastal subregion, summer air temperatures rarely rise above 0 °C or thereabouts. In the ice slope subregion, mean monthly air temperatures during summer (December to March) are generally less than about −5 °C, while the central ice plateau is substantially colder, less than −15 °C. In the maritime zone, during at least one month in summer, the mean air temperature rises above 0 °C at sea level while mean monthly winter temperatures rarely fall below −10 °C.

Discussion

Maritime Antarctic

Cryptogams dominate the vegetation of the maritime antarctic although there are small closed stands of the only two phanerogams, *Deschampsia antarctica* and *Colobanthus quitensis* (Fig. 1), present on the antarctic continent. Mosses form closed stands in wetter habitats with local accumulations of peat. Lichens predominate in exposed and higher sites. There are several species of macrofungi and snow algae which may be abundant in summer months.

Signy Island, South Orkney Islands (60°S, 45°W) falls within the maritime antarctic region (Holdgate, 1964). It lies 640 km from the Antarctic Peninsula, the nearest land mass, 900 km south-west of the island of South Georgia, and 1 440 km south-east of Tierra del Fuego. The climate is cold oceanic. The topography is generally rugged. There are a number of freshwater lakes and a small ice cap (Holdgate, 1967; Heywood, 1970). The vegetation of the island is similar to that found elsewhere in the maritime antarctic (Longton, 1967; Gimingham & Smith, 1970; Lindsay, 1971; Smith, 1972; Alison & Smith, 1972; Smith & Corner, 1973; Longton & Holdgate,

Fig. 1. Signy Island: *Deschampsia-Colobanthus* association. Typical maritime community in Maritime Antarctica.

1979). Intensive interrelated biological and microclimatic studies have been conducted at two reference sites in contrasting bryophyte systems (Tilbrook, 1973).

The macrovegetation of Signy Island is predominantly cryptogamic, with approximately 200 species of lichen, 75 mosses, 12 hepatics and only two vascular plants. Floristic studies of the algae are incomplete although Broady (1977a, 1977b, 1979a, 1979b) reported an algal flora of at least 162 species.

The most extensive closed stands of vegetation have developed in the more stable areas of scree slopes and valley floors, rocky ridges, plateaux and headlands. More open communities occur in areas subjected to cryoturbic disturbance. The more exposed rock surfaces are colonized almost solely by lichens. *Caloplaca, Verrucaria* and *Xanthoria* species are abundant on rocks directly influenced by sea spray. Nitrophilous species of *Biatorella, Buellia, Haematomma, Lecidea, Physcia, Rinodina* and *Xanthoria* are abundant near bird nesting sites on cliffs and rocky slopes. On inland rocks, stands of *Lecidea, Ochrolechia, Pertusaria, Placopsis* and *Rhizocarpon* species are abundant. On the more stable lithosols and rock surfaces, communities dominated by fruticose lichens, particularly *Usnea* and *Himmantornia,* and cushion-forming mosses such as species of *Andreaea, Grimmia* and *Dicranoweisia* are common. The mosses are usually dominant where there is a more stable substratum and abundant water supply. Only 17 of the moss species produce sporophytes (Webb, 1973). In the wetter depressions mosses such as *Calliergon, Calliergidium* and *Drepanocladus* species predominate and lichens are usually absent. Annual growth increments of species in these wetter depressions may be as much as 4 cm (Collins, 1973), contrasting with the lesser increments of only 3–5 mm for mosses in drier and unstable habitats (Webb, 1973).

The green alga *Prasiola crispa* (Lightf.) Menegh. is locally abundant in permanently wet areas and in pools or runnels enriched by seal excrement or near penguin colonies. The cyanophyte *Nostoc commune* Vaucher is also locally abundant on bare, wet solifluction slopes supplied with slightly alkaline meltwater (Horne, 1972).

Snow algae, chiefly *Chlamydomonas, Rhaphidonema* and *Ochromonas* species are commonly found on the surface of firm snow during summer (Fogg, 1977).

Data from the many studies that have been conducted in the dry moss turf community and wet moss carpet community that comprise the Signy Island Terrestrial Reference Sites have been synthesized by Davis (1981). Such intensive integrated studies have, as yet, no parallel in continental antarctic communities.

Continental Antarctic

A much less diverse flora with a narrow range of cryptogamic communities is found in the extreme cold and arid environment of continental antarctic localities than in the maritime antarctic. Bare ground predominates. Under such conditions water availability assumes great importance in determining plant distribution (Llano, 1965; Rudolph, 1971).

1. Southern Victoria Land.
Longton (1973, 1979) reported few plant associations from Ross Island. The most widespread vegetation consisted of sparsely developed communities of turf- and cushion-forming mosses in habitats ranging from dry cinder slopes to sand and gravel in seepage areas and at the side of and in meltwater streams. The bryophyte colonies seldom exceeded 4 cm in depth and total plant cover was frequently under 5%. *Sarconeurum* was generally found in drier habitats than *Bryum* species. Exposed rock habitats usually supported only small scattered thalli of crustaceous lichens such as *Xanthoria elegans* (Link) Th. Fr., *Candelaria concolor* var. *antarctica* Murray, *Caloplaca darbishieri* Dodge & Baker and *Lecanora melanophthalma* (Ram.) Ram. These communities were similar to those described for the Syowa Station area, Prince Olav Coast (Nakanishi, 1977) and the Mawson Station area, Mac. Robertson Coast (Seppelt & Ashton, 1978) although species dominance differed. Although no attempt was made to classify the vegetation the communities at Cape Hallett, nothern Victoria Land, also appear to be similar (Rudolph, 1963).

Detailed studies of the terrestrial algae of continental antarctic localities have been attempted only in recent years. Broady (1981d, 1981e, 1983) has studied the ecology and taxonomy of the terrestrial algae of southern Victoria Land. Other studies on soil, epilithic and chasmolithic algae (Broady, 1981a, 1981b) and on chasmolithic algae (Broady, 1981c) for other coastal localities have also been made. The role of endolithic cyanobacteria in the dry valley desert environment of southern Victoria Land has been discussed by Friedmann & Ocampo (1976) and Friedmann (1977, 1980, 1982).

The dry valleys of southern Victoria Land (Fig. 2) have been studied as examples of extremely arid, cold environments where the importance of the endolithic habitat occupancy as a niche for microbial life could be investigated (Friedmann, 1977, 1980, 1982; Friedmann & Ocampo, 1976).

Fumarole vegetation has been studied on Mt Erebus by Broady (1982, 1983, 1984). Heated soils around fumaroles (Fig. 3) provide considerable zonation of algal growth and, in addition, unidentified moss protonema have been recovered. A well-developed fumarole vegetation has also been reported for the South Sandwich Islands (Longton & Holdgate, 1979; Smith, 1984a, 1984b) and, following recent eruptions, on Deception Island (Collins, 1969) where areas of fumarole activity and their associated vegetation are ephemeral (Smith, 1984b). Other geothermically heated areas on the island currently support a community of short turf-forming mosses (Smith, 1984b, 1984c).

2. Vestfold Hills, Ingrid Christensen Coast

The Vestfold Hills 'oasis' is a 400 km^2 area of ice-free rock mainly of low relief. Much of the landscape is covered by moraines and tills (Fig. 4). The physiography and biology of the region have been generally described by Johnstone *et al.* (1973) while Burton & Campbell (1980) have discussed the climate. A detailed discussion of the geology, geomorphology, Cainozoic history, climate, vegetation, terrestrial invertebrates and significance of the Vestfold Hills is given in a recent book (Pickard, 1986a).

Davis Station (68°35′S, 77°58′E) is situated approximately 25 km from the continental ice plateau. Consequently, katabatic winds, which play a significant role in determining local plant distribution at Mawson Station (Seppelt & Ashton, 1978), dissipate before reaching the coastal perimeter near the Station, and therefore have an effect only close to the ice plateau.

Fig. 2. Taylor Valley: looking towards McMurdo Sound, with Lake Fryxell and Canada Glacier. Valley is approximately 5 km wide.

Fig. 3. Mt Erebus: steaming ground near the summit.

Fig. 4. Vestfold Hills: typical coastal desert habitat with glacial till.

The flora of the Vestfold Hills comprises at least 82 species of terrestrial algae (Broady, 1986), six moss species (Seppelt, 1986a) and at least 23 lichen species (Seppelt, 1986b). The lichens and mosses are distributed chiefly in the eastern or inland sector and their distribution patterns reflect the availability of drift snow, time since exposure of the substrate from the ice plateau and time since the last glaciation, elevation, and proximity to saline waters (Seppelt *et al.*, this volume). Very few occurrences of lichens or mosses have been noted towards the coastal margin where the low terrain is densely clothed with moraine and till. Mobile sands in these areas may actively inhibit colonization by the extremely slow growing lichens and mosses. In contrast, sublithic algae are widespread in these till-covered areas (Broady, 1981a; Seppelt *et al.*, this volume).

Notwithstanding sampling deficiencies in the south-eastern and margins of the oasis, epilithic algae (Broady, 1981b; Seppelt *et al.*, this volume) appear to reflect closely the generalized patterns of lichen and moss distribution. The epilithic algae are chiefly encrusting growths of cyanobacteria on rocks which receive fresh meltwater percolations. The community does not appear to be salt tolerant.

Chasmoendolithic algae, growing in capillary-sized cracks in boulders and *in situ* country rock outcrops, are no doubt far more widespread than preliminary searches indicate (Broady, 1981c; Seppelt *et al.*, this volume). There are indications that such algae are more abundant in coastal locations than further inland. At least in coastal regions these chasmoendoliths appear not to be associated with chasmoendolithic lichens, such as *Lecidea phillipsiana* Filson, *Biatorella cerebriformis* (Dodge) Filson and *Acarospora gwynnii* Dodge & Baker, which have so far been reported only from inland localities and not the coastal margin.

Schofield & Ahmadjian (1972) have implicated the absence of a freely available nitrogen source, together with other environmental factors, in accounting for the paucity of vegetation in the ice-free areas of Antarctica. *Nostoc* may be an important source of nitrogen fixation in the poorly developed antarctic cold desert soils.

Although there are many communities of bryophytes in the Vestfold Hills which have no obvious algal associations, Davey (1982) reported *Nostoc* was always associated with bryophytes. *Nostoc* may contribute significant amounts of nitrogenous compounds for bryophyte growth.

Communities of *Nostoc commune* are found in association with bryophyte stands in areas of meltwater runoff. Davey (1982) discussed *in situ* nitrogen fixation by *Nostoc commune* in a stand of moss and algae at a site near Crooked Lake. Submerged colonies of *Nostoc* had the highest fixation rates with decreasing rates of fixation as dryness of the substrate or habitat increased. Fixation of nitrogen by cyanobacteria is maximal during periods of optimal microclimatic conditions. Davey (1983) reported diurnal fluctuations of nitrogenase activity and increasing rates of fixation with increased temperature. Moisture levels also affect fixation rates (Davey, 1982). *In situ* determination of nitrogenase activity showed that summer activity ceased when vegetation surface temperatures fell below $-7\,^{\circ}C$ (Davey & Marchant, 1983).

While there are no large colonies of snow petrels, nesting sites are widely scattered in many parts of the Vestfold Hills (Brown, 1966). Rich lichen growths, particularly *Xanthoria mawsonii* Dodge, *X. elegans*, *Physcia caesia* (Hoffm.) Hampe and *Candelariella antarctica* Filson occur where meltwater runoff is enriched by nutrients derived from such nesting sites.

There are few localities with extensive moss beds and these are generally restricted to inland areas (Seppelt *et al.*, this volume). As in many other continental localities, *Bryum* species predominate. The taxonomy of this genus in Antarctica is confused but recent studies (Seppelt & Kanda, 1986) have shown that only two species – *B. argenteum* Hedw., and *B. pseudotriquetrum* (Hedw.) Meyer, Scherb., & Gaertn., – are represented in continental localities, the latter species including all material formerly identified as *B. algens* Card. (Seppelt, 1984). *Bryum* species exhibit a wide range of morphological variation in response to environmental factors. When inundated by meltwater, shoots of *Bryum* may etiolate conisderably (Seppelt, 1983). The aquatic *B. korotkeviczíae* Savich-Lyubitskaya & Smirnova, which has been found in a number of water bodies in the Vestfold Hills, has been shown to be an aquatic habitat-induced variant of the terrestrial *B. algens* (Seppelt, 1983), although this species should now be considered as *B. pseudotriquetrum* (Seppelt & Kanda, 1986). Etiolated shoots of this species have been recovered from Holocene deposits near several water bodies (Pickard & Seppelt, 1984b).

The terrestrial flora of the Vestfold Hills is representative of most of continental Antarctica.

Thus, the oasis is not unique floristically. However, the presence on intertidal rocks near Davis Station of the lichen *Verrucaria* (Seppelt, 1986), the only record of this genus on the antarctic continent beyond the Antarctic Peninsula, and the deposits of subfossil moss (see above) are of particular interest.

The Vestfold Hills, because of the area of ice-free rock, geomorphic history and range of habitats, provide an excellent opportunity for the study of terrestrial life in continental Antarctica. Much remains to be studied of the taxonomy, ecology and physiology of the biota. Soils and soil-forming processes have not been addressed.

A complex relationship exists between plant distribution patterns and topography, proximity to the continental ice cap, and past geomorphic history (Seppelt *et al.,* this volume). Intensive field studies will need to be undertaken in order to understand these relationships. It is time to develop a fully integrated approach to the study of the terrestrial biology of the region.

References

Allison, J. S. & R. I. L. Smith, 1973. The vegetation of Elephant Island, South Shetland Island. Bull. br. antarct. Surv. 33 & 34: 185–212.

Broady, P. A., 1977a. The Signy Island terrestrial reference sites, VII. The ecology of the algae of Site 1, a moss turf. Bull. br. antarct. Surv. 45: 47–62.

Broady, P. A., 1977b. The Signy Island terrestrial reference sites, IX. The ecology of the algae at Site 2, a moss carpet. Bull. br. antarct. Surv. 47: 13–29.

Broady, P. A., 1979a. Feeding studies on the collembolan Cryptopygus antarcticus Willem at Signy Island, South Orkney Islands. Bull. br. antarct. Surv. 48: 37–46.

Broady, P. A., 1979b. The terrestrial algae of Signy Island, South Orkney Islands. Br. antarct. Surv. Sci. Rep. 98, 117 pp.

Broady, P. A., 1981a. The ecology of sublithic terrestrial algae at the Vestfold Hills, Antarctica. Br. Phycol. J. 16: 231–240.

Broady, P. A., 1981b. Ecological and taxonomic observations on subaerial epilithic algae from Princess Elizabeth Land and Mac. Robertson Land, Antarctica. Br. Phycol. J. 16: 257–266.

Broady, P. A., 1981c. The ecology of chasmolithic algae at coastal locations in Antarctica. Phycologia 20(3): 259–272.

Broady, P. A., 1981d. Algae of southern Victoria Land, a taxonomic and ecological study, 1980-81. Melb. Univ. Program Antarct. Stud. Rep. 31, 26 pp.

Broady, P. A., 1981e. Non-marine algae of Cape Bird, Ross Island and Taylor Valley, Victoria Land, Antarctica. Melb. Univ. Program Antarct. Stud. Rep. 37, 97 pp.

Broady, P. A., 1982. Algae from fumarolic soils on Mt Erebus, Ross Island, Antarctica. Melb. Univ. Program Antarct. Stud. Rep. 51, 48 pp.

Broady, P. A., 1983. Botanical studies at Ross Island, Antarctica in 1982-83. Preliminary report. Melb. Univ. Program Antarct. Stud. Rep. 58, 40 pp.

Broady, P. A., 1984. Taxonomic and ecological investigations of algae on steam-warmed soil at Mt Erebus, Ross Island, Antarctica. Phycologia 23: 257–271.

Broady, P. A., 1986. Ecology and taxonomy of the terrestrial algae of the Vestfold Hills. In J. Pickard (ed.), Antarctic Oasis. Terrestrial environments and history of the Vestfold Hills. Academic Press, Lond.: 165–202.

Brown, D. A., 1966. Breeding biology of the Snow Petrel, Pagodroma nivea Forster. ANARE Sci. Rep. Publ. 89. Antarct. Div., Sci., Melbourne, 64 pp.

Burton, H. R. & P. J. Campbell, 1980. The climate of the Vestfold Hills, Davis Station, Antarctica, with a note on its effect on the hydrology of hypersaline Deep Lake. ANARE Sci. Rep. Publ. 129. Aust. Govt. Publ. Service, Canberra, 50 pp.

Collins, N. J., 1969. The effects of volcanic activity on the vegetation of Deception Island. Bull. br. antarct. Surv. 21: 79–94.

Collins, N. J., 1973. Productivity of selected bryophyte communities in the maritime Antarctic. In L. C. Bliss & F. E. Wiegolaski (eds), Primary Production and Production Processes, Tundra Biome. Tundra Biome Steering Committee, Stockholm: 177–183.

Davey, A., 1982. In situ determination of nitrogen fixation in Antarctica using a high sensitivity portable gas chromatograph. Aust. J. Ecol. 7: 395–402.

Davey, A., 1983. Effects of abiotic factors on nitrogen fixation by blue-green algae in Antarctica. Polar Biol. 2: 95–100.

Davey, A. & H. J. Marchant, 1983. Seasonal variation in nitrogen fixation by Nostoc commune Vaucher at the Vestfold Hills, Antarctica. Phycologia 22(4): 377–385.

Davis, R. C., 1981. Structure and function of two antarctic terrestrial moss communities. Ecol. Monogr. 51(2): 125–143.

Fogg, G. E., 1967. Observations on the snow algae of the South Orkney Islands. Phil. Trans. R. Soc. Lond. Ser. B. 252: 279–287.

Friedmann, E. I., 1977. Microorganisms in antarctic desert rocks from dry valleys and Dufek Massif. Antarct. J. U.S. 12: 26–30.

Friedmann, E. I., 1980. Endolithic microbial life in hot and cold deserts. Proc. 4th College Park Colloquium on Chemical Evolution. Limits of Life. University of Maryland, 1978. Origins of Life 10: 223–235.

Friedmann, E. I., 1982. Endolithic microorganisms in the antarctic cold desert. Science 215: 1045–1053.

Friedmann, E. I. & R. Ocampo, 1976. Endolithic blue-green algae in the dry valleys: primary producers in the antarctic desert ecosystem. Science 193: 1247–1249.

Gimingham, C. H. & R. I. L. Smith, 1970. Bryophyte and lichen communities in the Maritime Antarctic. In M. W. Holdgate (ed.), Antarctic Ecology 2, Academic Press, Lond.: 752–785.

Heywood, R. B., 1970. Ecology of the fresh-water lakes of Signy

Island, South Orkney Islands, I. Catchment areas, drainage systems and lake morphology. Bull. br. antarct. Surv. 14: 25–43.

Holdgate, M. W., 1964. Terrestrial ecology in the Maritime Antarctic. In R. Carrick, M. W. Holdgate & J. Prevost (eds.), Biologie Antarctique. Hermann, Paris: 181–194.

Holdgate, M. W., 1967. Singy Island. Phil Trans R. Soc. Lond. Ser. B. 252: 173–177.

Holdgate, M. W., 1970. Vegetation. In M.W. Holdgate (ed.), Antarctic Ecology 2. Academic Press, Lond.: 729–733.

Holdgate, M. W., 1977. Terrestrial ecosystems in the Antarctic. Phil. Trans. R. Soc. Lond. Ser. B. 279: 5–25.

Horne, A. J., 1972. The ecology of nitrogen fixation on Signy Island, South Orkney Islands. Bull. br. antarct. Surv. 27: 1–18.

Johnstone, G. W., D. J. Lugg & D. A. Brown, 1973. The biology of the Vestfold Hills, Antarctica. ANARE Sci. Rep. Publ. 123. Antarct. Div., Dept. Sci., Melbourne, 62 pp.

Korotkevich, Y. S., 1966. Gyebotanicheskoye raeonirobaniye. (Geobotanical regionalization). Map 135. In Anon. (1966) Atlas Antarktiti (Antarctic Atlas). Glavnoe upravlenie geodezii i kartografii Ministerstra geologii USSR: Moskva & Leningrad. (Soviet Geography Review & Translation 8: 262–507). [Russian]

Lindsay, D. C., 1971. Vegetation in the South Shetland Islands. Bull. br. antarct. Surv. 25: 59–83.

Lindsay, D. C., 1977. Lichens of cold deserts. In M. R. D. Seaward (ed.), Lichen Ecology. Academic Press, London.: 183–209.

Llano, G. A., 1965. The flora of Antarctica. In T. Hatherton (ed.), Antarctica: A New Zealand Antarctic Society Survey. Methuen, Lond.: 331–350.

Lognton, R. E., 1967. Vegetation in the Maritime Antarctic. Phil. Trans. R. Soc. Lond. Ser. B. 252: 213–235.

Longton, R. E., 1973. A classification of terrestrial vegetation near McMurdo Sound, continental Antarctica. Can. J. Bot. 51: 2339–2346.

Longton, R. E., 1979. Vegetation ecology and classification in the Antarctic Zone. Can. J. Bot. 57: 2264–2278.

Longton, R. E. & M. W. Holdgate, 1967. Temperature relationships of Antarctic vegetation. Phil. Trans R. Soc. Lond. Ser. B. 252: 237–250.

Longton, R. E. & M. W. Holdgate, 1979. The South Sandwich Islands: IV. Botany. Br. Antarct. Surv. Sci. Rep. 94.

Markov, K. K., V. I. Bardin, V. L. Lebedev, A. I. Orlov & I. A. Suetova, 1970. The geography of Antarctica. Israel prog. Sci Translation, Jerusalem (English translation of Markov, K. K., et al. 1968. Geografiya Antarktidy. Izdatel'stvo Mysl, Moskva).

Nakanishi, S., 1977. Ecological studies of the moss and lichen communities in the ice-free areas near Syowa Station, Antarctica. Antarct. Rec. 59: 68–96.

Pickard, J. (ed.), 1986a. Antarctic Oasis. Terrestrial environments and history of the Vestfold Hills. Academic Press, Lond., 367 pp.

Pickard, J., 1986b. Spatial relations of the vegetation of the Vestfold Hills, Antarctica. In J. Pickard (ed.), Antarctic Oasis. Terrestrial environments and history of the Vestfold Hills, Antarctica. Academic Press, Lond.: 275–308.

Pickard, J. & R. D. Seppelt, 1984a. Phytogeography of Antarctica. J. Biogeog. 11: 83–102.

Pickard, J. & R. D. Seppelt, 1984b. Holocene occurrence of the moss Bryum algens Card. in the Vestfold Hills, Antarctica. J. Bryol. 13: 209–217.

Rudolph, E. D., 1963. Vegetation of Hallett Station area, Victoria Land, Antarctica. Ecology 44: 585–586.

Rudolph, E. D., 1971. Ecology of land plants in Antarctica. In L. O. Quam (ed.), Research in the Antarctic. Am. Soc. Adv. Sci., Wash. D.C.: 191–211.

Schofield, E. & V. Ahmadjian, 1972. Field observations and laboratory studies of some Antarctic cold desert cryptogams. Antarct. Res. Ser. 20: 97–141.

Seppelt, R. D., 1983. The status of the Antarctic moss Bryum korotkevicziae. Lindbergia 9: 21–26.

Seppelt, R. D., 1984. The bryoflora of the Vestfold Hills and Ingrid Christensen Coast, Antarctica. ANARE Res. Notes 20. Antarct. Div., Dept. Sci. Hobart, 30 pp.

Seppelt, R. D., 1986a. Mosses of the Vestfold Hills. In J. Pickard (ed.), Antarctic Oasis. Terrestrial environments and history of the Vestfold Hills. Academic Press, Lond.: 221–245.

Seppelt, R. D., 1986b. Lichens of the Vestfold Hills. In J. Pickard (ed.), Antarctic Oasis. Terrestrial environments and history of the Vestfold Hills. Academic Press, Lond.: 247–274.

Seppelt, R. D. & D. H. Ashton, 1978. Studies on the ecology of the vegetation at Mawson Station, Antarctica. Aust. J. Ecol. 3: 373–388.

Seppelt, R. D., P. A. Broady, J. Pickard & D. A. Adamson, this volume. Plants and landscape in the Vestfold Hills. Hydrobiologia.

Seppelt, R. D. & H. Kanda, 1986. Morphological variation and taxonomic interpretation in the moss genus Bryum in Antarctica. Mem. Natl. Inst. Polar Res., Ser. E., No. 37: 27–42.

Smith, R. I. L., 1972. Vegetation of the South Orkney Islands with particular reference to Signy Island. Br. Antarct. Surv. Sci. Rep. 68.

Smith, R. I. L., 1984a. Terrestrial plant biology. In R. M. Laws, (ed.), Antarctic Ecology, I. Academic Press, Lond.: 61–162.

Smith, R. I. L., 1984b. Colonization by bryophytes following recent volcanic activity on an antarctic island. J. Hattori bot. Lab. 56: 53–63.

Smith, R. I. L., 1984c. Colonization and recovery by cryptogams following recent volcanic activity on Deception Island, South Shetland Islands. Bull. br. antarct. Surv. 62: 25–51.

Smith, R. I. L. & R. W. M. Corner, 1973. Vegetation of the Arthur Harbour, Argentine Islands region of the Antarctic Peninsula. Bull. br. antarct. Surv. 33 & 34: 89–122.

Tilbrook, P. J., 1973. The Signy Island terrestrial reference sites. I. An introduction. Bull. br. antarct. Surv. 33 & 34: 65–76.

Walton, D. W. H., 1984. The terrestrial environment. In R. M. Laws (ed.), Antarctic Ecology. I. Academic Press, Lond.: 1–60.

Webb, R., 1973. Reproductive behaviour of mosses on Signy Island, South Orkney Islands. Bull. br. antarct. Surv. 36: 61–77.

Weyant, W. S., 1966. The Antarctic climate. Antarct. Res. Ser. 8: 47–59.

Hydrobiologia 165: 185–196 (1988)
J. M. Ferris, H. R. Burton, G. W. Johnstone & I. A. E. Bayly (eds) Biology of the Vestfold Hills, Antarctica
Kluwer Academic Publishers

Plants and landscape in the Vestfold Hills, Antarctica

R. D. Seppelt[1], P. A. Broady[2], J. Pickard[3] & D. A. Adamson[3]
[1]*Antarctic Division, Department of Science, Kingston, Tasmania 7050, Australia;* [2]*Department of Botany, University of Canterbury, Christchurch, New Zealand;* [3]*School of Biological Sciences, Macquarie University, North Ryde, NSW 2113, Australia*

Key words: Antarctica, algae, mosses, lichens, ecology

Abstract

Maps are presented showing the recorded distribution and species density of terrestrial plants in the Vestfold Hills. The distribution, biomass and species diversity of terrestrial lithic algae, mosses and lichens is influenced positively by availability of meltwater from drift snow and by additional nutrient supply (probably N and P) near bird nest sites. The terrestrial plants are affected negatively by exposure (including sand blast) and salinity. These four factors are probably the most important environmental characteristics exercising local control over plant distribution and abundance within the limits set by temperature.

Large changes in salinity, degree of exposure, water supply and nutrient supply occur across the Vestfold Hills, with the most favourable conditions generally occurring in the eastern half fairly close to the ice sheet.

Plant distribution and abundance are also discussed in relation to the length of time that particular areas have been exposed as a result of ice retreat. With increasing time of exposure, plant diversity and abundance rise but subsequently fall sharply as conditions become more arid, saline, or both. This temporal sequence can be explained by considering changes in the important factors that control plant growth.

Introduction

The Vestfold Hills comprise about 400 km^2 of ice-free rock. The relief is generally low, the highest point being only about 160 m above m.s.l. The hills and valleys are draped with a variable cover of glacial till and support a flora of at least six moss species (Seppelt, 1986a), a minimum of 23 lichen species (Seppelt, 1986b), and at least 82 species of terrestrial algae, in which Cyanobacteria (about 34 species) and the Chlorophyta (about 36 species) predominate (Broady, 1986).

The eastern interior is generally more rugged and higher than the western half, with numerous hills and steep-sided valleys. Towards the coast there is a broad band of more gentle relief (up to 10 km in breadth) and wide valleys and open plains covered with moraines and tills.

The geology of the region has been described by Crohn (1959), Law (1959), McLeod (1963) and Collerson & Sheraton (1986). A general summary was given by Johnstone *et al.* (1973). The country rock is chiefly gneissic with pyroxene-quartz-feldspar rocks predominating. Numerous basaltic dykes transect the whole area. The geological relationships between the Vestfold Hills and the neighbouring localities of Princess Elizabeth Land have been discussed by Wellman & Williams (1982), while a detailed discussion of the ancient metamorphics of the Vestfold Hills – Prydz Bay region has been given by Sheraton & Collerson (1983).

The climate is cold, dry and windy. Mean summer maximum temperatures rise to about +2 °C but, as is typical of continental antarctic localities, temperatures are below 0 °C for most of the year, falling to as low as −40 °C in winter. Climatic data have been summarized by Burton & Campbell (1980) and a general discussion of the climate is given by Streten (1986).

The area of exposed country rock in the Vestfold Hills has a significant effect on the climate. The low albedo of rock (compared with ice and snow) results in a mean annual temperature for Davis (68°35′S, 77°58′E) of −10.2 °C, compared with −11.2 °C for Mawson (67°36′S, 62°53′E) which, in contrast with the Vestfold Hills, is situated on a small coastal rock outcrop at the edge of the ice plateau in Mac.Robertson Land. Davis is, on average, warmer than other antarctic stations at similar latitudes.

The wind at Davis, 25 km from the ice plateau, has an annual mean velocity of 5 m s^{-1}, which is low for antarctic stations (Streten, 1968). Katabatic winds, present near the edge of the continental ice sheet, dissipate over the intervening hills and have little or no effect near the station (Lied, 1963). This contrasts with the marked katabatic component of the winds at Mawson (Seppelt & Ashton, 1978).

Although cloud cover and snow precipitation are greater at coastal Davis than at the plateau edge, there is a higher incidence of drift show nearer the plateau. Drift snow is of considerable importance to the terrestrial vegetation in providing a source of moisture supplementing snow fall, which occurs mainly in winter. Snow accumulates only in the lee of ground relief where shelter from wind also reduces loss by ablation (Fig. 9a). As the antarctic climate is extremely arid, meltwater is a major factor governing plant distribution. On a local scale many distribution patterns reflect almost exactly the first accumulations of drift snow.

There is a general belief that antarctic terrestrial plant communities have a simple structure, e.g. Usher (1983), this belief presumably referring to the small stature, low biomass, low productivity and low species diversity of the vegetation. Antarctic terrestrial algal, moss and lichen communities lack the complex spatial structure and the many trophic levels of consumers in, for example, forest communities. However, there is a wide range of antarctic plant communities from endolithic algae and lichens (Friedmann, 1977, 1982; Friedmann & Ocampo, 1976; Broady, 1981c) to epiphytes (Longton, 1979; Seppelt & Ashton, 1978; Broady, 1981b), including multispecies assemblages displaying both vertical and horizontal complexity. Given the extent of plant abundance, some of the visually striking moss and lichen communities with their associated fungal, algal and invertebrate populations may be as complex as other physically larger communities elsewhere.

Methods

The distribution of lichens and mosses in the Vestfold Hills was determined initially by recording species presence in 1 km grid squares during the summer of 1978/79 with supplementary records obtained during 1980/81. Voucher specimens were collected for the verification of records and also for taxonomic studies. Specimens are currently housed in the herbarium of the Antarctic Division (ADT). Further analyses of lichen and moss distributions were undertaken on 70, 2 km squares. A total of 181, of a possible 635, 1 km grids were traversed for the lichen and moss survey.

Algal distribution was determined similarly with samples of epilithic, sublithic and chasmolithic algae, initially by direct microscopic observations of samples followed by culture of some of the taxa at the University of Melbourne.

Results and discussion

Lithic algal distribution

Lithic algae, or those associated with rock (Golubic *et al.*, 1981) may be categorized into five major groups (Friedmann & Galun, 1974): epilithic (on the exposed surfaces), chasmoendolithic (in cracks and fissures), cryptoendolithic (in structural cavities within porous rocks), euendolithic (organisms actively boring into rocks and forming tunnels) and sublithic (on the undersurfaces of transculent stones embedded on the soil surface). Chasmoendolithic and cryptoendolithic algae from continental Antarctica have been discussed by Friedmann & Ocampo (1976), Friedmann (1977) and Thompson (1979). Broady (1979, 1981b, 1981c) has studied chasmoendolithic algae from the South Orkney Islands and coastal locations in Mac.Robertson Land and Princess Elizabeth Land as well as subaerial epilithic algae from these continental localities.

There are clear patterns in the distribution of terrestrial algae between habitats (Broady, 1986). Only seven species have been recovered from epilithic and nine species from chasmoendolithic habitats, reflecting the harshness of these habitats. Where there is a greater availability of water a wider diversity of algae is encountered, with a maximum of 48 species recovered from unenriched soils. From soils enriched in nutrients by animals 23 species have been recorded, while 24 sublithic species have so far been identified. Forty-one species have been identified as epiphytes.

Numerous taxonomic difficulties remain, however, and a great deal of further study of algal ecology is needed.

Figure 1 shows the distribution of epilithic algae, mostly Cyanobacteria, on rocks receiving percolation of meltwater. These epilithic communities do

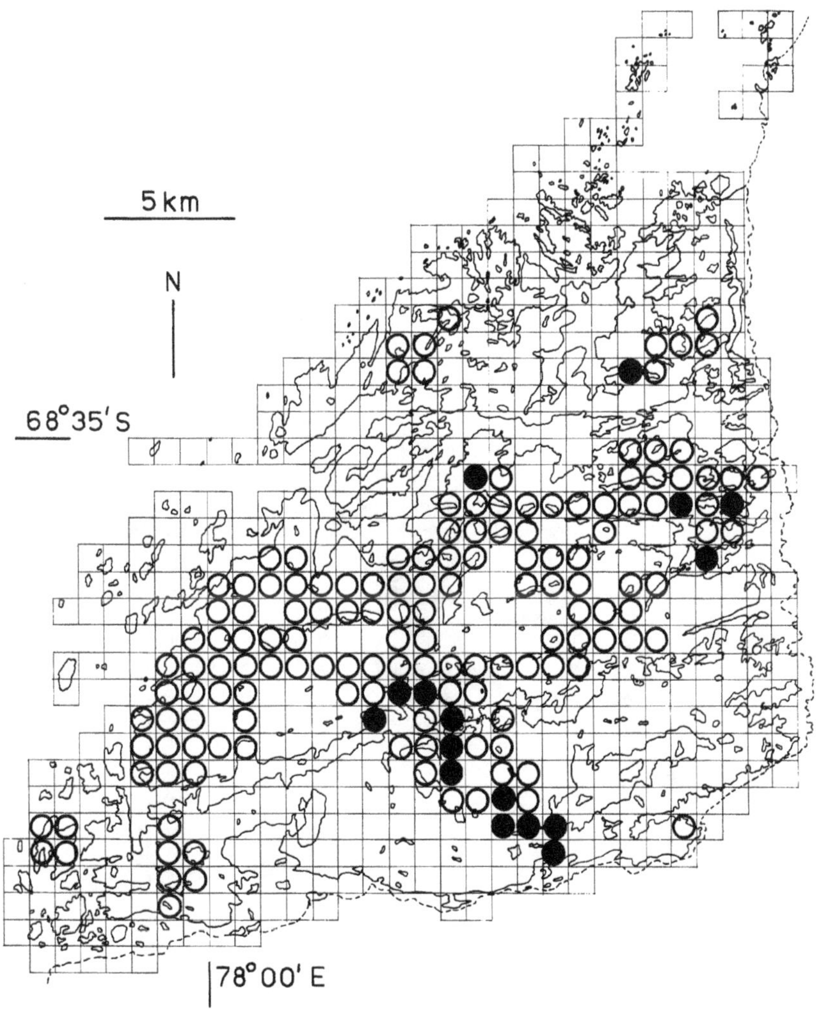

Fig. 1. Distribution of epilithic algae in 1 km squares. Presence is indicated by closed circles. Open circles indicate that epilithic algae were not observed. Blank squares were not examined. The organisms were mostly Cyanobacteria which formed black encrustations on rocks receiving percolations of fresh meltwater.

not appear to be salt tolerant and have been found only in areas where snow drifting is frequent. They are probably more abundant in the north-eastern corner and along the interior margin of the Vestfold Hills than indicated in Fig. 1; but we are unable to verify this due to the limited number of field observations in these localities.

Chasmoendolithic algae (Fig. 2), although obviously widespread, have not been searched for comprehensively in the Vestfold Hills. Such algae may be more abundant in coastal regions than further inland but firm conclusions cannot yet be drawn.

Wind may have both direct and indirect effects on the local distribution patterns of these communities. Broady (1981c) found that in coastal chasmoendolithic communities *Prasiococcus calcareus* (Boye Petersen) Vischer occurred only in areas subjected to wind blown salt while a *Desmococcus* species occurred only in areas not affected in this way. Abrasion of the windward surfaces of rock substrata by

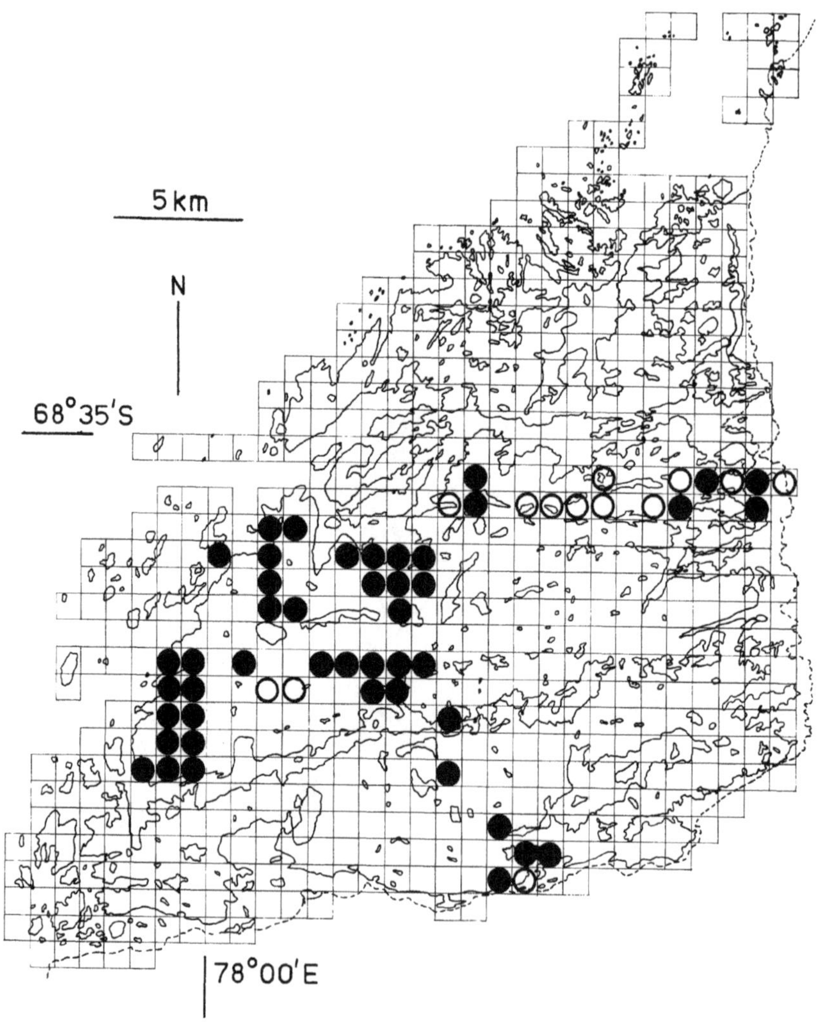

Fig. 2. Distribution of chasmoendolithic algae. Symbols as for Fig. 1. More extensive sampling is required but a wide distribution in the western part of Broad Peninsula is clear.

windborne sand and ice crystals inhibit colonization. Chasmoendoliths have generally been found only on leeward rock faces in the Vestfold Hills, the exposed northern to eastern aspects being both wind-affected and generally free of snow drifts. In contrast, Friedman (1977) reported that in the dry valleys of Victoria Land, cryptoendolithic communities were generally found on faces with northerly aspects where there is greater insolation.

Sublithic algal growths are widespread in the Vestfold Hills in nearly all areas visited (Fig. 3), even on the driest soils (Broady, 1981a). Absence of sublithic algae from grid squares may reflect sampling error rather than real absence. Abundance in grid squares varies widely. Twenty-four taxa have been identified (Broady, 1986) but only four are abundant: *Chroococcidiopsis* sp., and *Plectonema* sp., (Cyanophyta); *Desmococcus* sp., and *Prasiococcus calcareus* (Chlorophyta). Habitat selection by the different algae resulted in *Chroococcidiopsis* and *Desmococcus* favouring dry, raw mineral soils, *Plectonema* moist, raw mineral soils and *Prasiococcus*

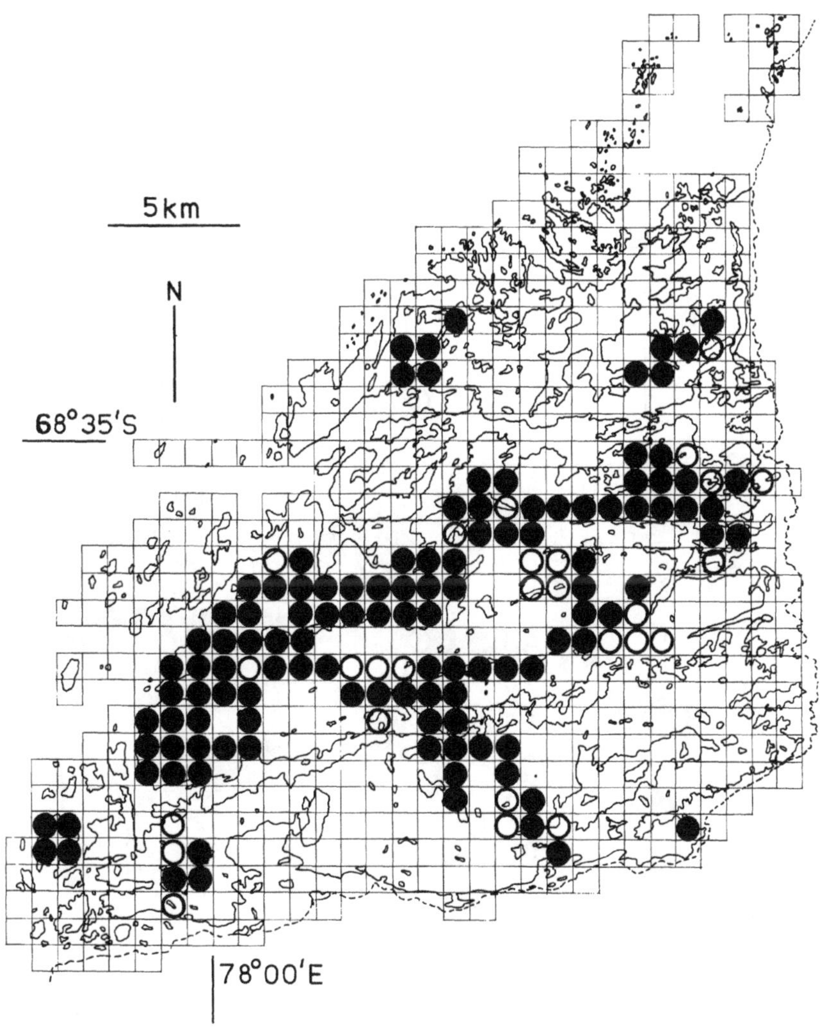

Fig. 3. Distribution of sublithic algae on the undersurfaces of transculent quartz stones and the rare erratics of marble. Symbols as for Fig. 1.

calcareus seal and bird influenced soils. Chlorophyll *a* estimates showed sublithic algal growths to be considerably more abundant than growths in adjacent exposed soil surfaces but less abundant than superficial algal growths in favourable wet microenvironments (Broady, 1981a). Although not widely studied in continental antarctic regions, sublithic algae have also been reported from Ongul Island (Fukushima, 1959) and from the dry valleys of southern Victoria Land (Cameron, 1972; Thompson, 1979). Their distribution right across the Vestfold Hills reflects the sheltered sublithic environment which allows algal growth even in the harsh conditions on the western part of the hills.

Lichen and moss distribution

Figures 4 & 5 show the distribution of mosses and lichens in the Vestfold Hills. In the initial surveys a total of 24 species was recorded. Further taxonomic studies have revealed the presence of an additional moss, *Ceratodon purpureus* (Hedw.) Brid (*C. minutifolius* Card., *sensu* Horikawa & Ando, 1963),

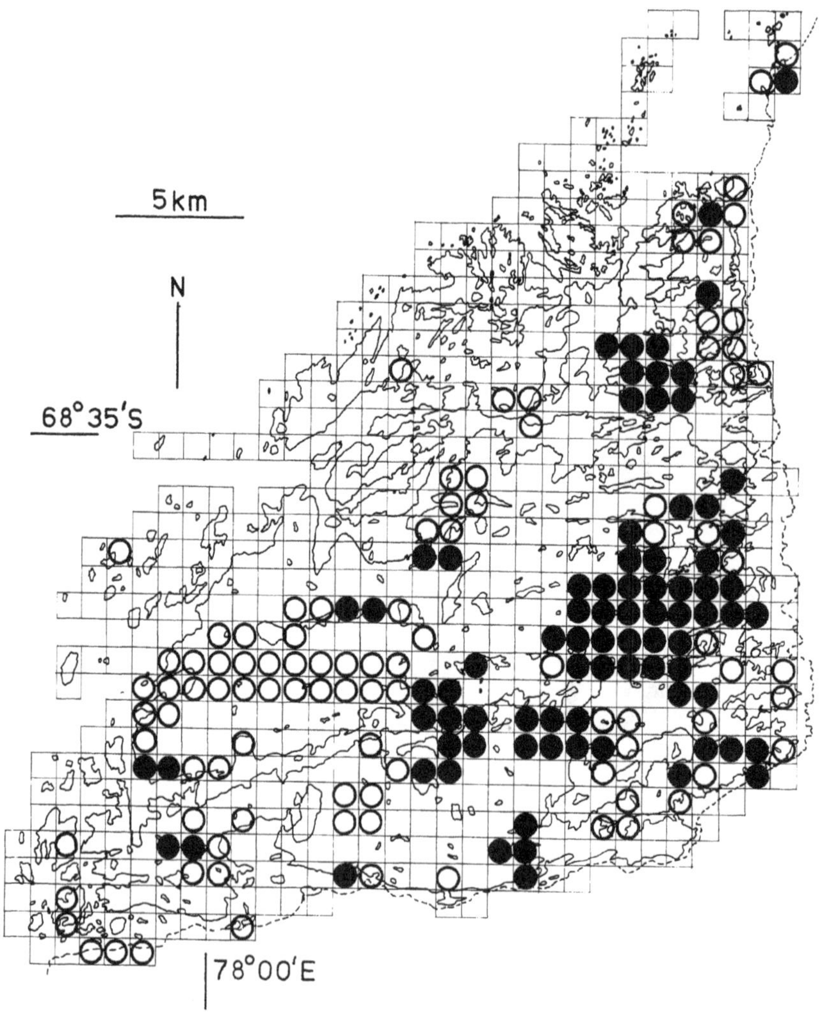

Fig. 4. Distribution of moss. Symbols as for Fig. 1.

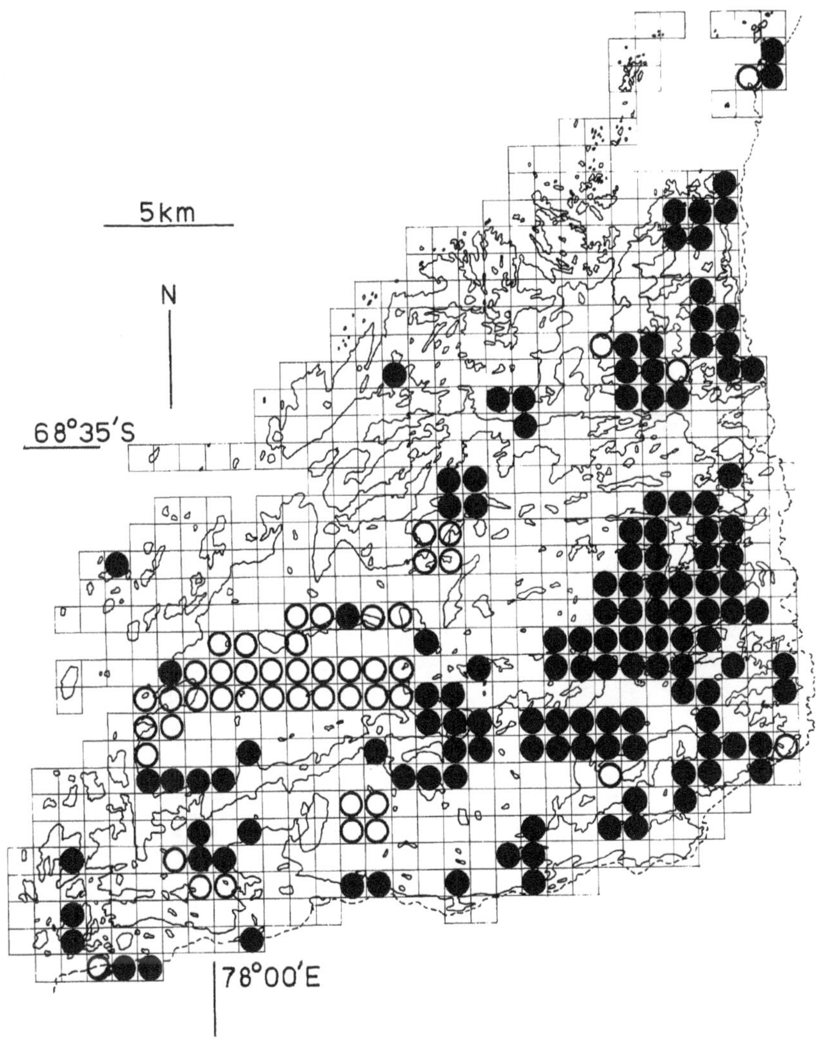

Fig. 5. Distribution of lichen. Symbols as for Fig. 1.

and several additional lichens (*Verrucaria* sp., *Buellia* sp., *Lecidea* sp.) have been identified but none of these have been included in the analyses. Such taxonomic difficulties emphasize the necessity for the collection of voucher specimens not only for studies of taxonomy and phytogeography but also for ecological and physiological studies.

Analysis of distribution patterns has also been undertaken on 70, 2 km squares, indicating densities of between one and 21 species per square (Figs. 6, 7 & 8). Figure 6 shows concentrations of species in some localities, such as Ellis Rapids, Lake Zvezda and Lichen Valley, while in others, particularly in the seaward and southern perimeters and south-eastern corner, there are comparatively few species. Only one grid had more than 20 species; five grids had 16 – 20 species; 14 grids had 11 – 15 species; 12 grids had 6 – 10 species; 35 grids had between one and five species present. While these 67 grids represent a biased sampling which reflected collecting intensities, a reasonable effort was made to examine less favourable habitats by specific searches and by vary-

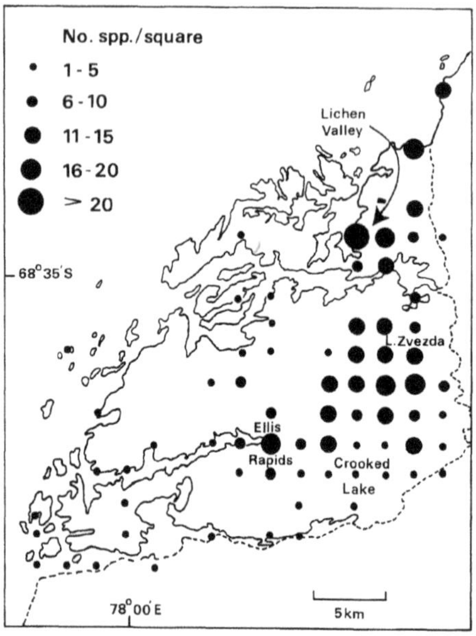

Fig. 6. Number of moss and lichen species in 2 km squares. Each spot is centered within each square rather than over sites.

ing foot traverse routes across the hills. The pattern of distribution shows clear variation between moss and lichen in numbers of species and in geographic range (Figs. 7 & 8). Analysis of coarse distribution data in maps as well as by principal coordinate ordinations and canonical correlation analysis of the ordination vectors (Pickard, 1986) reveals several patterns worthy of closer investigation.

First, is the reduction in number of species close to the ice sheet and Sørsdal Glacier (as seen for example in Figs. 4 & 6) real or an artefact? The reduction may, in fact, be real, particularly in the area affected by the Chelnock Glaciation (Pickard & Adamson, 1983), when at about 1000–2000 a B.P., ice surged north from the Sørsdal Glacier, crossing Crooked Lake before it halted and retreated to its present position. As with the narrow easternmost strip of the Hills close to the ice sheet, there may have been insufficient time for complete colonization of the area and conditions may still be unsuitable for colonization.

Second, wind direction has remained essentially constant from the north-east over the last 4000–6000 a (Pickard, 1982) so that salt from the

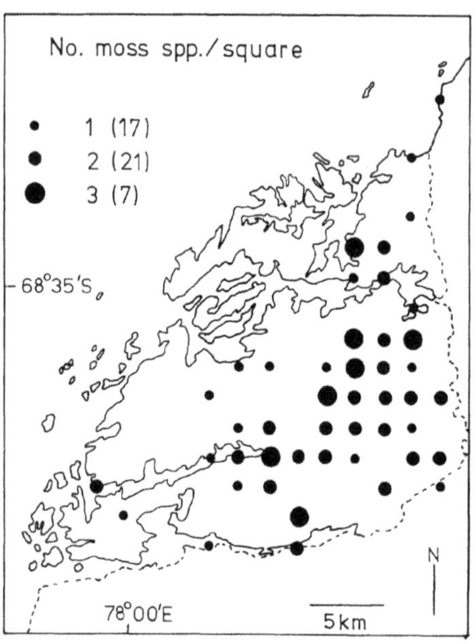

Fig. 7. Number of moss species in 2 km squares, spots centrally located.

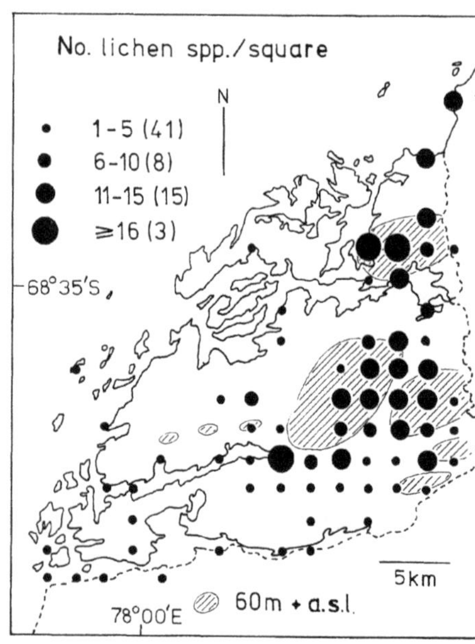

Fig. 8. Number of lichen species in 2 km squares, spots centrally located.

open sea, marine embayments and saline lakes in the northern, central and western sectors of the Hills has been blowing over Long Peninsula, western Broad Peninsula and western Mule Peninsula for some thousands of years. Salt impregnated sediments have also been elevated above sea level by isostatic uplift along the western edge of the Hills.

Third, there is a clear east-west gradient of species

a.

b.

Fig. 9. a. Remains of ephemeral snow drift in lee situations at Ellis Rapids. Algae, lichen, and moss encrusted with algae, form a prominent dark zone. *b.* Crustose and foliose lichens on a sheltered ledge in a cleft near the eastern margin of the Hills. Remnants of drift snow lie on the foliose lichens along the back of the ledge. *c.* Moss turf with convoluted surface showing fragmentation into polsters. This is as luxurious a community of terrestrial plants as occurs in the Hills. *d.* Moss community at Lake Stinear near the limit of moss survival. It is one of the most westerly moss communities on the central part of Broad Peninsula. The plants are affected by wind-blown sand.

richness and complexity of communities. In the east, lichen assemblages close to the ice sheet (Fig. 9b) give way to moss beds with epiphytes and generally richer lichen assemblages (Fig. 9c). In the western half there is a rapid decline to simple lichen assemblages of only one or two species. Interpretation of this complex gradient as a classical succession following ice retreat is unsatisfactory. Species richness does increase with distance from the present edge of the ice sheet to reach a peak at about 1.5 km from the ice (Fig. 9c). Over this distance and at any particular site, species richness increases with the time elapsed since deglaciation. This pattern conforms to the classical succession model. However, the model breaks down with increasing distance from the ice sheet or, what is essentially the same thing, increasing time since deglaciation. Beyond 5–10 km from the ice sheet there is a sharp reduction in abun-

c.

d.

dance/richness. The land here has been deglaciated for 3000 – 5000 a or even longer, thus allowing accumulations of both salt and loose sand to reach toxic or detrimental levels (Fig. 9d). The climax crashes.

Fourth, areas relatively rich in mosses and lichens are frequently associated with elevated land. The main areas over 60 m above msl are shown in Fig. 8. This correlation may be related to two factors. In the eastern areas of higher relief there is an abundance of suitable nesting sites for birds, particularly snow petrels. As a result there are many localized areas rich in plant nutrients, particularly phosphorous. The higher land also traps abundant snow from the nearby ice sheet. In this way favourable nutrient and moisture conditions are provided for small sites which lie outside the more saline zone to the west and, in lee situations, are also sheltered from sand and snow blast.

The terrestrial flora of the Vestfold Hills may be considered as representative of much of continental Antarctica. Differing taxonomic interpretations make overall biogeographic comparisons difficult, particularly in regard to the moss genus *Bryum* and lichen genera such as *Buellia*, *Lecidea*, *Caloplaca*, *Xanthoria* and *Rinodina* (Pickard & Seppelt, 1984; Seppelt, 1986a, 1986b). Of particular interest is the occurrence of a species of the marine lichen *Verrucaria*, the only record of this genus from continental antarctic regions (Seppelt, 1986b).

The Vestfold Hills, while not possessing a unique terrestrial flora, provide a wide range of contrasting habitats which, coupled with the Holocene history of deglaciation of the oasis, provide an ideal opportunity for the study of terrestrial polar ecosystems. At scales ranging from kilometres to fractions of a metre, extreme environmental gradients exist in important variables such as moisture, nutrients, salinity and exposure. However, exploiting the ecological and physiological potential of the area requires the prior resolution of important taxonomic problems.

References

Broady, P. A., 1979. The terrestrial algae of Signy Island, South Orkney Islands. Br. antarct. Surv. Sci. Rep. 98.

Broady, P. A., 1981a. The ecology of sublithic terrestrial algae at the Vestfold Hills, Antartica. Br. Phycol. J. 16: 231 – 240.

Broady, P. A., 1981b. Ecological and taxonomic observations on subaerial epilithic algae from Princess Elizabeth Land and Mac.Robertson Land, Antartica. Br. Phycol. J. 16: 257 – 266.

Broady, P. A., 1981c. The ecology of chasmolithic algae at coastal locations of Antarctica. Phycologia 20(3): 259 – 272.

Broady, P. A., 1986. Ecology and taxonomy of the terrestrial algae of the Vestfold Hills. In J. Pickard (ed.), Antarictic Oasis. Terrestrial environments and history of the Vestfold Hills. Academic Press, Lond.: 165 – 202.

Burton, H. R. & P. J. Campbell, 1980. The climate of the Vestfold Hills, Davis Station, Antarctica, with a note on its effect on the hydrology of hypersaline Deep Lake. ANARE Sci. Rep., Publ. 129. Aust. Govt. Publ. Service, Canberra, 50 pp.

Cameron, R. E., 1972. Ecology of blue-green algae in antarctic soils. In T. V. Desikachary (ed.), Taxonomy and Biology of Blue-Green Algae. Centre for Advanced Studies in Botany, University of Madras, Madras: 353 – 384.

Collerson, K. D. & J. W. Sheraton, 1986. Bedrock geology and crustal evolution of the Vestfold Hills. In J. Pickard, (ed.), Antarctic Oasis. Terrestrial environments and history of the Vestfold Hills. Academic Press, Lond.: 21 – 62.

Crohn, P. W., 1959. A contribution to the geology and glaciology of the western part of Australian Antarctic Territory. ANARE Rep. Publ. 49. Antarct. Div., Dept. External Affairs, Melbourne, 104 pp.

Friedmann, E. I., 1977. Microorganisms in antarctic desert rocks from dry valleys and Dufek Massif. Antarct. J. U.S. 12: 26 – 30.

Friedmann, E. I., 1982. Endolithic microorganisms in the antarctic cold desert. Science 215: 1045 – 1053.

Friedmann, E. I. & M. Galun, 1974. Desert algae, lichens and fungi. In G. W. Brown (ed.), Desert Biology, 2. Academic Press, Lond.: 165 – 212.

Friedmann, E. I. & R. Ocampo, 1976. Endolithic blue-green algae in the dry valleys: primary producers in the antarctic desert ecosystem. Science 193: 1247 – 1249.

Fukushima, H., 1959. General report on fauna and flora of the Ongul Island, Antarctica, especially on freshwater algae. J. Yokohama munic. Univ. Ser. C Nat. Sci. 31: 1 – 20.

Golubic, S., E. I. Friedmann & J. Schneider, 1981. The lithobiontic ecological niche, with special reference to microorganisms. J. Sediment. Petrol. 51(2): 475 – 478.

Horikawa, Y. & H. Ando, 1963. A review of the antarctic species of Ceratodon described by Cardot. Hikobia 3(4): 275 – 280.

Johnstone, G. W., D. J. Lugg & D. A. Brown, 1973. The biology of the Vestfold Hills, Antarctica. ANARE Sci. Rep. Publ. 123. Antarct. Div., Dept. Sci., Melbourne, 62 pp.

Law, P., 1959. The Vestfold Hills. ANARE Rep. Publ. No. 47. Antarct. Div., Dept. External Affairs, Melbourne, 50 pp.

Lied, N. T., 1963. Notes on sastrugi and sand dune observations, A.N.A.R.E. Satellite Station, Vestfold Hills, 1961. Aust. Meteorol. Mag. 40: 35 – 46.

Longton, R. E., 1979. Vegetation ecology and classification in the antarctic zone. Can. J. Bot. 57: 2264 – 2278.

McLeod, I. R., 1963. An outline of the geology of the sector from

45°E to 80°E, Antarctica. Sci. Comm. Antarct. Res. Bull. 15: 262–275.

Pickard, J., 1982. Holocene winds of the Vestfold Hills, Antarctica. N.Z. J. Geol. Geophys. 25: 353–358.

Pickard, J., 1986. Spatial relations of the vegetation of the Vestfold Hills, Antarctica. In J. Pickard (ed.), Antarctic Oasis. Terrestrial environments and history of the Vestfold Hills. Academic Press, Lond.: 275–308.

Pickard, J. & D. A. Adamson, 1983. Perennially frozen lakes at glacier/rock margins, East Antarctica. In R. L. Oliver, P. L. James & J. B. Jago (eds.), Antarctic Earth Science. Aust. Acad. Sci., Canberra: 470–472.

Pickard, J. & R. D. Seppelt, 1984. Phytogeography of Antarctica. J. Biogeog. 11: 83–102.

Seppelt, R. D., 1984. The bryoflora of the Vestfold Hills and Ingrid Christensen Coast, Antarctica. ANARE Res. Notes 20. Antarct. Div., Dept. Sci., Hobart, 30 pp.

Seppelt, R. D., 1986a. Mosses of the Vestfold Hills. In J. Pickard (ed.), Antarctic Oasis. Terrestrial environments and history of the Vestfold Hills. Academic Press, Lond.: 221–246.

Seppelt, R. D., 1986b. Lichens of the Vestfold Hills. In J. Pickard (ed.), Antarctic Oasis. Terrestrial environments and history of the Vestfold Hills. Academic Press, Lond.: 247–274.

Seppelt, R. D. & D. H. Ashton, 1978. Studies on the ecology of the vegetation at Mawson Station, Antarctica. Aust. J. Ecol. 3: 373–388.

Sheraton, J. W. & K. D. Collerson, 1983. Archaean and Proterozoic geological relationships in the Vestfold Hills – Prydz Bay area, Antarctica. J. Aust. Geol. Geophys. 8: 119–128.

Streten, N. A., 1968. Some features of mean annual wind speed data for coastal East Antarctica. Polar Rec. 14: 315–322.

Streten, N. A., 1986. Climate of the Vestfold Hills. In J. Pickard (ed.), Antarctic Oasis. Terrestrial environments and history of the Vestfold Hills. Academic Press, Lond.: 141–164.

Thompson, K., 1979. Plants in antarctic rocks: adaptation to the most extreme environments. 49th Aust. N.Z. Assoc. Adv. Sci. Conf., 63 pp.

Usher, M. B., 1983. Pattern in the simple moss-turf communities of the subantarctic and Maritime Antarctic. J. Ecol. 71: 945–958.

Wellman, P. & J. W. Williams, 1982. Extent of Archaean and Late Proterozoic rocks under the ice cap of Princess Elizabeth Land, Antarctica, inferred from geophysics. J. Aust. Geol. Geophys. 7: 213–218.

Hydrobiologia 165: 197–208 (1988)
J. M. Ferris, H. R. Burton, G. W. Johnstone & I. A. E. Bayly (eds) Biology of the Vestfold Hills, Antarctica
Kluwer Academic Publishers

A survey of the terrestrial Tardigrada of the Vestfold Hills, Antarctica

J. D. Miller[1], P. Horne[2], H. Heatwole[1], W. R. Miller[1] & L. Bridges[1]
[1]*Department of Zoology, University of New England, Armidale, N.S.W. 2351, Australia;* [2]*Department of Zoology, La Trobe University, Melbourne, Victoria, 3083, Australia*

Key words: Tardigrada, Antarctica, Vestfold Hills, biogeography

Abstract

A survey of the terrestrial tardigrades inhabiting growths of algae, lichens and mosses in the Vestfold Hills, Antarctica, was carried out at 11 and 35 sites during the austral summers of 1980 and 1982, respectively. In all, 24 species of plants were collected from which four genera and four species of Tardigrada were recovered. A key to the tardigrades of the area is presented. The distribution and associational patterns of the tardigrades are discussed in the context of other studies of antarctic Tardigrada.

Introduction

Historically, research on the Tardigrada has emphasized classification, phylogeny, and description of genera and species (Ramazzotti & Maucci, 1982). Contemporary research has approached the Tardigrada in a more comprehensive and integrated manner. Recent symposia (Higgins, 1975; Weglarska, 1979; Nelson, 1982a) provide a synthesis of current research into the biology of the phylum and the review by Nelson (1982b) provides a comprehensive summary of the developmental biology of the group. There have been a few investigations into the population and community ecology of antarctic tardigrades (Jennings, 1979; Davis, 1981; Miller, 1981), and into their cultivation in the laboratory (Ellsworth *et al.*, 1961; Dougherty, 1964).

Richters (1904) was the first to describe a species of tardigrade from Antarctica. Later, he described other species and considered the distribution of antarctic tardigrades (Richters, 1907, 1908, 1909). Murray (1910) contributed four species to the list, bringing the total to 17. Presently, only 28 species representing seven genera are known from the region (Jennings, 1976a; Sudzuki, 1979).

Early workers in the Vestfold Hills, East Antarctica, did not find tardigrades. They were not found in collections made by Korotkevich (1964) or Johnstone *et al.* (1973); nor were they reported by Sudzuki (1979) who examined a few samples collected by H. R. Burton from Tryne Fjord and Middle Tarn. However, tardigrades were among the microfauna reported in a study of an antarctic mite but they were not systematically studied (Rounsevell, 1981). Everitt (1981) reported on the ecology of two tardigrades, *Hypsibius arcticus* (Murray) and *Isohypsibius* sp., from Deep Lake Tarn.

Description of the Vestfold Hills

A detailed description of the Vestfold Hills is provided by Johnstone *et al.* (1973) and other papers in this volume. In brief, the area is an oasis comprised of dry, low hills interspaced with saline lakes that are remnants of fjords. The lakes were formed 'by isostatic uplift' following the retreat of the ice cap; 'evaporation in the dry and windy environment' has concentrated the salts to many times the original strength (Barker, 1981). The primary source of moisture in the Vestfold Hills is meltwater from snowdrifts (Rounsevell, 1981). For at least seven months of the year the moisture in the soil does not exceed 0.5% and the temperature remains below 0 °C (Rounsevell, *op. cit.*). Data on solar radiation, temperature, relative humidity, wind and snow have been summarized by Burton & Campbell (1980).

Methods

During the 1980 survey, core samples or small pieces of moss were collected from 11 sites. Although a more extensive collection was made during the year (Horne, unpub. data), only the results of the initial survey are presented here. During the 1982/83 survey, collections were made from 35 locations in the Vestfold Hills and adjacent islands. In total, samples were collected from 40 sites including six that were sampled in both surveys (Fig. 1). Locations follow Natmap (NMP 80/154), Vestfold Hills, Princess Elizabeth Land (1982, scale 1:50000). Collection sites were distributed in three general regions: along the edge of the permanent ice field and glacier,

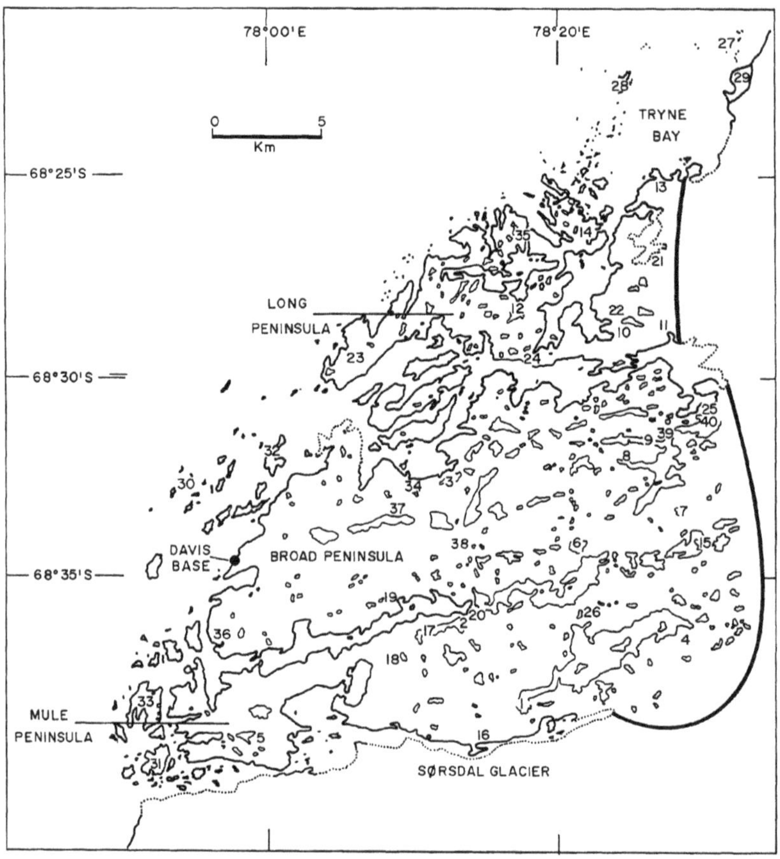

Fig. 1. Location of collection sites in the Vestfold Hills (68°35′S, 78°00′E). Collection sites are identified by numbers. * sampled in 1980; ** sampled in both 1980 & 1982/83; unmarked, sampled in 1982/83 only. Note: not all of the names used here have been accepted by the Antarctic Names Committee of Australia. 1 ** Mossel Hut, 2 Watts Lake Hut, 3 Brookes Hut, 4 Crooked Lake Depot, 5 Mule Peninsula, 6 Depot Lake, 7 ** Trajer Ridge, 8 Zvezda Lake, 9 ** Lake Cowen, 10 ** Lichen Valley, 11 Taaffe Ridge, 12 Long Peninsula, 13 Grace Lake, 14 B Depot, 15 Pauk Lake, 16 Sørsdal Glacier, 17 Watts Lake/Anderson Lake, 18 Poseidon Basin, 19 Learner's Way, 20 ** Ellis Rapids, 21 Plateau Lake, 22 Moss Cirque, 23 Long Peninsula (near Rookery Lake), 24 Tryne Crossing, 25 ** Platcha Hut, 26 Crooked Lake/Druzhby Lake, 27 Wyatt Earp I., 28 North Tryne I., 29 Walkabout Rocks, 30 Magnetic I., 31 Zolotov I., 32 Plough I., 33 Hawker I., 34 Drainage to Shirokaya Bay, 35 Channel to Taynaya Bay, 36 * Lookout Hill, 37 * Lake Stinear, 38 * Club Lake, 39 * Stalker Hill, 40 * Bisernoye Lake.

throughout the central lakes area, and along the coastal fringe and adjacent islands.

At collection sites, plants were selected casually; only visually dominant species of algae, lichens and mosses were taken. Two categories of samples were collected. The first was a core (1 cm diameter in 1980, 2 cm in 1982) extracted to the bottom of the vegetation, usually moss. The second sample (about 2 × 2 cm) was taken to the bottom of the vegetation, usually lichens and algae. In all cases, the site of the sample was repaired by filling holes or rearranging the edges of the surrounding plant material in order to re-establish the continuity of the microclimate of the vegetation.

Treatment of samples followed the methods of Jennings (1976a, 1976b). In 1980, samples were processed at Davis and were not subjected to freezing or centrifugation. In 1982/83, plant samples were dried at room temperature (approximately 25 °C) to a constant mass over a 48–96 h period and then frozen for transport to Australia. In the laboratory, samples were thawed, rehydrated in distilled water for 48 h then washed at least six times through a 100 × 200 μm mesh and the vegetation squeezed. The resulting fluid and debris were centrifuged at 1 500–2 000 rpm for 5 min and the supernatant discarded. Concentrated formalin was added slowly to make a final concentration of 5% in each sample. The preserved mixture of fluid and debris was recentrifuged and concentrated to a volume of 3 ml.

The resulting centrifugates were examined for tardigrades at 80× magnification under a binocular dissection microscope fitted with a movable stage. Specimens were extracted using a microcapillary tube, stained in alcoholic iodine and mounted in Hoyer's medium. The taxonomic criteria and terminology follow Ramazzotti (1972), Horning *et al.* (1978) and Schuster *et al.* (1980).

The mosses collected during the 1980 survey were not identified. However, in 1982/83 one species of alga, 20 lichens and three mosses were identified on the basis of growth morphology, colour and selected chemical tests (Lamb, 1962, 1968; Kashiwadani, 1970). The collections do not represent the complete flora of the area because only the visually dominant plants were collected.

Results

One species from each of four genera were recovered: *Hypsibius allisonii* Horning *et al.*; *Diphascon puniceum* (Jennings)?; *Macrobiotus fuciger* Murray?; *Pseudechiniscus novaezeelandiae* (Richters). *Hypsibius allisonii* is similar to *Hypsibius dujardini* (Doyère) except that in the former the buccal tube is shorter and wider. The species was described from a sample of moss from one locality in the South Island, New Zealand. *Diphascon puniceum*? was originally described from Signy Island as *Hypsibius (Diphascon) puniceus* which was previously unknown beyond the Island. Jennings (1976b) provided descriptions, illustrations and arguments for separate species status; Schuster *et al.* (1980) re-elevated *Diphascon* to generic level. Briefly, this species is separated from other species of *Diphascon* by possessing a granular posterior area, a small pharyngeal bulb and by the relative size of the macroplacoids. *Macrobiotus fuciger*? was illustrated and described by Horning *et al.* (1978) and Jennings (1976b). It was originally described from the South Orkney Islands and has a cosmopolitan distribution (Table 1). Species of *Pseudechiniscus* are identified by the number, length and location of dorsal and lateral spines. Cirrus A is slightly longer in these specimens than in other examples (Nelson, pers. commun.).

Both *Diphascon puniceum* and *Macrobiotus fuciger* are followed by '?' throughout the text because identification of these species is not certain. *Macrobiotus fuciger* was identified from adult morphology without the aid of eggs. *Diphascon puniceum* closely resembles the specimens found. Generic assignment is not questioned.

The following key to the known tardigrades of the Vestfold Hills is derived from keys and descriptions by Schuster & Grigarick (1965), Jennings (1976b), Morgan & King (1976) and Horning *et al.* (1978) as well as from specimens collected in the Vestfold Hills.

1. Dorsal plates absent; lateral cirrus A absent . 2
   Dorsal plates present; lateral cirrus A present on head plate on posteriolateral margin; pseudosegmental plate present between dorsal plate and terminal plate .... *Pseudechiniscus*
2. Two double claws on each leg equal in size and

shape and situated symmetrically on the foot .... *Macrobiotus*
Two double claws on each leg dissimilar in size and shape and situated symmetrically on the foot 3
3. Pharyngeal tube between the stylet supports and the bulb about equal to 1/2 that of the pharyngeal bulb 4
Pharyngeal tube between stylet supports and the bulb about equal to the length of the pharyngeal bulb ....*Diphascon*
4. Secondary branch of external claw and base continuously and evenly arched 5
Secondary branch of external claw at right angle to base of claw ....*Isohypsibius*
5. Principal arm of the posterior double claw highly curved at distal end; pharyngeal tube narrow (1.5 μm); pharyngeal bulb oval with round-oval macroplacoids ....*Hypsibius arcticus*
Principal arm of the posterior double claw only slightly curved at distal end; pharyngeal tube wider (2–2.5 μm); pharyngeal bulb oval with apophysis and two macroplacoids about equal in length ....*Hypsibius allisonii*

A comparison of both survey results is limited because only six sites were sampled in common and because only moss was collected in the 1980 survey. In only one case (Mossel Hut, Fig. 1) were samples taken from the same moss bed; at the other five sites, collections were made less than 100 m apart. In

Table 1. Antarctic Tardigrades found elsewhere in the World. The data were compiled from: Argue, 1971, 1972; Dastych, 1976; Gustavson, 1963; Horning *et al.*, 1978; King *et al.*, 1981; Miheleie, 1967; Miller, 1975; Morgan, 1980; Morgan & King, 1976; Murray, 1907, 1910; Nelson & Horning, 1979; Peterson, 1951; Puglia, 1959; Ramazzotti, 1972; Riggin, 1962; Schuster & Grigarick, 1965, 1970; Schuster & Toftner, 1982; Sudzuki, 1979.

Tardigrade Species		Arctic	Europe & Britain	Asia	North America	Greenland/Iceland	Scandinavia	Africa	Australia	New Zealand	South America	Oceanic Islands
Echiniscus	*arctomys* Ehrbg	*	*	*	*			*	*	*		*
	capillatus Ramazzotti		*									
	wendti Richters	*	*		*	*						
	kergeuelensis		*									
Pseudechiniscus	*novaezeelandiae* Richters								*	*	*	
Diphascon	*alpinus* (Murray)	*	*	*	*					*		
	chilenensis (Plate)	*	*	*	*	*	*		*	*	*	*
	pinguis Marcus		*		*	*						
	puniceum (Jennings)									*		
	scoticus (Murray)	*	*	*	*	*	*		*	*	*	*
Hypsibius	*antarcticus* (Richters)		*									
	arcticus (Murray)	*	*	*	*			*		*		
	convergens (Urbanowicz)		*		*			*		*	*	
	allisonii Horning *et al.*									*		
	oberhaeuseri (Doyére)	*	*	*	*	*	*	*		*	*	*
	dujardini (Doyére)				*			*		*	*	
Isohypsibius	*asper* (Murray)		*									
	papillifer (Murray)		*						*	*		*
	renaudi Ramazzotti											*
Milnesium	*tardigradum* Doyére	*	*	*	*	*	*	*	*	*	*	*
Orcella	*mollis* Murray								*			
Macrobiotus	*fuciger* Murray		*	*	*	*				*	*	
	ambigius Murray	*	*									

1982/83, an average of 14 samples (range 2–25) and an average of 4.6 species of plants (range 1–15) were collected from each site.

There was no clear association between tardigrades and plants (Table 2). *Hypsibius allisonii* exhibited the least specific associational pattern. It occurred in the alga *Prasiola crispa* (Lightf.) Menegh., crustose lichens, foliose lichens and mosses. *Diphascon puniceum*? was also recovered in algae and crustose and foliose lichens but not from any of the moss samples. *Macrobiotus fuciger*? was found in one crustose lichen and in moss; *Pseudechiniscus novaezeelandiae* was retrieved from crustose and foliose lichens and from moss samples but not from algae. In addition to the species of plants that yielded tardigrades (Table 2), 10 species of lichens were examined that did not yield any tardigrades. They were: *Biatorella cerebriformis* (Dodge) Filson, *Buellia lignoides* Filson, *Buellia soredians* Filson, *Caloplaca citrina* (Hoffm.), *Caloplaca* (*Xanthoria*) *elegans* (Link.), *Lecidea phillipsiana* Filson, *Lecanora expectans* Darbshire, *Physica caesia* (Lightf.) Menegh., *Umbilicaria aprina* Nyl., *Umbilicaria crestata* Dodge & Baker.

Tardigrades occurred in only 25% of the 491 samples collected in 1982/83. Rotifers were contained in 78% and nematodes were found in 59% of the samples. In samples that contained them, tardigrades occurred at the lowest relative abundance (1–25) in 61% of the samples; nematodes and rotifers also occurred at their lowest relative abundance (1–200) in the majority of samples (64% and 50%, respectively) (Table 3).

The four species of tardigrades were recovered from samples collected at 18 of the 40 sites (45%); these were located in the more inland portions of the Vestfold Hills in association with freshwater seepage areas. The most widely distributed tardigrade was *Hypsibius allisonii* which was found at 16 of the 18 positive sites in at least one of the surveys (Table 4). The only two sites at which it was not recovered were Club Lake and Plateau Lake (Fig. 1). The most restricted distribution was shown by *Diphascon puniceum*? which was found at only two sites (Plateau Lake

Table 2. Association of tardigrades and plants in the Vestfold Hills, based on the 1982/83 survey.

	Hypsibius allisonii	*Diphascon puniceum?*	*Macrobiotus fuciger?*	*Pseudechiniscus novaezeelandiae*
ALGAE				
Prasiola crispa (Lightf.) Menegh.	+	+	–	–
LICHENS				
Buellia frigida Darbshire	+	+	–	–
Buellia grammiae Filson	+	+	–	+
Caloplaca athallina Darbshire	+	–	–	–
C. (Xanthoria) mawsoni Dodge	+	–	–	–
Candelariella antarctica Filson	+	–	+	+
Lecanora melanophthalma (Th. Fr.) Zahl.	+	–	–	+
Rhizocarpon flavum Dodge & Baker	–	+	–	–
Rinodina olivaceobrunnea Dodge & Baker	+	–	–	–
MOSSES				
Bryum sp. Hedwig.	+	–	–	+
Ceratodon sp. Bridel	+	–	+	–
Sarconeurum sp. Bryhn	+	–	–	+

Table 3. Percentage occurrence of tardigrades, nematodes and rotifers in the total, tardigrade positive and tardigrade negative samples. Scale of relative abundance for tardigrades is 0 = 0, 1 = 1–25, 2 = 26–50, 3 = 51–75, 4 = 76–100, 5 = 100+. Scale of relative abundance for nematodes and rotifers is 0 = 0, 1 = 1–200, 2 = 201–400, 3 = 401–600, 4 = 601–800, 5 = 801+.

Samples	Scale of relative abundance					
	0	1	2	3	4	5
Total						
Tardigrades	75	15	2	3.2	0.8	2
Nematodes	41	52	4.4	0.4	0.6	0
Rotifers	22	57	12	3	2.3	1.6
Tardigrade Positive						
Tardigrades	–	61	11.2	12.9	3.2	9.6
Nematodes	29	64.5	2.4	1.6	1.6	0
Rotifers	5.6	50	23	8	10	4.8
Tardigrade Negative						
Tardigrades	–	–	–	–	–	–
Nematodes	46	49	5	0	0.2	0
Rotifers	27	60.9	8.1	1.3	0.8	0.8

and Ellis Rapids, Fig. 1). *Macrobiotus fuciger*? was found at four sites and *Pseudechiniscus novaezeelandiae* was recovered from six sites; only one site (Mossel Hut) had both. Tardigrades were recovered from samples collected at Mossel Hut and Lake Cowen in both surveys (Table 4). In contrast, in 1980 tardigrades were retrieved from samples collected at Lichen Valley, Platcha Hut area and Trajer Ridge (Fig. 1) but not from these general vicinities in 1982/83. Two species of tardigrades were recovered from samples taken from Ellis Rapids in 1982/83 but none were found in 1980.

All sites in the vicinity of the head of Ellis Fjord yielded tardigrades (Learner's Way, Poseidon Basin, Ellis Rapids, Watts Lake Hut and the area between Watts Lake and Anderson Lake, Fig. 1). Three species inhabited the moss bed located between Mossel Hut and Mossel Lake (Fig. 1). Although all positive samples were collected from moist freshwater seepage areas, two sites (Laternula Lake and the area between Anderson Lake and Watts Lake, Fig. 1) were in close proximity to saline lakes. Two species were found at Walkabout Rocks (site 29, Fig. 1) at the northern end of Tryne Bay. This is an isolated outcrop surrounded by ice and sea. The occurrence of tardigrades at the latter three sites is a strong indication thay they will be retrieved from the coastal areas in other collections, almost certainly from *Prasiola crispa*, a widely distributed terrestrial alga, and from lichens.

At present the occurrence and distribution of tardigrades in the Vestfold Hills cannot be adequately explained. The paucity of species stands in sharp contrast to the 16 species reported from Signy Island (Jennings, 1976b) and 11 species from the Antarctic Peninsula (Jennings, 1976a). However, the number of species reported from other localities closer to the Antarctic Circle is also low (Morikawa, 1962; Murray, 1910; Richters, 1904, 1909; Sudzuki, 1964, 1979).

Table 4. Distribution of tardigrades among the various positive collection sites in the two surveys.

Site	Site Number (Fig. 1)	*Hypsibius allisonii*	*Diphascon puniceum?*	*Macrobiotus fuciger?*	*Pseudechiniscus novaezeelandiae*
1980 Survey					
Mossel Hut	1	–	–	+	+[a]
Trajer Ridge	7	+	–	–	–[b]
Lake Cowen	9	+	–	–	+[a]
Lichen Valley	10	+	–	–	+[b]
Platcha Hut	25	+	–	–	–[b]
Club Lake	38	–	–	–	+
Stalker Hill	39	+	–	–	–
1982/83 Survey					
Mossel Hut	1	+	–	+	+[a]
Watts Lake Hut	2	+	–	–	–
Crooked Lake Depot	4	+	–	–	–
Mule Peninsula	5	+	–	–	–
Lake Cowen	9	+	–	–	+[a]
Taaffe Ridge	11	+	–	–	+
Watts Lake/ Anderson Lake	17	+	–	+	–
Poseidon Basin	18	+	–	–	–
Learner's Way	19	+	–	–	–
Ellis Rapids	20	+	+	–	–[b]
Plateau Lake	21	–	+	–	+
Moss Cirque	22	+	–	+	–
Walkabout Rocks	29	+	–	+	–

[a] Positive in both surveys; [b] Positive in one survey but not in the other.

Discussion

Distribution of plants

East Antarctica (20°E to 170°E) lies parallel to the Antarctic Circle and hosts a flora of over 70 species, nearly half of which are restricted to the area (Dodge, 1965). The phytogeography of Antarctica was recently reviewed by Pickard & Seppelt (1984).

The history of antarctic freshwater algal studies and a review of contemporary work including a list of species and their distribution was presented by Hirano (1965). Since then other species have been found and a new species described from the antarctic region (Broady, 1976; Parker *et al.*, 1972). Two genera and one species of terrestrial alga (*Nostoc* Vaucher, *Oscillatoria* Vaucher, and *Prasiola crispa*) were reported by Rounsevell (1981) from the Vestfold Hills. A brief account of the distribution and taxonomy of terrestrial algae of the Vestfold Hills is given in Seppelt *et al.* (this volume).

Approximately 58 genera of lichens occur in the antarctic region (31 crustose genera, 14 foliose genera, 12 fructicose genera) (Dodge, 1965).

The taxonomy and distribution of antarctic mosses have been reviewed several times (Greene, 1967, 1968a, 1968b; Greene *et al.*, 1970; Robinson, 1972). Few species are endemic (Robinson, 1972). The genus *Bryum* Hedwig. contains more species in the Antarctic than any other bryophyte group.

In the Vestfold Hills, a collection of at least 17 species of 12 genera of lichens and one bryophyte was made by Johnstone *et al.* (1973); fewer lichens occurred along the seaward portions; the numbers increased 'particularly near fjords and at the edge of the ice-cap'. The single bryophyte, *Bryum antarcticum* Hook., Fil. & Wils., has been reported from five localities by Johnstone *et al.* (*op. cit.*) who speculated that it probably 'occurs at many sites throughout the Vestfold Hills where there is an adequate water supply from melting snow'. The distribution of lichens and mosses in the Vestfold Hills is more fully discussed in Seppelt *et al.* (this volume) and placed in the broader context of antarctic terrestrial ecosystems by Seppelt & Broady (this volume). In general, the distribution of tardigrades reflects that of the plants; however, more thorough investigation is required.

Although studies have addressed the ecology and physiology of antarctic bryophytes (Matsuda, 1968; Rastorfer, 1972; Gimingham & Smith, 1971; Gimingham, 1967; Longton & Holdgate, 1967), not enough is known to allow speculation about the interactions and internal environment required for survival of tardigrades. Factors which limit the distribution and growth of plants, including a short growing season, low temperatures and restricted free water (Wace, 1965; Ugolini, 1972; Longton, 1979; Taylor, 1955), must also influence the distribution of the microfauna. Certainly, the growth form in which the interspaces between stems is filled with sandy soil alters the microclimate in the turf (Longton, 1974). The valuable comments concerning the ecology of the vegetation at Mawson Station by Seppelt & Ashton (1978) may be applicable to those species of plants which also occur in the Vestfold Hills and may eventually provide insight into the plant-animal ecological association.

Distribution of tardigrades

Tardigrades, as a group, show several patterns of distribution. Many species have very wide ranges. Of 40 species reported from the arctic region, eight are also found in the Antarctic (Murray, 1907); nearly 25% (10 of 42) of Icelandic tardigrades (Morgan, 1980) occur in the Antarctic. Because the distribution of tardigrades is incompletely known, surveys of defined areas provide valuable data; however, because of the sporadic nature of most collecting, a second survey in the same area might produce quite different results. A survey of the Scandinavian Peninsula recorded 60 species of tardigrades including 24 previously unknown from the region but of these only six were undescribed species (Pasa & Maucci, 1979). Even more interestingly, 13 species previously reported from the survey area were not found. This was not solely the result of faulty collecting procedures. Clearly, the tardigrade fauna, as shown by presently available surveys, must be an approximation of the total fauna of the Vestfold Hills and Antarctica.

Tardigrades are not evenly distributed throughout a habitat and are not necessarily represented in every sample (Hallas, 1977; Schuster & Grigarick, 1970).

Tardigrades were found in only 50% of 45 plant species collected in a survey of Santa Cruz Island (Schuster & Grigarick, 1970). Only 29 of 65 sites in New Brunswick, Canada, proved to host tardigrades (Argue, 1971). In New Zealand, only 42.6% of 1 354 samples derived from a wide survey contained tardigrades (Horning *et al.*, 1978); that survey demonstrated the presence of 55 species of tardigrades. Survey results from most parts of the world (Morgan & King, 1976; Schuster & Grigarick, 1965; Hallas, 1977; Riggin, 1962; Murray, 1910; Puglia, 1959) support the two points raised by Schuster & Grigarick (1970) concerning the distribution of tardigrades: (1) tardigrades are not equally or uniformly distributed and (2) host-plant associations do not explain the observed distributions. Certainly, other factors are involved.

Because many tardigrades are cosmopolitan (Morgan & King, 1976; Nelson, 1982b), it is not surprising that 60% of the species previously reported to occur in Antarctica also occur elsewhere in the world (Table 1). For example, *Hypsibius oberhaeuseri* (Doyére) has been reported from 15 countries. It is an easily recognized species having nine coloured bands and a papillate cuticle. Similarly, the predatory tardigrade, *Milnesium tardigradum* (Doyére) occurs in at least 24 countries; it too is easily identified. Nine species at present appear to be restricted to the antarctic region. In some cases, however, this may reflect intensity of collecting (Jennings, 1976a) rather than endemism. An array of species has been found wherever detailed systematic collecting has been conducted (Peterson, 1951; Morgan & King, 1976; Jennings, 1976a, 1976b; Hallas, 1977; Horning *et al.*, 1978; Pasa & Maucci, 1979). To a lesser extent, the restricted distribution of some species may result from inadequate identification. The taxonomy of several species is in doubt; members of the genus *Diphascon* are a good example (Peterson, 1951; Jennings, 1976a, 1976b).

In the Scott Glacier region, Queen Maud Land, no tardigrades were found in soil which yielded lichens, bacteria, moulds, algae, amoebae and rotifers (Claridge *et al.*, 1971). However, Dougherty & Harris (1963) reported four species of tardigrades (*Diphascon alpinus* (Murray), *Hypsibius arcticus, Hypsibius oberhaeuseri, Macrobiotus polarus* Murray) from Ross Island and adjacent areas of Victoria Land.

Heywood (1977) recovered 10 species of tardigrades (*Echiniscus* sp., *Macrobiotus ambigius* Murray, *Macrobiotus fuciger, Hypsibius dujardini, Hypsibius oberhaeuseri, Isohypsibius asper* (Murray), *Isohypsibius renaudi* Ramazzotti, *Diphascon alpinus/pinguis, Diphascon scoticus* (Murray), *Milnesium tardigradum*) from the vicinity of Ablation Point, Alexander Island, along the Antarctic Peninsula.

Tardigrades were reported but not identified from four locations in the area of McMurdo Sound (Armitage & House, 1962). An unidentified species of *Macrobiotus* was found on the Clark Peninsula near the abandoned Wilkes Base by Thomas (1965).

Although no tardigrades were recovered from two locations in the Vestfold Hills by a Soviet expedition, they were found in collections from the Bunger Hills and the Obruchev Hills; unfortunately these were not identified (Korotkevich, 1964).

The distributional patterns of tardigrades reflect the location of investigations rather than the biology of the animals. Most studies have focused on the South Orkney Islands, the Antarctic Peninsula and periphery of the Ross Sea (Fig. 2). Until the present report, only *Hypsibius arcticus* (a very widespread species) and one undescribed species of Tardigrada had been reported from near Davis Station (Everitt, 1981).

The antarctic continent is not a continuous habitat for terrestrial organisms but rather it consists of a series of thawed ‘islands’ scattered around the periphery of permanent ice. Consequently, it is possible that the biogeography of Antarctica should be examined from the standpoint of insular biogeographic theory in which immigration, local extinction and high species turnover may play dominant roles (for a review of insular biogeographic theory see MacArthur & Wilson, 1963, 1967; Heatwole *et al.*, 1981).

In summer the antarctic continent consists of a series of widely scattered ice-free areas and nunataks isolated from each other by broad expanses of permanent ice. For strictly terrestrial organisms, such as tardigrades and mosses, these isolated patches of thawed land take on insular characteristics. In biogeographic terms they, in combination with the true

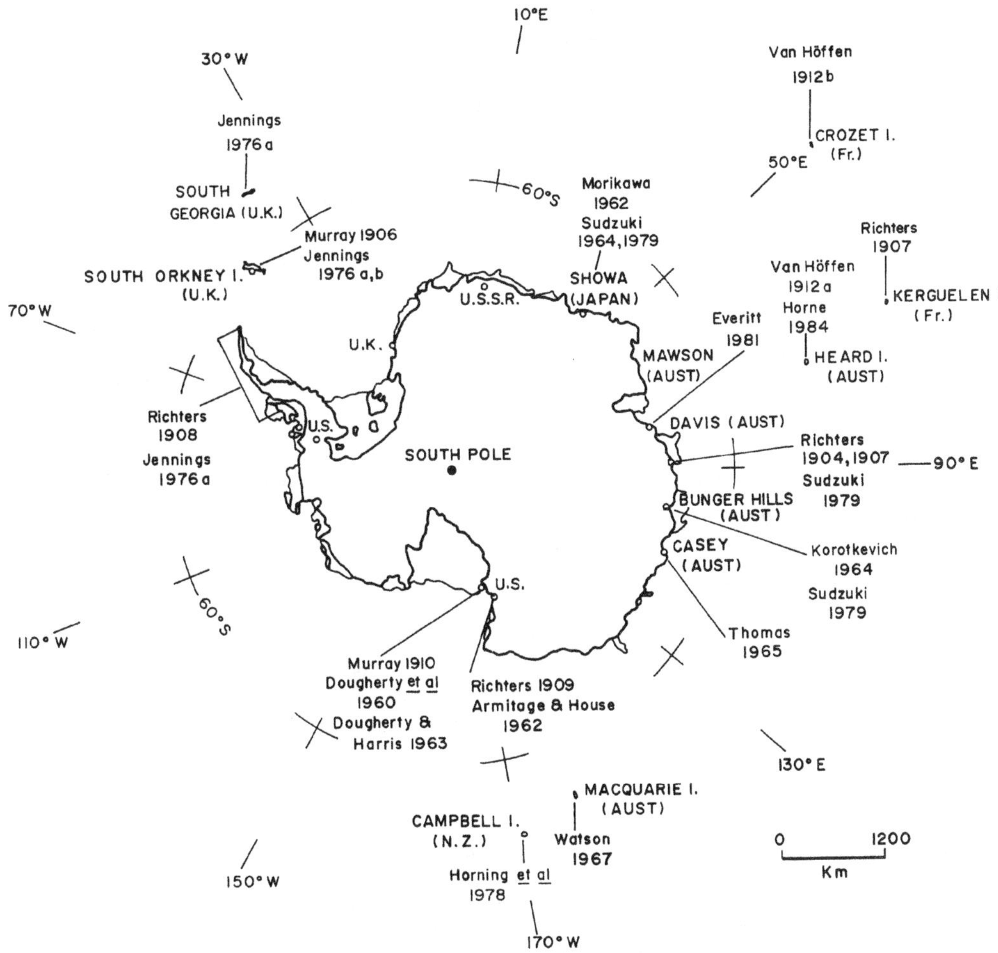

Fig. 2. Map of the greater antarctic region indicating where collections of tardigrades have been made.

antarctic offshore islands and the sub-antarctic islands, should probably be treated as a far flung archipelago embedded in a medium of ice and/or sea-water.

The insular analogy has been applied to isolated mountain tops, lakes and ponds, caves, vacant urban lots and a wide variety of other isolates (see Heatwole *et al.*, 1981 for a brief summary). Heatwole *et al.* (1981) have cautioned against interpreting the analogy too closely because the factors influencing dispersal may differ greatly across different kinds of dispersal barriers (land versus sea, etc.). They suggest that a theory of the biogeography of isolates, of which real islands are one type, should be developed and that the attributes of different kinds of isolates be compared. However, the insular analogy may be especially apt in the case of antarctic landmasses. Given the nature of the polar terrestrial biota, it is unlikely that dispersal by sea currents is as important as in milder climates where plant seeds may be dispersed in this way or terrestrial arthropods dispersed in flotsam. Thus, wind and birds are probably the major dispersal agents in the south polar region. Ice and sea-water would present similar barriers to dispersal by these agents and these two media can probably be assumed to be equivalent in the isolation of polar organisms. In this context, factors influencing migration of the propagules of terrestrial organisms among exposed land areas, their survival once there, and the resultant species turnover should be examined. The number of species of terrestrial taxa occupying islands should be related to such parameters as the area of those masses, the distances separating them from other such lands, and the directional qualities of dispersal agents (wind direction, migration paths of birds). Features influencing survival or extinction once there should be examined, e.g. mode and degree of resistance to prolonged periods of unfavourable conditions, biotic interactions among taxa, and reproductive strategies. Finally, long-term assessments of the community structure and species turnover should be made for selected land areas. We believe that the distribution of the antarctic terrestrial biota is most appropriately and most efficiently studied through a combination of these ecological and survey techniques.

Addendum

Since the present manuscript went to press in 1984, an important paper on the antarctic tardigrade fauna came to our attention. It is: Dastych, H. 1984. The Tardigrada from antarctic with descriptions of several new species. Acta Zool. Cracov. 27: 377–436. Dastych reports 25 species of tardigrades from antarctic mosses and lichens, and describes eight of them as new to science. This represents a considerable increase in the number of species known from the area. Of the 22 localities from which his material came, all but three were on the Antarctic Peninsula, South Shetland Islands, or South Georgia Island, i.e. areas with milder conditions than those of continental Antarctica. Thus, many of his species may only occur in those peripheral localities and the main part of the continent may be relatively impoverished. However, it is also possible that the tardigrade fauna of the entire antarctic region is richer than previously supposed. We have investigations under way that should ascertain which situation exists.

Acknowledgements

We gratefully acknowledge the help of Dr D. Nelson in determining the taxonomic status of the tardigrades. Logistic support was provided by the Australian National Antarctic Research Expeditions during the 1980/81 overwintering period for PH and during the austral summer of 1982–1983 for JDM. Other logistic support for JDM was financed by the Ian Potter Foundation and by Internal Research Grants from the University of New England, Armidale, NSW. Thanks are extended to the helicopter pilots (John Robertson and Colin Wiles) during the 1982–83 summer and to the men of Davis Station for their help with the fieldwork. D. Horning, R. Harden, M. Jones and M. McCoy critically read the manuscript which was typed by S. Higgins.

References

Argue, C. W., 1971. Some terrestrial tardigrades from New Brunswick, Canada. Can. J. Zool. 49: 401–415.

Argue, C. W., 1972. Tardigrades from New Brunswick, Canada. 2. Can. J. Zool. 50: 87–94.

Armitage, K. B. & H. B. House, 1962. A Limnological reconnaissance in the area of McMurdo Sound, Antarctica. Limnol. Oceanogr. 7: 36–41.

Barker, R. J., 1981. Physical and chemical parameters of Deep Lake, Vestfold Hills, Antarctica. ANARE Sci. Rep. Publ. 130. Aust. Govt. Publ. Service, Canberra, 73 pp.

Broady, P. A., 1976. Six new species of terrestrial algae from Signy Island, South Orkney Islands, Antarctica. Br. phycol. J. 11: 387–405.

Burton, H. R. & P. J. Campbell, 1980. The climate of the Vestfold Hills, Davis Station, Antarctica, with a note of its effect on the hydrology of hypersaline Deep Lake. ANARE Sci. Rep. Publ. 129. Aust. Govt. Publ. Service, Canberra, 50 pp.

Claridge, G. G., I. B. Campbell, J. D. Stout & M. E. Dutch, 1971. The occurrence of soil organisms in the Scott Glacier region, Queen Maud Range, Antarctica. N.Z. J. Sci. 14: 306–312.

Dastych, H., 1976. Tardigrada from the Himalaya. Bull. Acad. pol. Sci. 24: 521–523.

Davis, R. C., 1981. Structure and function of two antarctic terrestrial moss communities. Ecol. Monogr. 51: 125–143.

Dodge, C. W., 1965. Lichens. In J. van Mieghem & P. van Oye (eds), Biogeography and Ecology in Antarctica. Junk, The Hague: 194–200.

Dougherty, E. C., 1964. Cultivation and nutrition of micrometazoa. II. An antarctic strain of tardigrade Hypsibius arcticus (Murray, 1907) Marcus 1928. Trans. Am. microsc. Soc. 83(1): 7–11.

Dougherty, E. C. & L. G. Harris, 1963. Antarctic micrometazoa: fresh water species in the McMurdo Sound area. Science 140: 497–498.

Ellsworth, C., E. C. Dougherty, D. J. Ferral & B. Solberg, 1961. Xenic cultivation of antarctic micrometazoa I. The tardigrade Hypsibius arcticus (Murray, 1907) Marcus 1928. Amer. Zool. 1: 350–351.

Everitt, D. A., 1981. An ecological study of an antarctic freshwater pool with particular reference to Tardigrada and Rotifera. Hydrobiologia 83: 225–237.

Filson, R. B., 1966. The Lichens and Mosses of MacRobertson Land. ANARE Sci. Rep. Publ. 82. Antarct. Div. Dept. External Affairs, Melbourne, 169 pp.

Gimingham, C. H., 1967. Quantitative community analysis and bryophyte ecology on Signy Island. Phil. Trans r. Soc., Lond. Ser. B, 252: 251–259.

Gimingham, C. H. & R. I. L. Smith, 1971. Growth form and water relations of mosses in the maritime Antarctic. Br. antarct. Surv. Bull. 25: 1–27.

Greene, S. W., 1967. Bryophyte distribution. In V. C. Bushnell (ed.), Terrestrial Life of Antarctica. Antarctic Map Folio Ser. 5. Am. Geogr. Soc., N.Y.: 11–13.

Greene, S. W., 1968a. Studies in antarctic bryology. 1. A basic check list of mosses. Revue bryol. lichen. 36 (1–2): 132–135.

Greene, S. W., 1968b. Studies in antarctic bryology. 2. Andreaea, Neuroloma. Revue bryol. lichen. 36 (1–2): 136–146.

Greene, S. W., D. M. Greene, P. D. Brown & J. M. Pacey, 1970. Antarctic moss flora 1. The genera Andreaea, Pohlia, Polytrichum, Psilopilum, and Sarconeurum. Br. antarct. Surv. Sci. Rep. 64, 118 pp.

Gustavson, T. C., 1963. A report on collecting, preparing, and mounting tardigrades from central eastern North Dakota. N. Dakota Acad. Sci. 17: 28–31.

Hallas, T. E., 1975. Interstitial water and Tardigrada in a moss cushion. Ann. Zool. Fennici 12: 255–259.

Hallas, T. E., 1977. Survey of the Tardigrada of Finland. Ann. Zool. Fennici 14: 173–183.

Hallas, T. E., 1978. Habitat preference in terrestrial tardigrades. Ann. Zool. Fennici 15: 66–68.

Heatwole, H., T. Done & E. Cameron, 1981. Community ecology of a coral cay, a study of One Tree Island, Great Barrier Reef. Junk, The Hague, 375 pp.

Heywood, R. B., 1977. A limnological survey of the Ablation Point area, Alexander Island, Antarctica. Phil. Trans. r. Soc., Lond. Ser. B, 279: 39–54.

Higgins, R. P. (ed.), 1975. International Symposium on Tardigrades. Pallanza, Italy, 1974. Mem. Ist. ital. Idrobiol. 32 Suppl., 469 pp.

Hirano, M., 1965. Freshwater algae in the antarctic regions. In J. van Mieghem & P. van Oye (eds), Biogeography and Ecology in Antarctica. Junk, The Hague: 127–193.

Horne, P. A., 1984. Terrestrial invertebrates from McDonald Island and Heard Island, sub-antarctic. J. Aust. ent. Soc. 23: 38.

Horning, D. S., R. O. Schuster & A. A. Grigarick, 1978. Tardigrada of New Zealand. N.Z. J. Zool. 5: 185–280.

Jennings, P. G., 1976a. Tardigrada from the Antarctic Peninsula and Scotia Ridge region. Br. antarct. Surv. Bull. 44: 77–95.

Jennings, P. G., 1976b. The Tardigrada of Signy Island, South Orkney Islands, with a note on the Rotifera. Br. antarct. Surv. Bull. 44: 1–25.

Jennings, P. G., 1979. The Signy Island terrestrial reference sites: X. Population dynamics of Tardigrada and Rotifera. Br. antarct. Surv. Bull. 47: 89–105.

Johnstone, G. W., D. J. Lugg & D. A. Brown, 1973. The biology of the Vestfold Hills, Antarctica. ANARE Sci. Rep. Publ. 123. Antarct. Div., Dept. Sci., Melbourne, 62 pp.

Kashiwadani, H., 1970. Lichens of the Prince Olov Coast, Antarctica. Jap. antarct. Res. Exped. Sci. Rep. Ser. E, Publ. 30. Polar Research Center. Natn. Sci. Museum, Tokyo: 1–21.

King, P. E., M. R. Fordy & C. I. Morgan, 1981. The marine flora and fauna of the Isles of Scilly, Tardigrada. J. nat. Hist., Lond. 15: 145–150.

Korotkevich, V. A., 1964. Concerning the population of water bodies in the oasis of east Antarctica. Sov. Antarct. Exped. Info. Bull. 1: 154–161.

Lamb, M., 1962. Antarctic lichens I. The genera Usnea, Ramalina, Himantormia, Alectoria, Cornicularia. Br. antarct. Surv. Sci. Rep. 38: 1–34.

Lamb, M., 1968. Antarctic lichens II. The genera Buellia and Rinodina. Br. antarct. Surv. Sci. Rep. 61: 1–129.

Longton, R. E., 1974. Microclimate and biomass in communities of the Bryum association of Ross Island, Continental Antarctica. The Bryologist 77: 109–127.

Longton, R. E., 1979. Vegetation ecology and classification in the antarctic zone. Can. J. Bot. 57: 2264–2278.

Longton, R. E. & M. W. Holdgate, 1967. Temperature relationships of antarctic vegetation. Phil. Trans. r. Soc., Lond. Ser. B, 252: 237–250.

MacArthur, R. H. & E. O. Wilson, 1963. An equilibrium theory of insular zoogeography. Evolution 17: 373–387.

MacArthur, R. H. & E. O. Wilson, 1967. The Theory of Island Biogeography. Princeton Univ. Press, 203 pp.

Matsuda, T., 1968. Ecological study of the moss community and microorganisms in the vicinity of Syowa Station, Antarctica. Jap. Antarct. Res. Exped. Sci. Rep. Ser. E, 29: 1–57.

Miheleie, F., 1967. Ein Beitrag zur Kenntnis der Tardigrada Argentiniens. Verh. zool.–bot. Ges. Wien. 107: 43–56.

Miller, J. D., 1981. Ecological relationships of tardigrades and other microfauna from sub-antarctic islands and selected areas of the Antarctic, ANARE News, suppl. 1. ANARE Field Reports 1982–83 season: 81.

Miller, W. R., 1975. The taxonomy and distribution of the phylum Tardigrada in Montana west of the continental divide. M.A. Thesis, Univ. of Montana, Missoula, Montana.

Morgan, C. I., 1980. A systematic survey of the Tardigrada from Iceland. Acta Naturalia Islandica 27: 1–25.

Morgan, C. I. & P. E. King, 1976. British Tardigrades. Synop. Br. Fauna 9. The Linnean Soc. Lond. Academic Press, Lond., 133 pp.

Morikawa, K., 1962. Notes on some Tardigrada from the antarctic region. Jap. antarct. Res. Exped. Sci. Rep. Ser. E, Publ. 17. Dept. Polar Research. Natn. Sci. Museum, Tokyo: 3–6.

Murray, J., 1906. Scottish National Antarctic Expedition: Tardigrada of the South Orkneys. Trans. r. Soc. Edinb. 45 (12): 323–334.

Murray, J., 1907. Arctic Tardigrada, collected by Wm. S. Bruce. Trans. r. Soc. Edinb. 45: 669–681.

Murray, J., 1910. Tardigrada. Br. antarct. Exped. 1907–1909 1 (5): 81–185.

Nelson, D. R. (ed.), 1982a. Proceedings of the 3rd International Symposium on the Tardigrada, 1980. East Tennessee State University Press, Johnson City, Tennessee.

Nelson, D. R., 1982b. Developmental biology of the Tardigrada. In F. W. Harrison & R. R. Cowden (eds), Developmental Biology of Freshwater Invertebrates. Alan Liss Inc., N.Y.: 363–398.

Nelson, D. R. & D. S. Horning, 1979. Tardigrada of the Kowhai Bush, Kaikoura, New Zealand. In B. Weglarska (ed.), 2nd International Symposium on tardigrades, Krakow, Poland, 1977. Zesz. Nauk. Uniw. Jagiell. Pr. Zool. Krakow 25: 125–142.

Parker, B. C., G. I. Samsell & G. W. Prescott, 1972. Freshwater algae of the Antarctic Peninsula. 1. Systematics and ecology in the U.S. Palmer Station area. In G. A. Llano (ed.), Antarctic Terrestrial Biology. Antarct. Res. Ser. 20. Am. Geophys. Union, Natn. Acad. Sci.: 69–81.

Pasa, M. V. D. & W. Maucci, 1979. Moss Tardigrada from the Scandinavian peninsula. In B. Weglarska (ed.), 2nd International Symposium on tardigrades, Krakow, Poland, 1977. Zesz. Nauk. Uniw. Jagiell. Pr. Zool. Krakow 25: 47–85.

Peterson, B., 1951. The tardigrade fauna of Greenland. A faunistic study with some ecological remarks. Meddr. Grønland. 150 (5): 1–94.

Pickard, J. & R. D. Seppelt, 1984. Phytogeography of Antarctica. J. Biogeogr. 11: 83–102.

Puglia, C. R., 1959. Some aspects of the taxonomy, ecology and distribution of the tardigrades with emphasis on the tardigrades of east central Illinois. Ph.D. Thesis, Univ. of Illinois, Urbana, Illinois.

Ramazzotti, G., 1972. Il phylum Tardigrada, seconda edizione. Mem. Ist. ital. Idrobiol. 28: 732 pp.

Ramazzotti, G. & W. Maucci, 1982. A history of tardigrade taxonomy. In D. R. Nelson (ed.), Proceedings of the 3rd International Symposium on the Tardigrada, 1980. East Tennessee State University Press, Johnson City, Tennessee: 11–30.

Rastorfer, J. R., 1972. Comparative physiology of four west antarctic mosses. In G. A. Llano (ed.), Antarctic Terrestrial Biology. Antarct. Res. Ser., 20. Am. Geophys. Union, Natn. Acad. Sci.: 143–161.

Richters, F., 1904. Vorläufiger Bericht über die antarktische Moosfauna. Verh. dt. zool. Ges. 14: 236–239.

Richters, F., 1907. Die Fauna der Moosrasen des Gaussbergs und einiger Südlicher Inseln. Deutsche Südpolar-Expedition 1901–1903. Zoologie 9 (4): 259–302.

Richters, F., 1908. Antarktische Tardigraden. Zool. Anz. 34: 915–916.

Richters, F., 1909. Tardigrada unter Südpol. Zool. Anz. 36: 604–606.

Riggin, T. G., 1962. Tardigrada of Southwest Virginia: with the addition of a description of a new marine species from Florida. Va. agric. Expt. Sta. Tech. Bull. 152: 1–145.

Robinson, H. E., 1972. Observations on the origin and taxonomy of the antarctic moss flora. In G. A. Llano (ed.), Antarctic Terrestrial Biology. Antarct. Res. Ser., 20. Am. Geophys. Union, Natn. Acad. Sci.: 163–177.

Rounsevell, D. E., 1981. A population of Nanorchestes antarcticus (Acari: Prostigmata) at the Vestfold Hills, Antarctica. ANARE Sci. Rep. Publ. 131. Aust. Govt. Publ. Service, Canberra, 100 pp.

Schuster, R. O. & A. A. Grigarick, 1965. Tardigrada from western North America with emphasis on the fauna of California. Univ. Calif. Publ. Zool. 76: 1–67.

Schuster, R. O. & A. A Grigarick, 1970. Tardigrada of Santa Cruz island, California. The Pan-Pacif. Ent. 46: 184–193.

Schuster, R. O. & E. C. Toftner, 1982. Dominican Republic Tardigrada. In D. R. Nelson (ed.), Proceedings of the 3rd International Symposium on the Tardigrada, 1980. East Tennessee State University Press, Johnson City, Tennessee.

Schuster, R. O., D. R. Nelson, A. A. Grigarick & D. Christenberry, 1980. Systematic criteria of the Eutardigrada. Trans. am. micros. Soc. 99 (3): 284–303.

Seppelt, R. D. & D. H. Ashton, 1978. Studies of the ecology of the vegetation at Mawson Station, Antarctica. Aust. J. Ecol. 3: 373–388.

Seppelt, R. D. & P. A. Broady, this volume. Antarctic terrestrial ecosystems: The Vestfold Hills in context. Hydrobiologia.

Seppelt, R. D., P. A. Broady, J. Pickard & D. A. Adamson, this volume. Plants and landscape in the Vestfold Hills, Antarctica. Hydrobiologia.

Sudzuki, M., 1964. On the microfauna of the antarctic region. I. Moss-Water Community at Langhovde. Jap. antarct. Res. Exped. Sci. Rep. Ser. E, Publ. 19. Dept. Polar Research, National Science Museum, Tokyo: 1–41.

Sudzuki, M., 1979. On the microfauna of the antarctic region III. Microbiota of the terrestrial interstices. In T. Matsuda & T. Hoshiai (eds), Proceedings of the Symposium on Terrestrial Ecosystems in the Syowa Station area. Natn. Inst. Polar Res. Memoirs. Spec. Iss.: 104–126.

Taylor, B. W., 1955. The flora, vegetation and soils of Macquarie Island. ANARE Rep. Publ. 19. Antarct. Div., Dept. External Affairs, Melbourne, 192 pp.

Thomas, C. W., 1965. On populations in antarctic meltwater pools. Pacif. Sci. 19: 515–521.

Ugolini, F. C., 1972. Ornithogenic soils of Antarctica. In G. A. Llano (ed.), Antarctic Terrestrial Biology. Antarct. Res. Ser. 20. Am. Geophys. Union, Natn. Acad. Sci.: 181–193.

Van Hoffen, E., 1912a. Tiere und Pflanzen der Heard-Insel. Deutsch Südpolar Exped. 1901–1903, 2: 257–271.

Van Hoffen, E., 1912b. Die Tiere und Pflanzen von Possession-Eiland, der Crozet gruppe. Deutsche Südpolar Exped. 1901–1903, 2: 335–343.

Wace, N. M., 1965. Vascular plants. In J. van Mieghem & P. van Oye (eds), Biogeography and Ecology in Antarctica. Junk, The Hague: 201–266.

Watson, K. C., 1967. The terrestrial Arthropoda of Macquarie Island. ANARE Sci. Rep. Publ. 99. Antarct. Div., Dept. External Affairs, Melbourne, 90 pp.

Weglarska, B. (ed.), 1979. 2nd International Symposium on tardigrades, Krakow, Poland, 1977. Zesz. Nauk. Uniw. Jagiell. Pr. Zool., Krakow 25, 197 pp.

Hydrobiologia 165: 209–212 (1988)
J. M. Ferris, H. R. Burton, G. W. Johnstone & I. A. E. Bayly (eds) Biology of the Vestfold Hills, Antarctica
Kluwer Academic Publishers

Cuticle structure and habitat in the Nanorchestidae (Acari: Prostigmata)

D. E. Rounsevell[1,3] & Penelope Greenslade[2]
[1]*Antarctic Division, Department of Science, Kingston, Tasmania, 7050, Australia;* [2]*C/– CSIRO Division of Entomology, GPO Box 1700, Canberra, ACT, 2601, Australia;* [3]*Present address: Department of Lands Parks and Wildlife, GPO Box 44a, Hobart, Tasmania, 7001, Australia*

Key words: Acari, deserts, Antarctica, soil, cuticle, respiration

Abstract

Mites of the cosmopolitan family Nanorchestidae may numerically dominate soil faunas of both hot and cold deserts. The genera *Nanorchestes* and *Speleorchestes* differ in their distributions with the former more abundant in cold and the latter in hot regions. We suggest that this is related to differences in cuticular structure in the two genera. The cuticle of *Nanorchestes* spp. is elaborated with regularly spaced granulations which are absent from *Speleorchestes* spp. These granulations retain a layer of air over the body which may facilitate cuticular respiration in polar soils that are seasonally waterlogged and increase chances of survival by reducing freezing through direct contact with ice. We discuss the biology of the two genera in terms of the ecological strategies that are selected in hot and cold deserts.

Introduction

Mites of the family Nanorchestidae inhabitat the soil and soil surface and appear to numerically dominate mite faunas of arid areas. We have sampled them in the cold desert of Antarctica (Rounsevell, 1974, 1981) and in the temperate–subtropical desert of Australia (P. Greenslade, unpubl. data) and are able to compare aspects of the biology of members of two genera; *Speleorchestes* and *Nanorchestes.* Both genera occur on most continents but only *Nanorchestes* spp. are found in polar regions where they are the most abundant and widespread soil arthropods (Strandtmann, 1982). *Speleorchestes* spp. are abundant in hot deserts in Africa and North America (Den Heyer & Ryke, 1966; Wallwork, 1972; Whitford & Santos, 1980). In parts of the Great Victoria Desert in Australia, *Speleorchestes* spp. are sympatric with a species of *Nanorchestes* but the latter are mainly restricted to drainage systems. In Antarctica, *Nanorchestes* spp. occur mainly in soils that are seasonally exposed to seepage from melting snow and ice.

Block (1979, 1980) and Block & Somme (1982) have investigated the biochemical and physiological adaptations that allow free-living arthropods to 'super-cool', so avoiding freezing damage and permitting survival at sub-zero soil temperatures. However, a wide range of arthropods have this ability and some other attribute may account for the success of *Nanorchestes* spp. in the Antarctic.

Topsent & Trouessart (1890) first observed granulations on the cuticular striae of *Nanorchestes* using light microscopy. It is well known that *Speleorchestes* lack these granulations (Grandjean, 1939; Theron & Ryke, 1969), although at least some species possess tracheae (Grandjean, 1939) which are absent from *Nanorchestes*, but the adaptive significance of the structural differences between genera has not been discussed. Here we describe the granulations of a species of *Nanorchestes* and consider their possible role in respiration.

Methods

Live specimens of *Nanorchestes antarcticus* Strandtmann were collected from the Vestfold Hills, Antarctica and maintained in frozen soil at temperatures above $-5\,^{\circ}C$ for subsequent electron microscopy. In arid Australia, several undescribed species of *Nanorchestes* and *Speleorchestes* were collected in pitfall traps at Kunoth Paddock, near Alice Springs, and at Serpentine Lakes in the Great Victoria Desert.

Specimens of *N. antarcticus* and of undescribed Australian species of *Speleorchestes* and *Nanorchestes* were prepared for scanning electron microscopy by freeze-drying and then coating with gold under vacuum. Transverse and sagittal sections of a single *N. antarcticus* were prepared for transmission electron microscopy with an ultramicrotome after fixing in glutaraldehyde, post-fixing in osmium tetroxide and embedding in methacrylate.

Results

The *Nanorchestes* and *Speleorchestes* specimens that were examined all had flexible cuticles which were regularly folded over the entire body surface. In *Nanorchestes* spp. these folds or striations were elaborated by the addition of regularly spaced granulations aligned with the crests of the striations (Figs. 1 & 2). The average density of these granulations in *N. antarcticus* was 1×10^6 mm^{-2} and in an undescribed Australian *Nanorchestes* sp. from the Great Victoria Desert, 0.8×10^6 mm^{-2}.

When *N. antarcticus* was immersed in water the cuticle appeared irridescent. This appearance is consistent with the reflection of light from a meniscus and the presence of a thin film of air over the cuticle. The apparent meniscus disappeared when detergent was added to the water. Examination of transverse and sagittal section of *N. antarcticus* under the

Fig. 1. Scanning electron micrograph (3.5 K) of striations and granulations on the cuticle of *Nanorchestes antarcticus.*

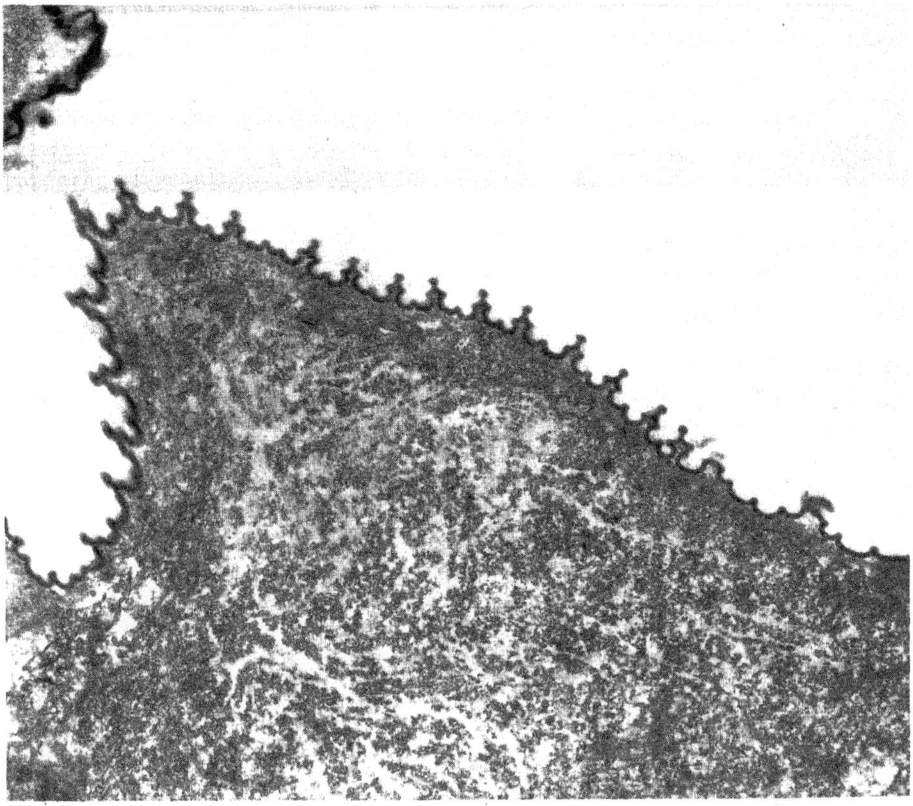

Fig. 2. Transmission electron micrograph (11 K) of *Nanorchestes antarcticus* cuticle showing cross-section of striations.

transmission electron microscope using magnifications up to 64 K failed to show any trachea-like structures. The cuticle was uniformly thin (approximately 0.2 μm) over most of the body surface including the granulations (Fig. 2).

Discussion

The densities of granulations in both Antarctic and Australian *Nanorchestes* spp. lie within the range of $0.5-1.8 \times 10^6$ mm^{-2} recorded by Hinton (1971) for the microtrichia of insect plastrons or spiracular gills. These plastrons support a meniscus and hold a thin layer of air so allowing gaseous transfer between the cuticle and surrounding water and enabling certain insects to respire in water (Hinton, 1969). In Antarctica *Nanorchestes* spp. occur in habitats that are periodically saturated with seepage water during the spring thaw. They do not possess tracheae, which function in air but which cannot function in water unless the stigmata are associated with a plastron. Our observations suggest that respiration in *Nanorchestes* spp. is cuticular and is facilitated by cuticular structures (granulations) analogous to the microtrichia of an insect plastron but covering most of the body. Because mature *Nanorchestes* individuals are very small (length 0.15 – 0.3 mm), cuticular respiration is probably adequate. *Speleorchestes* spp. are larger (length 0.3 – 0.6 mm) and some, at least, possess tracheae while none have a granulate cuticle.

We propose that antarctic *Nanorchestes* spp. and temperate – subtropical desert *Speleorchestes* spp.

are also, to a varying degree, **A**-selected. In both cases the environment is undoubtedly a harsh one for soil invertebrates; both are predominantly arid but they differ in the predictability of inputs of water. A highly predictable, seasonal water supply is characteristic of the habitat of *N. antarcticus*. Accordingly this species, with few if any interacting micro-arthropods, is an extreme **A**-strategist: it is long-lived, with low fecundity and very limited powers of dispersal (Rounsevell, 1974, 1981).

In contrast, the habitat of *Speleorchestes* spp. in central Australia is severe (high temperatures and evaporation rates) but it is much less constant in that the incidence of rain is highly unpredictable. Little is known of the biology of *Speleorchestes* but desert species appear to have some exploitative features. However, they live on or near the soil surface and are saltatorial and so may be sufficiently mobile to avoid immersion in water. Fecundity is higher in *Speleorchestes* spp. (Theron & Ryke, 1969) and they are more mobile than *Nanorchestes* spp.

During the antarctic winter, *N. antarcticus* is dormant in cold, dry soil, just above the permafrost, to depths of 5 cm (Rounsevell, 1981). Here the cuticular structure may confer an additional selective advantage by reducing the risk of freezing through direct contact with ice during initial flooding in the spring.

Although Block (1980) suggested that antarctic micro-arthropods were *K*-strategists in that they were adapted to a stable environment, he found subsequently (Block, 1985) that they display a number of the correlates of **A** or adversity selection (Greenslade, 1983), notably the poor migratory ability, extended life cycles and low reproductive rates that are associated with physically severe but very predictable environments supporting species-poor communities which suggest adaptation to somewhere between the extremes of **A** - and *r*-selection on the habitat templet (Greenslade, 1983).

In addition to life cycle differences between the two genera, we suggest that simple morphological differences in cuticular structure have contributed to different strategies for survival in extreme environments.

Acknowledgements

We thank Ms Sue Eckhert of the Baker Medical Research Institute, J. Nailon of the Department of Microbiology, Monash University and D. Claugher of the British Museum (Natural History) who prepared the specimens and took the photographs.

References

Block, W., 1979. Cold tolerance of micro-arthropods from Alaskan taiga. Ecol. Ent. 4: 103–110.

Block, W., 1980. Survival strategies in polar terrestrial arthropods. J. Biol. 14: 29–38.

Block, W., 1985. Arthropod interactions in an antarctic terrestrial community. In W. R. Siegfried, P. R. Condy & R. M. Laws (eds), Antarctic Nutrient Cycles and Food Webs. Springer-Verlag, Berlin: 614–619.

Block, W. & L. Somme, 1982. Cold hardiness of terrestrial mites at Signy Island, maritime Antarctica. Oikos 38: 157–167.

Den Heyer, J. & P. A. J. Ryke, 1966. A mesofaunal investigation of the soil in a thorntree (Acacia karoo) biotope. Rev. Biol. 5: 309–364.

Grandjean, F., 1939. Quelques genres d'acariens apportenant au groupe des Endeostigmata. Ann. Sci. nat. Zool. Biol. Anim. 2: 1–22.

Greenslade, P. J. M., 1983. Adversity selection and the habitat templet. Am. Nat. 122: 352–365.

Hinton, H. E., 1969. Some structures of insects as seen with the scanning electron microscope. Micron 1: 84–108.

Hinton, H. E., 1971. Plastron respiration in the mite Playtseius italicus. J. Insect Physiol. 17: 1185–1199.

Rounsevell, D. E., 1974. The ecology of the pan-antarctic mite Nanorchestes antarcticus Strandtmann. In G. A. Llano (ed.), Adaptations within Antarctic Ecosystems. Smithsonian Institution, Wash., D.C.: 1023–1033.

Rounsevell, D. E., 1981. A population of Nanorchestes antarcticus (Acari: Prostigmata) at the Vestfold Hills, Antarctica. ANARE Sci. Rep. Publ. 131. Antarct. Div., Dept. Sci., Hobart, 100 pp.

Strandtmann, R. W., 1982. Notes on Nanorchestes. II. Four species from Victoria Land, Antarctica. Pacif. Insects 24: 60–68.

Theron, P. D. & P. A. J. Ryke, 1969. The family Nanorchestidae, Grandjean, (Acari: Prostigmata) with descriptions of new species from South African soils. J. ent. Soc. Sth. Afr. 32: 31–60.

Topsent, F. & E. L. Trouessart, 1890. Sur un nouveau genre d'Acarien sauteur de la Manche. C. r. Séances Soc. Biol. Paris 3: 891–892.

Wallwork, J. A., 1972. Distribution patterns and population dynamics of the micro-arthropods of a desert soil in Southern California. J. Anim. Ecol. 41: 291–310.

Whitford, W. G. & P. F. Santos, 1980. Arthropods and detritus decomposition in desert ecosystems. In D. Dindal (ed.), Soil Biology as related to Land Use Practices. Environment Protection Authority, Washington: 770–778.

Hydrobiologia 165: 213–220 (1988)
J. M. Ferris, H. R. Burton, G. W. Johnstone & I. A. E. Bayly (eds) Biology of the Vestfold Hills, Antarctica
Kluwer Academic Publishers

Foraminiferida from Neogene sediments, Vestfold Hills, Antarctica

P. G. Quilty
Antarctic Division, Department of Science, Channel Highway, Kingston, Tasmania, 7050, Australia

Key words: Neogene, Antarctica, Foraminiferida, Vestfold Hills

Abstract

Sediment samples from several locations in the southern part of the Vestfold Hills have been examined for their foraminiferid content. Those from Early Pliocene sediments of Marine Plain immediately east of Burton Lake contain little foraminiferid evidence of marine deposition. Those from mid-late Holocene sections in and near the lakes contain good open marine faunas of species now known to be living on the continental shelf west of Cape Darnley.

Foraminiferid faunas characterized by calcareous forms are most common, the most prevalent dominated by *Globocassidulina crassa.*

The faunas examined contrast markedly with those in the present Prydz Bay which is characterized by agglutinated forms.

The results, in relation to planktonic faunas, conflict with an earlier record of a diverse fauna.

Introduction

The Vestfold Hills (68°30′S, 78°15′E) is an ice-free oasis of approximately 400 km^2. The area has been the subject of considerable study in recent years and many aspects are thoroughly reviewed in Pickard (1986). It is ice-free because the Sørsdal Glacier, near the edge of the antarctic ice sheet and immediately south of the region, flows to the west, and occupies a trench 750 m deep which controls its direction of flow. The darkness of the rocks helps ensure that it remains ice-free in summer (Adamson & Pickard, 1983).

The area seems to have been glaciated at the peak of the last glaciation (10000–25000 a B.P.). As glaciation declined, the Vestfold Hills became ice-free and what had been glacier-filled valleys evolved to their present condition of marine valleys penetrating the continental margin. Orientation is structurally controlled (Adamson & Pickard, 1983) along zones of relatively softer metamorphic rocks of the 'layered grey gneiss' and 'layered paragneiss' of Oliver *et al.* (1982).

Many 'fjords' are controlled in this way, but the rock type and trend also controls the Death Valley Lake System of Adamson & Pickard (1983), an incipient 'fjord'.

As deglaciation proceeded, a series of lakes was isolated. Local conditions ensured that some have had all saline water removed so that they now collect only molten ice and have become freshwater lakes. At the other extreme, some lakes have had very little freshwater input and have been desiccated by wind from the ice cap. Some of these now have water surfaces well below sea level (50 m in the case of Deep Lake) and salinities as high as 280‰ (eight to nine times normal marine salinity).

Evolution of the landscape through glacial, shallow marine to dry oasis has meant that sediments were initially widespread but have been reduced to a series of isolated pockets; the source of the material for this paper.

While modern sediments in Antarctica which yield foraminiferid faunas are common, Neogene sediments are rare. Apart from those reported here, most other records are from the Ross Sea area (e.g. Kellogg *et al.,* 1979). Perhaps the apparently relict faunas of Lützow-Holm Bay noted by Uchio (1960) would be included in the Quaternary as are those documented by Nomura (1983) and Tai (1984).

Foraminiferida were first described from sediments in the Vestfold Hills by Crespin (1960) who noted the presence of over 30 species, describing six as new. The samples had been collected by the earliest Australian expedition in the area, and came from two sediment sections, originally deposited under marine conditions but now occurring as isolated exposures in the vicinity of lakes isolated from the sea.

The sections providing Crespin's material are:

1. South Terrace, Lake Dingle (C1 on Fig. 1).
2. Terrace about 40 m below sea level, north-west corner, Deep Lake (C2 on Fig. 1).

Voronov (1964) made a passing reference to foraminiferids and other invertebrates from marine terraces 5 m and 15 m above sea level on the Vestfold Hills.

The most important paper to appear to date is that by Setty *et al.* (1980) who examined 72 samples, 36 from each of two transects from approximately lake level in Deep Lake (50.4 m below MSL) to elevations of 54.5 m above lake level (4 m above MSL). The sections were on the western side of the lake (S2 on Fig. 1) and south-eastern corner (S1 on Fig. 1). They recorded 56 species, noting that diversity was high but abundance low.

Their records include 14 planktonic species, apparently in marked contrast with the one or two species recovered from the neighbouring ocean. Interestingly, *Neogloboquadrina pachyderma* (Ehrenberg), often regarded as the only planktonic form in the vicinity of Antarctica, was not recorded.

The distribution of Foraminiferida in modern seas around Antarctica is now reasonably well known following the works resulting from many earlier voyages of exploration such as Brady (1884, HMS *Challenger*), Earland (1933, 1934, 1936, HMS *Discovery,* SY *Scotia,* HMS *Terra Nova*), Parr (1950, BANZARE), Wiesner (1931, Deutsche Südpol Expedition) and so on. In addition, there have been many more recent syntheses such as that by Echols & Kennett (1973). The area encompassed in individual re-

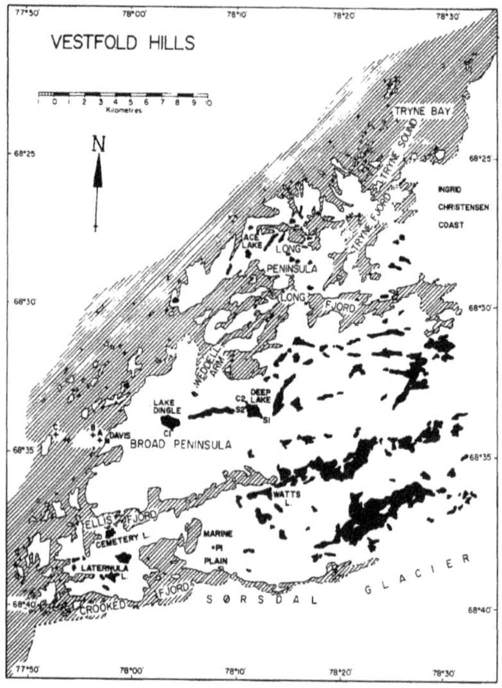

Fig. 1. Map of the Vestfold Hills with locations from which samples, mentioned in the text, were obtained.

cent studies has been less extensive (Anderson, 1975; Milam & Anderson, 1981; McKnight, 1962 and Pflum, 1966). More recently, Quilty (1985) has described the faunal associations in sediment samples from Prydz Bay (including those from the vicinity of Davis Station) and as far west as 80 km west of Mawson Station. Included in that paper are summaries of the results of papers by many authors which have examined foraminiferid associations and their control, especially in relation to carbonate compensation depth/lysocline (CCD) phenomena.

Adamson & Pickard (1983) divided the sediment regimes of the Vestfold Hills region into a series of genetically and temporally related groups.

- the Marine Plain immediately east of Burton Lake – incorporating sediments of age greater than 24000 a, considerably older than any other samples studied (now known to be Early Pliocene, Pickard *et al.,* 1986).
- the Watts Lake System, consisting of depocentres in the south-eastern part of the Vestfold Hills be-

tween Ellis and Crooked Fjords incorporating sediments which mostly date 5 000 – 8 000 a B.P.
- the Death Valley System, consisting of an irregular series of lakes just south of Long Fjord, containing sediments radiocarbon dated at *c.* 5 000 a B.P.
- Laternula Lake – an isolated lake at the western extremity of Mule Peninsula (the peninsula between Ellis and Crooked Fjords). For this paper Cemetery Lake is included in this system.

Zhang *et al.* (1983) briefly recorded foraminiferids from several samples in Watts and Mud Lakes.

For this study, samples were taken from each group; many of those discussed here are localities from which foraminiferids had not previously been studied.

In 1980, J. Pickard collected extensively from Marine Plain (P1 on Fig. 1) and 12 of his samples were examined in this study. Additional samples from Cemetery Lake, Laternula Lake (Laternula Lake System) and Watts Lake (Watts Lake System) were collected by the author in company with H. R. Burton, who also collected samples from a short core in Ace Lake and from the Mirabilite Valley (Death Valley System, South-west of Deep Lake and clearly related to the section examined by Setty *et al.*, 1980).

Undoubtedly more localities will be found to yield valuable information. It is hoped to report on these on an *ad hoc* basis.

All samples were taken as a standard volume (equivalent to about 12 g wet weight of sediment) and washed through 1 mm, 125 μm and 63 μm sieves. Only that portion resting on the 1 mm and 125 μm sieves was studied in detail. The samples were collected as part of the Antarctic Division's continuing program in the Vestfold Hills region.

Foraminiferid number is taken here to be the number of specimens per g of sediment. Diversity is expressed as the number of species constituting 95% of the fauna (Walton, 1964).

Results

Marine Plain

Pickard (1984) sampled Marine Plain extensively. I have examined 12 samples (for locality details see Pickard) from three transects. Transect 1 (down the main scarp of the Marine Plain) yielded seven samples; Transect 2, four samples, and holes on a northern extension of Transect 2, one sample. The localities are shown in Table 1.

During preliminary examination of the samples a slightly overgrown test of *Neogloboquadrina pachyderma* (Ehrenberg) was noted. Most samples contained siliceous sponge spicules, some shell material and a few diatoms. Enough evidence is therefore available to show that the sediment was deposited in a fully marine environment, and that calcium carbonate skeletons have not been dissolved. Deposition occurred above the CCD and in an environment deep enough to allow an occasional planktonic foraminiferid to accumulate.

Adamson & Pickard (1983) showed that the Marine Plain sediments are significantly older than the others, with a minimum age of 24 000 a B.P. and probably older than 30 000 a B.P. Later work (Pickard *et al.*, 1986) indicates that the Marine Plain sequence is much older, Early Pliocene and this is also referred to by Pickard (1986).

Table 1. Marine Plain sample details (after Pickard, 1984).

Sample No[a]	Section No[b]	Position (m)[c]	Depth (cm)[d]
Transect No. 1			
MP 1	1	1.55	15 – 25
MP 6	6	3.85	50 – 60
MP 10	8	8.90	20
MP 12	9	11.3	20 – 30
MP 15	12	18.05	20
MP 26	14	22.4	5 – 10
MP 30	15	25.6	40
Transect No. 2 (T2)			
MP 50	1	0.5	20 – 25
MP 69	5	10	32 – 40
MP 69B	5	10	55
MP 100	7	14	5
T2 Northern Extension			
MP 91	2	?	20 – 25

[a] Pickard (1984) sample No.; [b] Pickard (1984) section No.; [c] Distance from start of section; [d] Depth from surface in hole or section.

This locality is now known to be a source of a significant marine Pliocene vertebrate fauna including a new genus and species of dolphin.

Ace Lake core samples

Three samples were examined from Ace Lake hole at depths of 0–5 cm, 50–55 cm and 100–105 cm. All were barren of foraminiferids, because either none were initially present, or acidic conditions had led to dissolution of their skeletons.

Mirabilite Valley section T2

The base of this section lies about 40 m below sea level and thus represents almost the oldest sediments in the system. Ten samples were collected between the base and a height of 2.4 m by H. R. Burton. They are from the Death Valley Lake System of Adamson & Pickard (1983) and an uncorrected radiocarbon date of some 5 000 ± 500 a B.P. could be expected.

Within this section there can be recognized a trend reflecting change of environment with time (Table 2). Samples between the base of the section and a height of 0.9 m contain abundant indices of fully marine shallow conditions of deposition with abundant calcareous organisms including serpulids, echinoids, bryozoans, foraminiferids and ostracods. In some samples, this calcareous fauna is supplemented by siliceous sponge spicules and diatoms (*Triceratium* sp. in the base sample).

Table 2. Mirabilite Valley Section T2.

Height above base of Section (m)	Comments
2.42	Barren
2.40	Rare serpulids, sponges
2.05	Abundant sponge, spicules, *Triceratium* sp.
2.0	Abundant sponge spicules, diatoms agglutinated forams
0.9	Serpulids, bryozoans, echinoid spines, rare sponge spicules, forams, ostracods
0.8	Serpulids, sponge spicules, forams, echinoid spines, ostracods, polychaete jaws
0.6	Sponge spicules, diatoms
0.5	Abundant serpulids, echinoid spines, bryozoans, sponge spicules, forams, ostracods
0.3	Serpulids, echinoid spines, forams
Base	Bryozoans, forams, ostracods, *Triceratium* sp.

All are benthic organisms. The lack of a significant planktonic element suggests shallow inshore conditions.

The upper part of the section (2.0–2.42 m) shows quite marked contrast. Other than serpulids at 2.4 m it is apparently devoid of calcareous organisms, being dominated by sponge spicules and diatoms towards the base. The top sample is barren of fossils. The sample at 2.0 m contains agglutinated foraminiferids, one of the very few such samples in fossil remains from the Vestfold Hills.

Although foraminiferid numbers are low (generally 2–7 g^{-1} where present), they are high by the standards applying to Vestfold Hills samples. They are also quite diverse. While all calcareous faunas are dominated by *Globocassidulina crassa* (d'Orbigny) the additional fauna includes several miliolids and polymorphinids as well as *Eponides bradyi* Earland, *Cibicides refulgens* Montfort, *Patellina corrugata* Williamson, *Rosalina* spp. and *Astrononion antarcticum* Parr. Ostracod valves are approximately as abundant as foraminiferids and diversity and quality of preservation are high. They are worthy of study.

No trend is obvious in grain size or composition.

In addition to these samples at the base of the Death Valley System sediments, H. R. Burton has collected at the top of the sequence, about 4.5 m above sea level. Three samples were examined at this level. One (∝) is at a terrace edge, one on the terrace above and one just below. That above the terrace is barren of foraminiferids, possibly due largely to wind removal. The marine indices include bivalves and echinoids. The other samples both have a foraminiferid number of about one. Sample ∝ contains a few foraminiferids of two species, *Globocassidulina crassa* and *Cibicides refulgens*. The sample below the terrace is dominated by miliolids, especially *Miliolinella subrotunda* (Montagu), and with lower content of the two species listed above. Bryozoans, ostracod valves and serpulids are additional shallow marine indices.

Watts Lake

Two samples were collected in Watts Lake and represent two very different environments of marine deposition.

The first, from a serpulid bed (probably that dated by Adamson & Pickard at 6150 ± 95 a or 7680 ± 120 a, Fig. 2c) is dominated by *Cibicides refulgens, Globocassidulina crassa* and *Rosalina vilardeboana* d'Orbigny and has a foraminiferid number of eight. Also present were quite abundant ostracods, bryozoans, echinoids, bivalves and gastropods. This was a shallow water, fully marine environment.

The other sample is from an algal bed (Fig. 2d). Marine indices include echinoids and very rare gastropods. No foraminiferids were present. All are benthic organisms. The lack of a significant planktonic element suggests shallow inshore conditions.

Laternula and Cemetery lakes

Although Adamson & Pickard referred to dates from Laternula Lake only, it is highly likely that the neighbouring Cemetery Lake is part of the same system.

The Laternula Lake sample was taken from sediments containing abundant *Laternula* remains (Fig. 2a, b). The fauna recovered included marine indices such as gastropods, polychaetes, ostracod valves, radiolaria and diatoms. *Globocassidulina crassa* is the dominant foraminiferid with *Cibicides refulgens* and *Cassidulinoides porrectus* (Heron-Allen & Earland) as minor constituents. Few foraminiferids were present, about 1 g^{-1}.

The Cemetery Lake sample is from inside the paired valves of *Laternula* still in its living position. The only specimen recovered was the foraminiferid *Globocassidulina crassa.* Apart from *Laternula,* other marine organisms include gastropods, echinoids and the diatom *Triceratium* sp.

Discussion

Planktonic content and diversity

Setty *et al.* (1980) identified 14 species of planktonic foraminiferids and illustrated 10 from the 72 samples they examined from the margins of Deep Lake. These sediments are part of a depositional regime that also included the Mirabilite Valley samples and the upper terrace sediments. The samples, part of the Death Valley System of Adamson & Pickard (1983), contain sediments which have an uncorrected radiocarbon date of 5000 ± 500 a B.P. There is no accepted evidence to suggest that globally there has been any significant change in climate between that time and now. Therefore, it is expected that the composition of the planktonic foraminiferid fauna will remain approximately the same, that is, dominated by *N. pachyderma* with lesser *Globigerina antarctica* Keany & Kennett and *G. bulloides* d'Orbigny.

The fauna identified by Setty *et al.* (1980) is difficult to reconcile with the above comment. For example, *Globigerinoides ruber* (d'Orbigny) and *G. trilobus* (Reuss) suggests significantly warmer conditions than currently exist. The absence of *N. pachyderma,* normally so dominant, is very difficult to explain. Also, the possibility of these being significantly older and reworked is low or it should be reflected in the radiocarbon dates.

No planktonic species were identified in this study even though the sediments examined are a continuation of those reported by Setty *et al.* (*op. cit.*).

Trends

Setty *et al.* (*op. cit.*) stated that no trends of change were observable in their faunas and presented as possible explanations the

- change in wind characteristics
- non *in situ* deposition of foraminiferids
- uneven physiography

The apparent trend in the Mirabilite Valley samples is in marked contrast to these statements. Only detailed restudy of the faunas can resolve this conflict.

Phases of sedimentation

It is becoming clear that four episodes of sedimentation can be recognized.

1. Marine Plain – Early Pliocene.

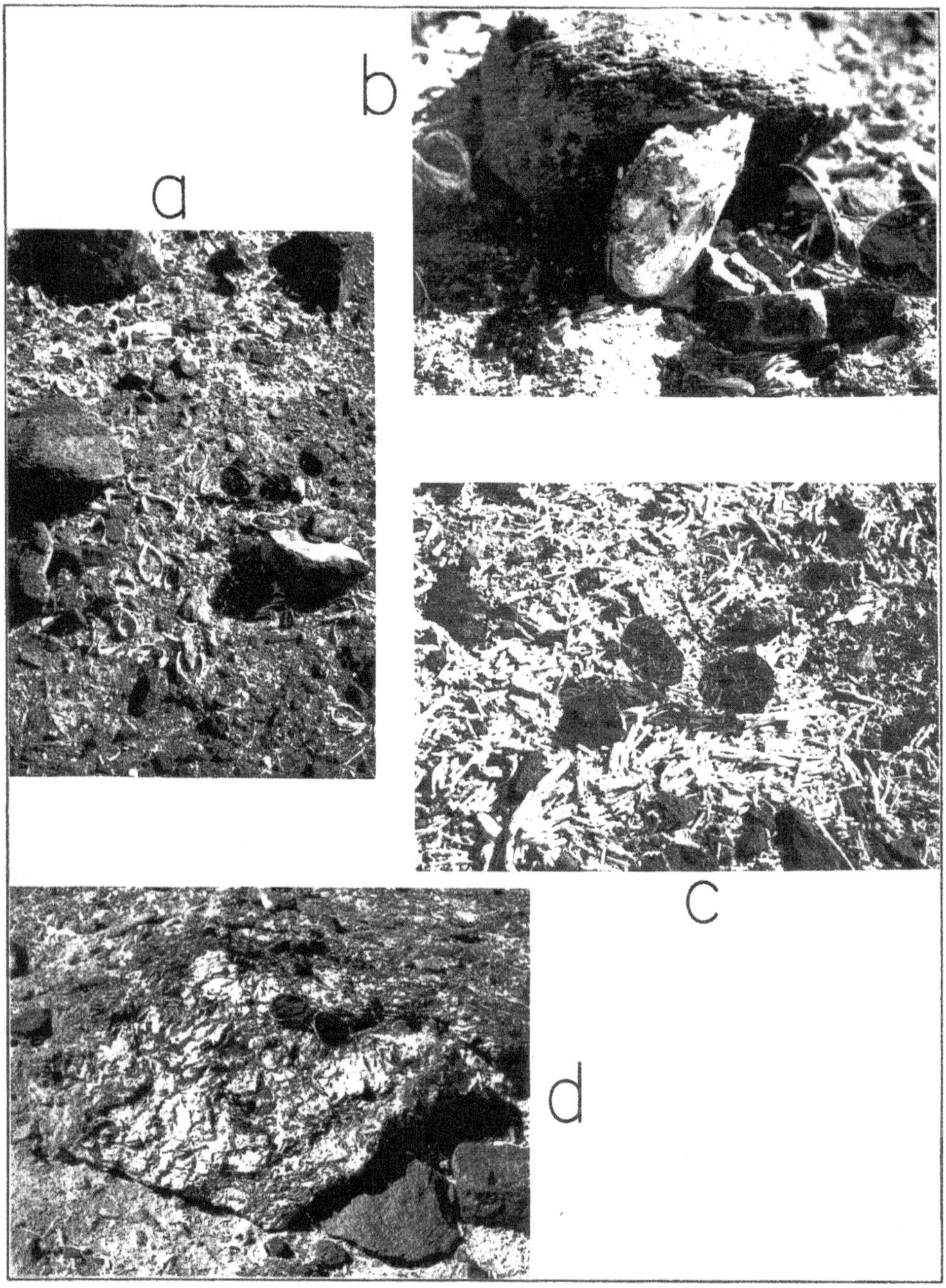

Fig. 2. In situ facies of some surface sediments, sunglasses are present for scale in each case. a. Horizontal surface showing truncated, articulated valves of *Laternula elliptica* (King & Broderip) in living vertical position, Cemetery Lake. b. Exhumed *L. elliptica* in living orientation, Cemetery Lake. c. Serpulid worm tubes, Watts Lake. d. Laminated algal mat, Watts Lake.

2. Death Valley Lake System – 5 000 ± 500 a B.P.
3. Laternula Lake System – 2 500 – 3 500 a B.P.
4. The present marine system.

The position of the Watts Lake System (5 000 – 8 000 a B.P. estimated by Adamson & Pickard, 1983) in this sequence is uncertain. Whether it embraces the Death Valley Lake System or is totally separate is not clear.

The Marine Plain 'fauna' is akin to that accumulating in Prydz Bay today with few foraminiferids and practically no calcareous species.

The other sections all contain benthic calcareous faunas with species identical with those now well known on the continental shelf west of Cape Darnley (Quilty, 1985), especially in the *Globocassidulina crassa* dominant fauna.

This paper does not take into account the recent papers by Li & He (1985) and Gou & Li (1985). The reason for excluding the papers is that the former claims to define many new species of foraminiferids which are named but not illustrated, thus constituting *nomina nuda.*

References

Adamson, D. & J. Pickard, 1983. Late Quaternary ice movement across the Vestfold Hills, East Antarctica. In R. L. Oliver, P. R. James & J. B. Jago (eds), Antarctic Earth Science. Aust. Acad. Sci., Canberra: 465 – 469.

Anderson, J. B., 1975. Ecology and distribution of foraminifera in the Weddell Sea of Antarctica. Micropaleontology 21: 69 – 96.

Brady, H. B., 1884. Report on the foraminifera dredged by HMS Challenger during the years 1873 – 1876. Rep. Voy. Challenger, Zool. 9: 1 – 814.

Crespin, I., 1960. Some recent foraminifera from Vestfold Hills, Antarctica. Spec. Vol. Sci. Rep. Geol. Dept., Tohoku Univ. 4: 19 – 31.

Earland, A., 1933. Foraminifera, Part II, South Georgia. Discovery Rep. 7: 27 – 138.

Earland, A., 1934. Foraminifera, Part III, the Falklands sector of the Antarctic (excluding South Georgia). Discovery Rep. 10: 1 – 208.

Earland, A., 1936. Foraminifera, Part IV, additional records from the Weddell Sea sector from material obtained by the S.Y. 'Scotia.' Discovery Rep. 13: 1 – 76.

Echols, R. J. & J. P. Kennett, 1973. Distribution of foraminifera in the surface sediments. In V. C. Bushnell (ed.), Marine Sediments of the Southern Ocean. Antarct. Map. Folio Ser., 17. Am. geogrl Soc., N.Y.: 13 – 17.

Gou, Y. & Y. Li, 1985. The Holocene ostracoda from section DW1 of Lake Watts in the Vestfold Hills, Antarctica. In A Collection of Antarctic Scientific Explorations 1980 – 1984, Vol. 1. Natn. Comm. Antarct. Res., People's Republic of China: 14 – 19.

Heron-Allen, E. & A. Earland, 1922. Protozoa, Part II, Foraminifera. Br. Antarct. ('Terra Nova') Exped., 1910. Nat. His. Rep., Zool. 6: 25 – 268.

Kellogg, T. B., L. E. Osterman & M. Stuiver, 1979. Late Quaternary sedimentology and benthic foraminiferal paleoecology of the Ross Sea, Antarctica. J. foramin. Res. 9: 323 – 335.

Li, Y. & X. He, 1985. Preliminary study on the Holocene Foraminifera assemblages and their ecological environment in the Vestfold Hills, Antarctica. In A Collection of Antarctic Scientific Explorations 1980 – 1984, Vol. 1. Natn. Comm. Antarct. Res., People's Republic of China: 20 – 21.

McKnight, W. M., 1962. The distribution of foraminifera off parts of the Antarctic coast. Bull. Am. Paleont. 44: 65 – 158.

Milam, R. W. & J. B. Anderson, 1981. Distribution and ecology of recent benthonic foraminifera of the Adelie-George V continental shelf and slope, Antarctica. Mar. Micropaleont. 6: 297 – 325.

Nomura, R., 1983. Foraminifera from the raised beach deposits on the east coast of Lützow-Holm Bay, Antarctica. Proc. 3rd Symposium Antarctic Geoscience, 1982 (N.I.P.R.): 219 – 230.

Oliver, R. L., P. R. James, K. D. Collerson & A. B. Ryan, 1982. Precambrian geologic relationships in the Vestfold Hills, Antarctica. In C. Craddock (ed.), Antarctic Geoscience. University of Wisconsin Press, Madison: 435 – 444.

Parr, W. J., 1950. Foraminifera. Rep. Br. Aust. N.Z. antarct. Res. Exped. 1929 – 1931. 5: 235 – 392.

Pflum, C. E., 1966. The distribution of foraminifera in the eastern Ross Sea, Amundsen Sea and Bellingshausen Sea, Antarctica. Bull. Am. Paleont. 50: 151 – 209.

Pickard, J., 1984. Landscape development and vegetation of the Vestfold Hills, East Antarctica. Ph.D. Thesis Macquarie Uni., Sydney.

Pickard, J. (ed.) 1986. Antarctic oasis. Terrestrial environments and history of the Vestfold Hills. Academic Press, 367 pp.

Pickard, J., A. Adamson, D. M. Harwood, G. H. Miller, P. G. Quilty & R. K. Dell, 1986. Early Pliocene marine sediments in the Vestfold Hills, East Antarctica: Implications for coastline, ice sheet and climate. South African J. Sci. 82: 520 – 521.

Quilty, P. G., 1985. Distribution of foraminiferids in sediments of Prydz Bay, Antarctica. Spec. Publ. S. Aust. Dept. Mines & Energy 5: 329 – 340.

Setty, M. G. A. P., R. Williams & K. R. Kerry, 1980. Foraminifera from the Deep Lake terraces, Vestfold Hills, Antarctica. J. foramin. Res. 10: 303 – 312.

Tai, Y., 1984. Benthic foraminifera from the Antarctic Ocean bottom sediments off Syowa and Mawson Stations, Antarctica. Proc. 4th Symposium Antarctic Geoscience, 1983 (N.I.P.R.): 93.

Uchio, T., 1960. Benthonic foraminifera of the Antarctic Ocean. Biol. Results Japan. Antarct. Res. Exped. 12: 3 – 20.

Voronov, P. S., 1964. On the geomorphology of East Antarctica. Sov. Antarct. Exped. Info. Bull. 1: 20 – 24.

Walton, W. R., 1964. Recent foraminiferal ecology and paleoecology. In J. Imbrie & N. Newell (eds), Approaches to Paleoecology, J. Wiley & Sons, N.Y.: 151–237.

Wiesner, H., 1931. Die Foraminiferen. Deutsche Südpolar-Exped. 1901–1903 20: 53–165.

Zhang, Q., Y. Xie & Y. Lie, 1983. A preliminary study of the evolution of the post Late Pleistocene Vestfold Hills environment, East Antarctica. In. R. L. Oliver, P. R. James & J. B. Jago (eds), Antarctic Earth Science, Aust. Acad. Sci., Canberra: 473–477.

Hydrobiologia 165: 221–226 (1988)
J. M. Ferris, H. R. Burton, G. W. Johnstone & I. A. E. Bayly (eds) Biology of the Vestfold Hills, Antarctica
Kluwer Academic Publishers

Changing distribution of late Quaternary terrestrial lacustrine and littoral environments in the Vestfold Hills, Antarctica

J. A. Peterson[1], B. L. Finlayson[2] & Zhang Qingsong[3]
[1]*Department of Geography, Monash University, Clayton, Victoria, 3168, Australia;* [2]*Department of Geography, University of Melbourne, Parkville, Victoria, 3052, Australia;* [3]*Institute of Geography, Academia Sinica, Beijing, China*

Abstract

Relict marine terraces and associated boulder pavements allow the mapping of coastal environments in the Vestfold Hills and adjacent islands approximately 6000 a B.P. These structures are best developed in what were restricted tidal waters and are ascribed to sea-ice dynamics. The retreat of glacier ice, the subsequent flooding of many areas by sea water and the drainage of most of these as isostatic uplift raised the Vestfold Hills in the Holocene, have produced an immense variety of lacustrine environments as well as a diverse marine littoral. Glacial advance and retreat have occurred in the mid-late Holocene and is particularly evident adjacent to the Sørsdal Glacier.

Introduction

During the last glacial stage, ice inundation of the Vestfold Hills is believed by most workers (but see Pickard, 1986) to have been total (e.g. Zhang *et al.*, 1983; Zhang & Peterson, 1984; Adamson & Pickard, 1986). In any case, there is little evidence of the nature and extent of local habitats (e.g. felsenmeer, cryovegetation fossils) during the last glacial stage and the immediate local history of habitats probably dates from ecesis which followed deglaciation and emergence during the post-glacial marine transgression (PMT). The nature and extent of marine, terrestrial and lacustrine habitats during the Holocene was governed by progressive changes in the spatial pattern of emergence due to variation in the relative magnitudes of eustatic sea level rise and glacio-eustatic rebound from place to place. The general trend was that islands became joined to the mainland, and arms of the sea were drained so that some are now landlocked and marked by lines of hypersaline lakes (Hand & Burton, 1981). As the relative significance of low albedo (rock) surfaces increased, an oasis (*sensu* Shumskiy, 1957) or heat island (Lied, 1963) effect established which today is reflected in a climate milder than around most of the Antarctic Ice Cap edge (Burton & Campbell, 1980). The positive feedback associated with antarctic oasis development augments deglaciation rates, and must be included in explanations for regional warming, and of changes in scope for ecesis.

Habitats

Geomorphological controls on the formation of lacustrine habitats determine that some lakes are older than others. The youngest are:

a) those most recently uncovered by shrinking of the northern margin of the Sørsdal Glacier (e.g. Chelnock Lake), and
b) those most recently cut-off by isostatic uplift from dominantly marine influence; for example Rookery Lake (Burton, 1981) and other lakes close to sea level.

The former are marked by the proximity of a series of ice-cored moraine including ice-cored shear moraine, (Pickard, 1983; Zhang *et al.*, 1983) attributed to ice margin advances during the second half of the Holocene (Zhang *et al.*, 1983; Zhang & Peterson,

1984; Adamson & Pickard, 1986). The latter are marked by diminishing and tenuous connection to the sea exercised during the highest tides of summer only. Burton Lake may also be placed in this category. Thus salinity is partly a function of lake age which in turn is a function of period since freedom from ice cover and/or tidal influence, and partly a matter of the extent to which evaporated water is replaced by snow melt.

The age of terrestrial habitats has been broadly inferred from the history of deglaciation and the dating of shallow-water marine molluscs, and the environmental history of the lakes and fjords as documented from the study of lake-bottom sediments, most of which remain to be sampled. Alternatively, the maximum ages for the hypersaline lakes will be inferred from studies of the distribution and rate of glacio-isostatic uplift, as revealed by precise levelling of their rims and now isostatically deformed shorelines, and by further dating.

Raised shorelines (Terraces)

These were mapped from air photo interpretations (see Fig. 1) and in the field work (see Fig. 2). The age of some of the terraces is indicated by Zhang *et al.* (1983, Fig. 1), Adamson & Pickard (1983) and Zhang & Peterson (1984). The terraces are prominent and for the most part well preserved (e.g. Pickard *et al.*, 1986).

In general it appears that, where dated, the most prominent and continuous of the terraces are approximately 6000 a old and 6 m above sea level, although the detailed pattern of any regional variation in this level has yet to be calculated. The distribution (Fig. 3) and morphological characteristics of these terraces, allows us to infer the Vestfold Hills coastal environments about 6000 a B.P. The terraces are surmounted by boulder pavements of appearance apparently not atypical of shores in cold climates where boulders are plentiful. Such pavements appear to result from several processes, the relative significance of which vary from one locality to another. Thus terrace slope, width, and the scope for allowing formation of a prominent terrace front seems to depend on the nature of overlooking cliffs, if any, as well as the frequency of wetting and drying and the intensity of frost action (e.g. see Pickard *et al.*, 1986). These latter factors have to be inferred

Fig. 1. Raised terraces are prominent on air photographs from the Vestfold Hills and formed the basis of geomorphological mapping. The examples given here are on the southern shores of large saline (ice-free) water bodies Club Lake (left) and Lake Jabs (right top), part of the 'Death Valley' system of lakes mentioned by Adamson & Pickard (1983).

Fig. 2. Raised shoreline with boulder pavement on top about 6 m asl around Watts Lake, a former arm of Ellis Fjord.

from the fossil landform and in such cases it is often useful to examine the distribution pattern. The well-developed raised terraces in the Vestfold Hills appear to relate to sheltered arms of the sea where floating sea-ice probably dominated the intertidal zone for nearly all months of the year. Many of the terraces carry boulder pavements. Well-developed modern shoreline boulder pavements have been described by Hansom (1983) who ascribes the packing and smoothing of their component cobbles to the way sea-ice can ride over them during the partial open water season.

The best-developed raised terraces in the Vestfold Hills occur around what would have been the most restricted tidal waters, suggesting that the terraces were moulded in summer by tidal heaving of sea-ice, which, although no longer land-fast, was too land-locked to float away to the open sea. At the same time, ice-push rampart formation, as occurs, for instance, around shallow littoral zones of large lakes in the modern sub-Arctic (e.g. Peterson, 1965) has apparently been inhibited, and this can be attributed to the steepness of the former shore profiles, and to lack of fetch.

The terraces, facing each other across what were narrow straits, today provide the clue for completing the sketch map of off-shore islands. There is a sharp contrast between the sheltered coasts (Fig. 4) with their terraces (Fig. 3) and the exposed coasts without them. It is, incidentally, on these inferred exposed coasts that fossils indicating the nature of any local mid-Holocene penguin rookeries are most likely to be found.

Coastal evolution

Late Pleistocene – early Holocene

The distribution of raised shorelines at the head of Long Fjord, around Tryne Fjord, and in Crooked Fjord, indicates that at the culmination of the post-glacial marine transgression (PMT) the coast in these areas included sectors of glacier ice. Some of the retreat since then would be accounted for by the time-lag until equilibrium between the new grounding line and the glacier margin was established. Further retreat can be accounted for by the 'heat island'

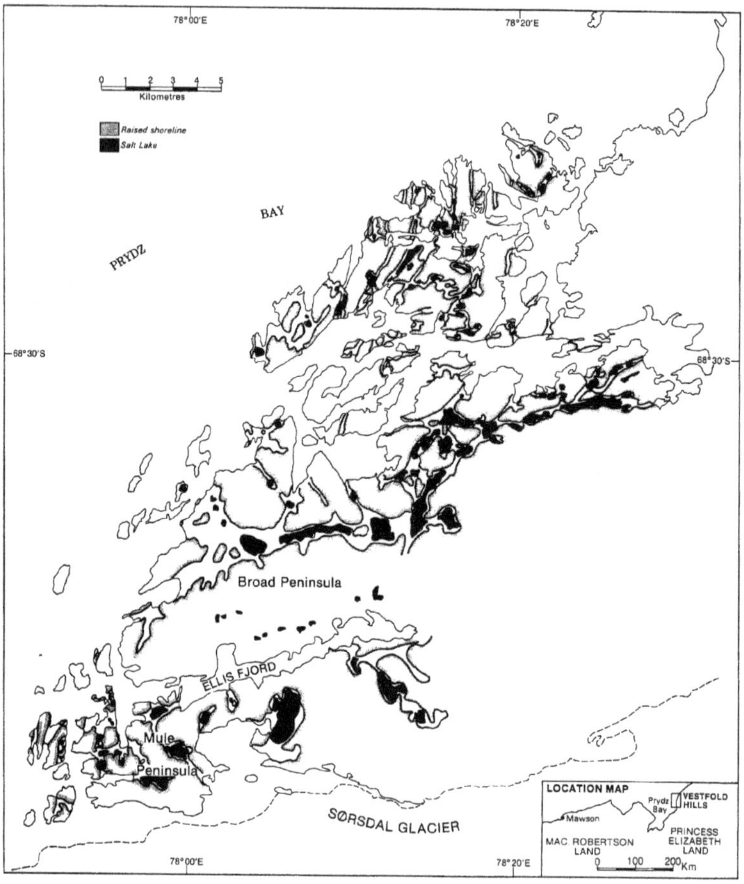

Fig. 3. The distribution of raised terraces, Vestfold Hills, Antarctica. This map was abstracted from the authors' *A. Geomorphic Map of the Vestfold Hills*, which is in preparation.

effect (Lied, 1963) mentioned earlier. In the case of Crooked Fjord however, the ice edge is the north-lateral margin of a major outlet glacier (the Sørsdal) some 700 m thick (Wellman & Williams, 1982). Antarctic continental ice flow rate is greatest on the outlet glaciers, and in their vicinity the 'heat island' effect would be of comparatively minor importance in accounting for ice retreat. The minimal deglaciation date for Clear Lake of about 8,350 a (Adamson & Pickard, 1983) demonstrates that ice margins at the mouth of Crooked Fjord have changed little since then. The disposition of raised shorelines suggests that tidal access to Crooked Fjord may have originally been across comparatively shallow sills, now occupied by Clear Lake and Laternula Lake, after ice-free areas were inundated by the PMT. The Sørsdal glacier margin near here floats in tidal waters and therefore major fluctuations in this area would not be expected. Moreover, with 700 m of ice nearby, the rate of isostatic recovery may not be as high as might otherwise be the case. Emergence will also be inhibited by the effect on sea level over the continental shelf of the Antarctic Ice Cap (Clark *et al.*, 1978).

Late Holocene

The distribution of ice-cored moraine and ice-cored shear moraine remnants (Zhang & Peterson, 1984) shows that the head of Crooked Fjord was closed during the most recent thickening of the Sørsdal glacier. Crooked Lake has formed since retreat from

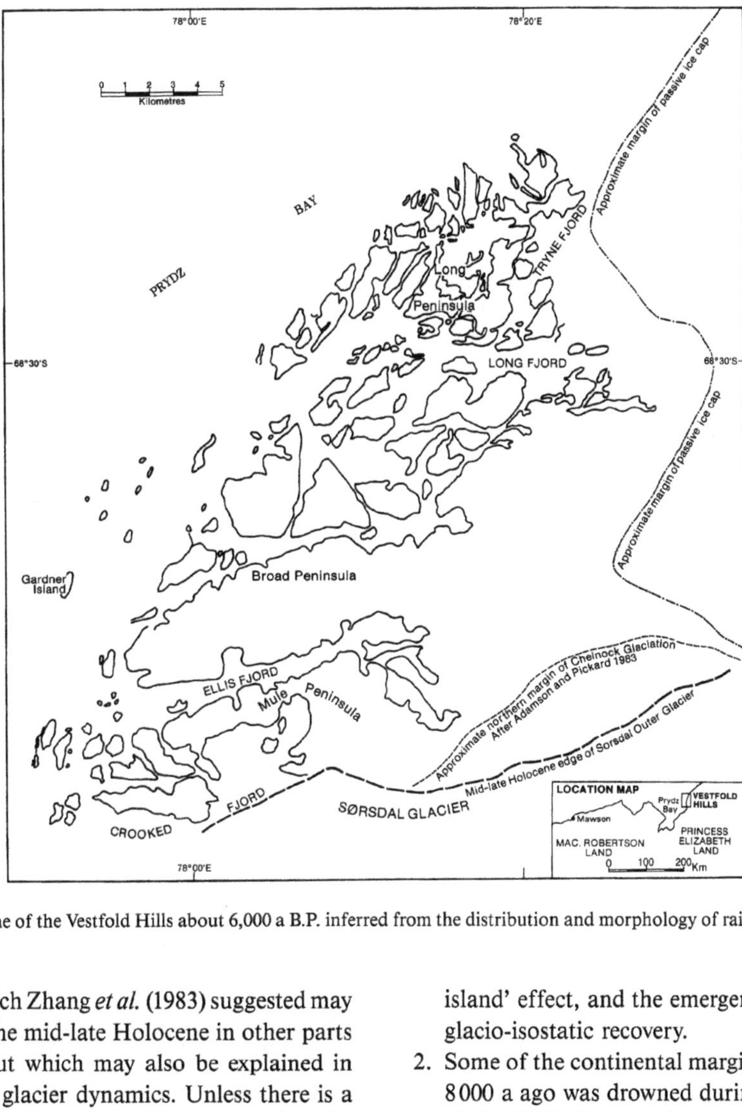

Fig. 4. The coastline of the Vestfold Hills about 6,000 a B.P. inferred from the distribution and morphology of raised shorelines (Fig. 3).

the position which Zhang *et al.* (1983) suggested may correlate with the mid-late Holocene in other parts of the world but which may also be explained in terms of outlet glacier dynamics. Unless there is a surge in the adjacent antarctic ice, or a marked decline in global ice volume, continued isostatic recovery will maintain emergence which will promote further diversity of lacustrine and shallow marine habitats.

Summary

1. There has been a net increase in terrestrial habitat and in the number of lakes over the last 8000 a or so. These changes have been due mainly to retreat of the ice margin in the face of the 'heat island' effect, and the emergence resulting from glacio-isostatic recovery.
2. Some of the continental margin that was ice free 8000 a ago was drowned during the final stages of the PMT. Some such areas may still be under water but most seem to have emerged since then.
3. Terrestrial, marine and lacustrine habitats close to the Sørsdal Glacier have been subject to glacier advance and retreat during the mid-late Holocene.
4. Lakes close to the coast and sea level have probably lost their marine connection comparatively recently and isolated arms of the sea with shallow thresholds are emerging at rates of about 2 mm a^{-1} (Zhang *et al.*, 1983; Zhang & Peterson, 1984).
5. The shallow and narrow arms of the sea at the

culmination of the PMT were perennially ice covered and their distribution can be inferred today from the distribution of raised terrace pairs that would have faced each other across straits too narrow and restricted for clearance of sea ice during the summer break-up period.

6. These terraces are characterized by boulder pavements and were formed by the tidal heaving of land-locked sea-ice during summer. The coasts of the Vestfold Hills at this time were dominated by the presence of off-shore islands. The exposed coasts of these islands were ice-free during summer and their distribution can be inferred by reference to a contour map marked with the position of the terraces.
7. The maximum age of the hypersaline lakes can be inferred from the dating (mainly radiocarbon) of the marine terraces that ring the valleys in which they are found, but only detailed levelling will evaluate further the chronology of basin isolation upon emergence. Detailed levelling of terraces is also required before spatial variation in the pattern of glacio-isostatic recovery can be documented, and more detailed palaeo-coastal maps drawn.

References

Adamson, D. A. & J. Pickard, 1983. Late Quaternary ice movement across the Vestfold Hills, East Antarctica. In R. L. Oliver, P. R. James & J. B. Jago (eds), Antarctic Earth Science. Aust. Acad. Sci., Canberra: 465–469.

Adamson, D. A. & J. Pickard, 1986. Cainozoic history of the Vestfold Hills, In J. Pickard (ed.), Antarctic Oasis. Terrestrial environments and history of the Vestfold Hills. Academic Press, Lond.: 63–97.

Burton, H. R. & P. J. Campbell, 1980. The climate of the Vestfold Hills, Davis Station, Antarctica, with a note on its effect on the hydrology of hypersaline Deep Lake. ANARE Sci. Rep., Publ. 129. Aust. Govt. Publ. Service, Canberra, 50 pp.

Burton, H. R., 1981. Chemistry, physics and evolution of antarctic saline lakes; a review. Hydrobiologia 82: 339–362.

Clark, J. A., W. E. Farrel & W. R. Peltier, 1978. Global changes in postglacial sea level: a numerical calculation. Quat. Res. 9: 265–287.

Hand, R. M. & H. R. Burton, 1981. Microbial ecology of an antarctic saline meromictic lake. Hydrobiologia 182: 363–374.

Hansom, J. D., 1983. Ice-formed intertidal boulder pavements in the sub-Antarctic, J. Sediment. Petrol. 53(1): 135–145.

Lied, N. T., 1963. Notes on sastrugi and snow dune observations. ANARE satellite station, Vestfold Hills, 1961, Australian Meteorological Magazine 40: 35–46.

Peterson, J. A., 1965. Ice-push ramparts in the George River basin. Labrador-Ungava Arctic 18(3): 189–193.

Pickard, J., 1983. Pingos in Antarctica. Quat. Res. 20: 105–109.

Pickard, J., 1986. The Vestfold Hills: a window on Antarctica. In J. Pickard (ed.), Antarctic Oasis. Terrestrial environments and history of the Vestfold Hills. Academic Press, Lond: 333–351.

Pickard, J., D. A. Adamson & C. W. Heath, 1986. The Evolution of Watts Lake, Vestfold Hills, East Antarctica, from marine inlet to freshwater lake. Palaeogeogr. Palaeoclimatol. Palaeoecol. 53: 271–288.

Shumskiy, P. A., 1957. Glaciological and geomorphological reconnaissance in the Antarctic in 1956. J. Glaciol. 3: 56–61.

Wellman, P. & J. W. Williams, 1982. The extent of Archaean and late Proterozoic rocks under the ice cap of Princess Elizabeth Land. BMR J. Geol. Geophys. 7: 217–228.

Zhang Qingsong, Zie Youyu & Li Yuanfang, 1983. A preliminary study of the evolution of the post-late-Pleistocene Vestfold Hills environment, East Antarctica. In R. L. Oliver, P. R. James & J. B. Jago (eds), Antarctic Earth Science. Aust. Acad. Sci., Canberra: 473–477.

Zhang Qingsong & J. A. Peterson, 1984. A geomorphology and Late Quaternary geology of the Vestfold Hills, Antarctica. ANARE Reports 133. Antarct. Div., Dept. Sci., Hobart, 84 pp.

Hydrobiologia 165: 227–237 (1988)
J. M. Ferris, H. R. Burton, G. W. Johnstone & I. A. E. Bayly (eds) Biology of the Vestfold Hills, Antarctica
Kluwer Academic Publishers

Birds of Prydz Bay, Antarctica: Distribution and abundance

T. L. Montague[1]
Antarctic Division, Department of Science, Kingston, Tasmania, 7050, Australia; [1]Present address: Department of Zoology, Monash University, Clayton, Victoria, 3168, Australia

Key words: Antarctica, birds, ecology, petrels, weather, ocean

Abstract

The distribution and abundance of 15 taxa of birds seen from a vessel in the region of Prydz Bay, Antarctica between November and December 1982 are presented. Antarctic petrels were by far the most frequent in the survey area while Adélie and emperor penguins made up 77% of the observed avian biomass. The distribution and abundance of birds was found to be related to ice conditions, air temperature and wind speed but not to water temperature or air pressure. The extent of species associations at sea are shown and the possibilities of diurnal fluctuations in bird numbers at sea are examined.

Introduction

In the austral summer of 1980–1981, the Australian Antarctic Division began a detailed physical and biological study of the Prydz Bay region (Fig. 1) with special reference to the ecology of antarctic krill *Euphausia superba* Dana. The programme was initiated in response to Australia's commitment to the international BIOMASS (Biological Investigations of Marine Antarctic Systems and Stocks) programme, investigating the distribution and abundance of krill. An ecological study of the avian predators of krill, involving the censusing of breeding colonies, monitoring the diet of birds and shipboard censusing, was undertaken as part of this study. While shipboard observations of the birds in Prydz Bay have been made in the past (Johnstone & Kerry, 1976), it was not until 1981 during the FIBEX (First International BIOMASS Experiment) cruise (January–March 1981) that bird observation data for the region were collected concurrently with oceanographic and other biological data.

This paper reports observations on distribution and abundance of birds in Prydz Bay, in November and December 1982.

Methods

Observation techniques

Three hundred and fifty-eight observations were made from the resupply vessel M.V. *Nella Dan* in the Prydz Bay region between 19 November and 19 December 1982. For this study, Prydz Bay was defined as the area within latitudes 60° and 70° South and longitudes 60° and 90° East. The cruise track is shown in Fig. 1. Each observation involved counting the number of individuals of each species seen during a 10 min period. The date, Greenwich Mean Time, ship's activity, position, heading and speed were also recorded, along with a number of environmental variables such as cloud cover, wind speed and direction, air temperature and pressure etc. Observations were made using Zeiss 10 × 40 binoculars from the boat deck 8.2 m above the water. This enabled a line of sight to the horizon, on calm days, of approximately 11 km although for this study only birds occurring within an estimated 1 km radius of the ship were recorded. Codes used to complete the sheets and the methods of observation are given in BIOMASS Working Party on Bird Ecology (1982).

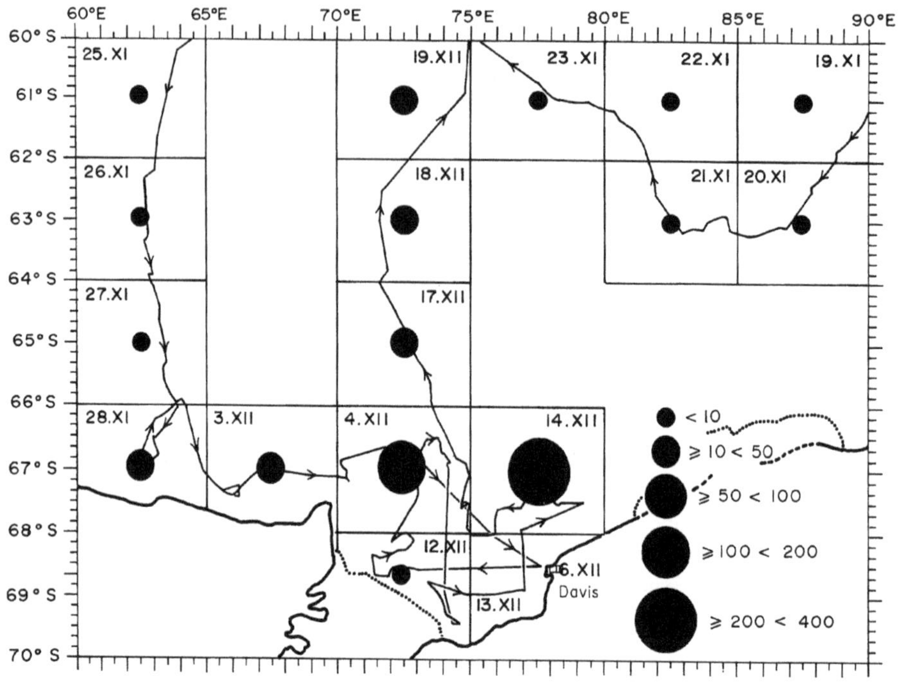

Fig. 1. The cruise track taken during the ABDEX I cruise, November and December 1982 and the mean total abundance of birds in the Prydz Bay region.

The validity of censusing sea birds using shipboard techniques and the problems and considerations involved are discussed in Routh (1949), Bailey (1966), Gill (1967), Bailey & Bourne (1972), Cooke & Mills (1972), Johnstone & Kerry (1976), Dixon (1977), Brown (1980) and Powers (1982). It is important to note that the abundance and biomass of penguins given below are under-estimates because penguins were usually visible only when they were present on ice.

Analyses

The nonparametric Kruskal-Wallis test (Sokal & Rohlf, 1969) was used to compare mean ranked data for total abundance, species diversity and biomass over an abitrary grid (2° Lat. × 5° Long.) to detect spatial variation of each of these parameters. The effect of selected meteorological and oceanographic conditions (air and water temperatures, wind speed, air pressure, wave height and ice cover) on total abundance, species diversity and biomass were similarly examined. Diurnal variation in numbers of birds observed was analysed by dividing the day into 4 h intervals, ranking these intervals according to total or species abundance, and applying the Kruskal-Wallis test.

The extent to which the various bird species were associated was estimated from cluster analysis of a dissimilarity matrix; the matrix was constructed using the Czekanowski dissimilarity coefficient and clustering was achieved using flexible sorting (Clifford & Stephenson, 1975). The groups obtained from this procedure were subsequently confirmed by principal coordinate analysis (Clifford & Stephenson, 1975).

Results

Distribution and abundance

Fifteen taxa were identified from a total of 15543 birds recorded during the 30 days (Table 1). The average number of birds seen per observation was 43.4 (n = 358, SD = 221.5, range 0 to 2319). This included 16 occasions (4%) when no birds were seen. The mean total abundance varied significantly with latitude and longitude (Lat. S: H = 57.23, $P < 0.001$, Long. E: H = 24.72, $P < 0.001$) where H is the statistic used in the Kruskal-Wallis test. Figure 1 shows that birds were most abundant (within the survey area) at latitudes 66° to 68° South but there was not a smooth change in numbers across the latitudes surveyed. The mean total abundance increased with longitude from 60° to 80° East and declined significantly between 80° and 90° East.

The frequencies and percent total abundance of each species seen in the study area are shown in Table 1. Antarctic petrels, Adélie penguins, snow petrels, southern fulmars, cape petrels and emperor penguins constituted 95% of all birds counted. Distribution and abundance plots for each of these species are shown in Figs. 2a–2f. Observations of prions, white-chinned petrels, shearwaters and both species of albatross were restricted to the ice-free, northern grid blocks of Prydz Bay. Cape petrels were seen in every block while antarctic petrels and snow petrels were seen in all but one. Both penguin species, south polar skuas and southern giant-petrels were not seen north of 62 °S and were most numerous south of 66 °S.

The mean species diversity per 10 min observation was 2.88 (n = 341, S.D. = 1.36, range 1–7). Mean species diversity generally increased with latitude and decreased with longitude in grids to the east of 80 °E (Fig. 3). The grid block just north of Davis station was found to have the greatest species diversity.

The mean biomass per observation was 111.8 kg (S.D. = 1047.75, range 0–19009). The two penguin species made up 76.8% of the biomass and were responsible for the increase in biomass with longitude across Prydz Bay from 60 °E to 80 °E and with latitude south to 68 °S (Figs. 2c, 2f & 4).

Table 1. Frequency, % occurrence, body weight, biomass and % biomass of each species seen in Prydz Bay. Except as noted, species body weights taken from Griffiths *et al.* (1982).

Species	Common name	Frequency	%	Body weight (kg)	Biomass (kg)	% biomass
Thalassoica antarctica	Antarctic petrel	7264	46.73	0.70	5804.8	14.25
Pygoscelis adeliae	Adélie penguin	2267	14.59	6.00[a]	13602.0	33.40
Pagodroma nivea	Snow petrel	2019	12.99	0.30	605.7	1.49
Fulmarus glacialoides	Southern fulmar	1964	12.64	1.00	1964.0	4.82
Daption capense	Cape petrel	702	4.52	0.45	315.9	0.78
Aptenodytes forsteri	Emperor penguin	595	3.83	30.00[a]	17850.0	43.83
Oceanites oceanicus	Wilson's storm-petrel	324	2.09	0.04	13.0	0.03
Procellaria aequinoctialis	White-chinned petrel	187	1.20	1.21	226.3	0.56
Pachyptila spp	Prion sp.	103	0.66	0.15	15.1	0.04
Macronectes giganteus	Southern giant-petrel	58	0.37	4.10	237.8	0.58
Stercorarius maccormicki	South polar skua	38	0.24	1.26	47.9	0.12
Phoebetria palpebrata	Light-mantled albatross	13	0.08	2.70	35.1	0.09
Sterna spp	Arctic/antarctic tern	6	0.04	0.14	0.8	0.002
Puffinus spp	Shearwater sp.	2	0.01	0.65[b]	1.3	0.003
Diomedea exulans	Wandering albatross	1	0.01	8.60	8.6	0.02
	Total	15543	100.00			

[a] Body weights taken from Stonehouse (1975).
[b] Weight taken from Montague *et al.* (1986).

(a)

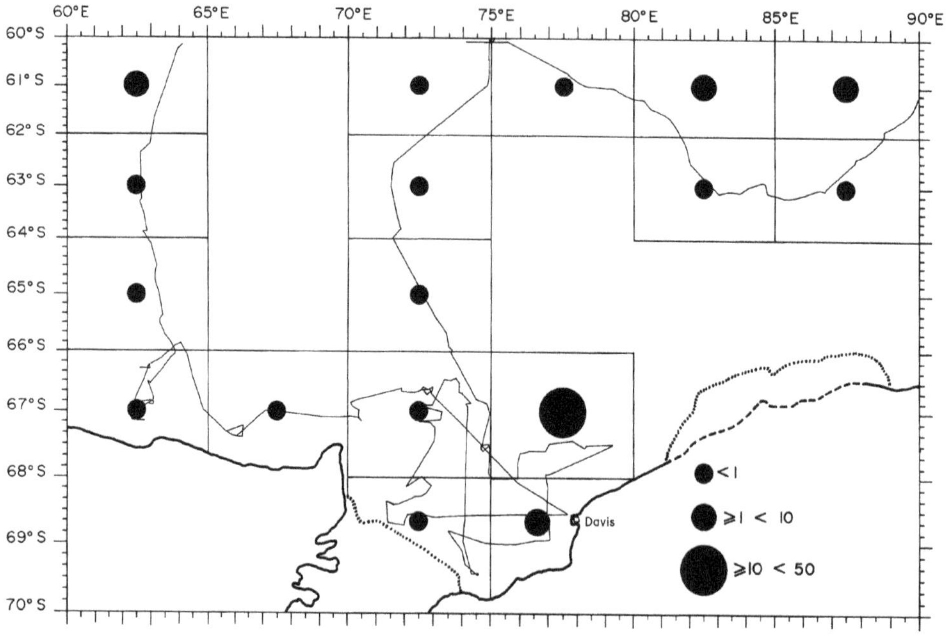

(b)

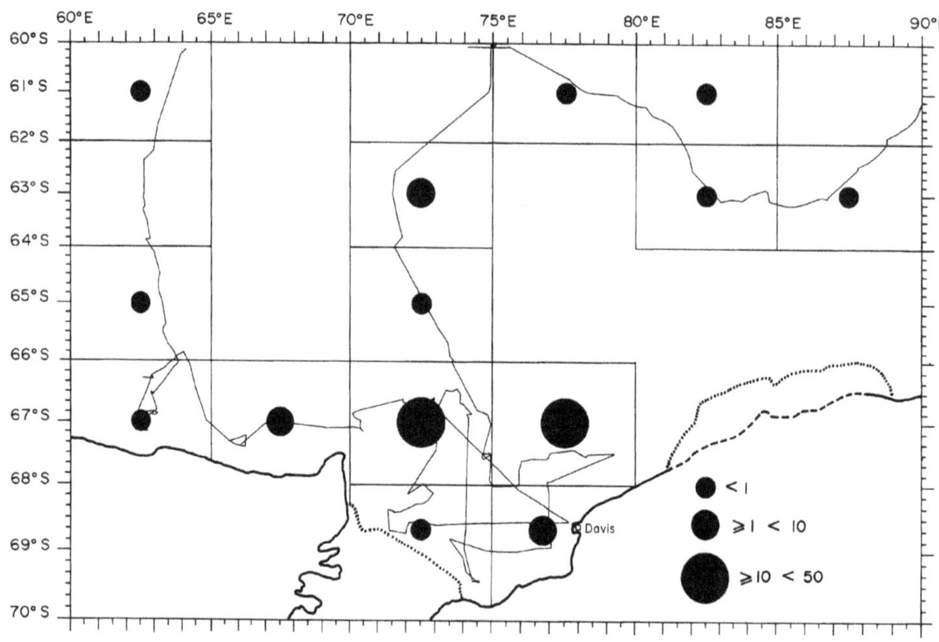

Fig. 2. The average number of birds seen per 10 min observation, (a) cape petrel, (b) southern fulmar (c) emperor penguin (d) antarctic petrel, (e) snow petrel, (f) Adélie penguin.

(c)

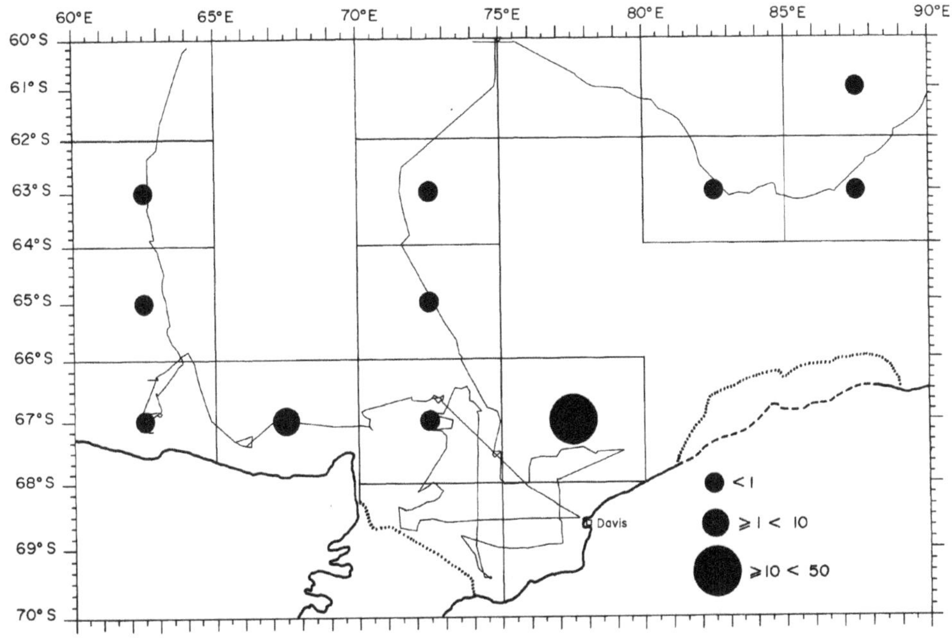
60°E
65°E
70°E
75°E
80°E
85°E
90°E
60°S
61°S
62°S
63°S
64°S
65°S
66°S
67°S
68°S
69°S
70°S
Davis
< 1
≥ 1 < 10
≥ 10 < 50

(d)

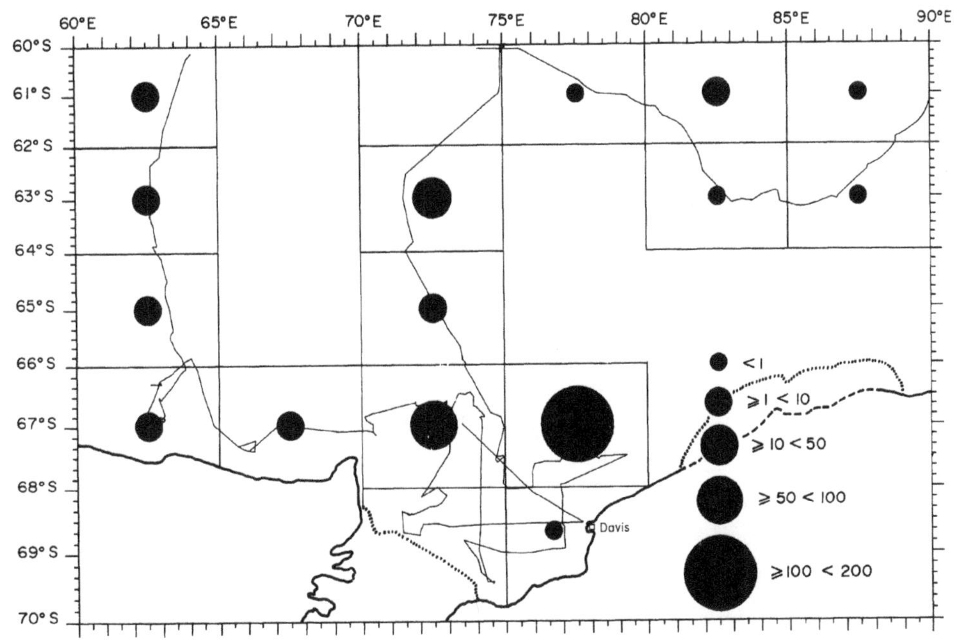
60°E
65°E
70°E
75°E
80°E
85°E
90°E
60°S
61°S
62°S
63°S
64°S
65°S
66°S
67°S
68°S
69°S
70°S
Davis
< 1
≥ 1 < 10
≥ 10 < 50
≥ 50 < 100
≥ 100 < 200

(e)

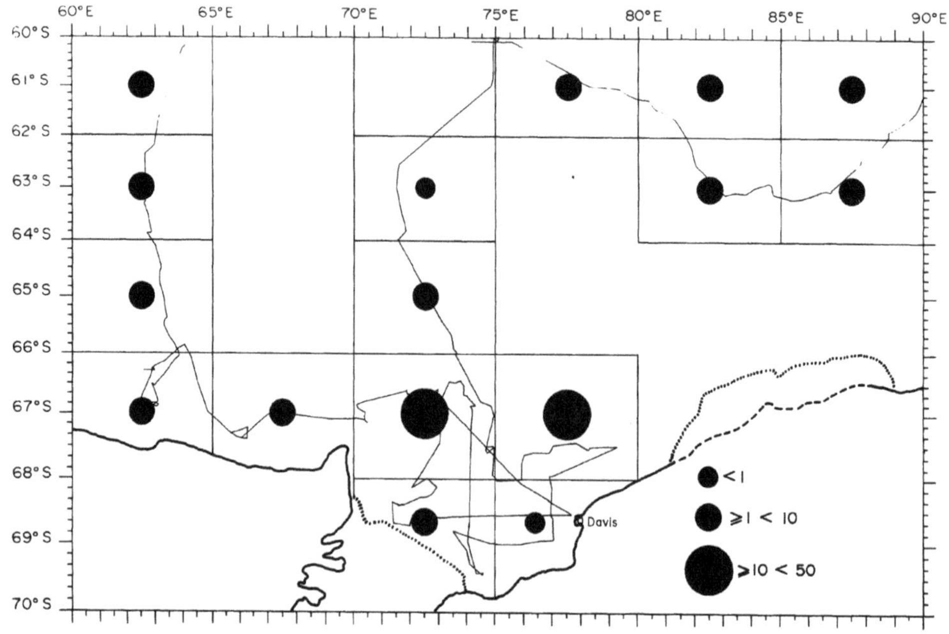
60°E
65°E
70°E
75°E
80°E
85°E
90°E
60°S
61°S
62°S
63°S
64°S
65°S
66°S
67°S
68°S
69°S
70°S
Davis
< 1
≥1 < 10
≥10 < 50

(f)

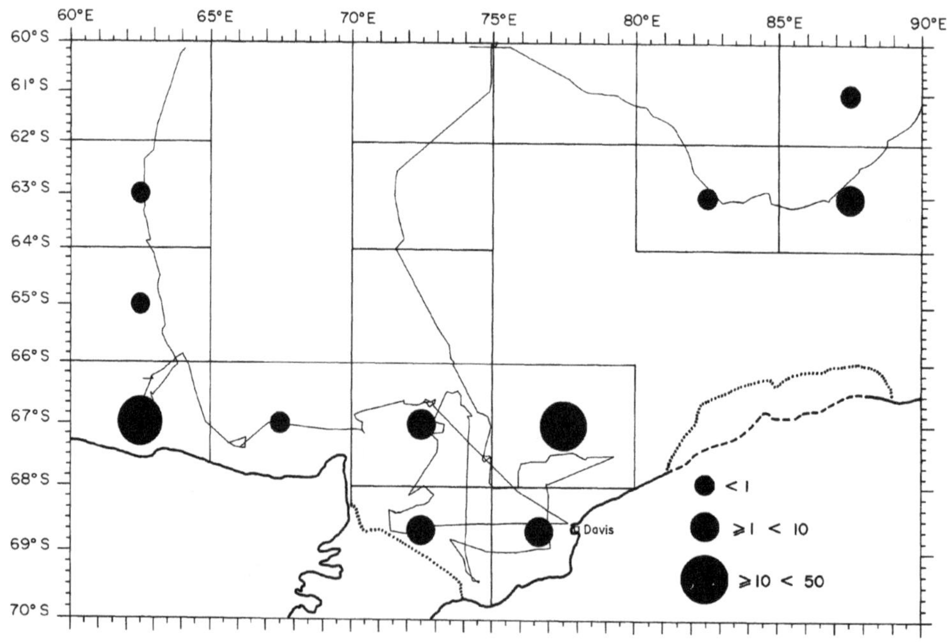
60°E
65°E
70°E
75°E
80°E
85°E
90°E
60°S
61°S
62°S
63°S
64°S
65°S
66°S
67°S
68°S
69°S
70°S
Davis
< 1
≥1 < 10
≥10 < 50

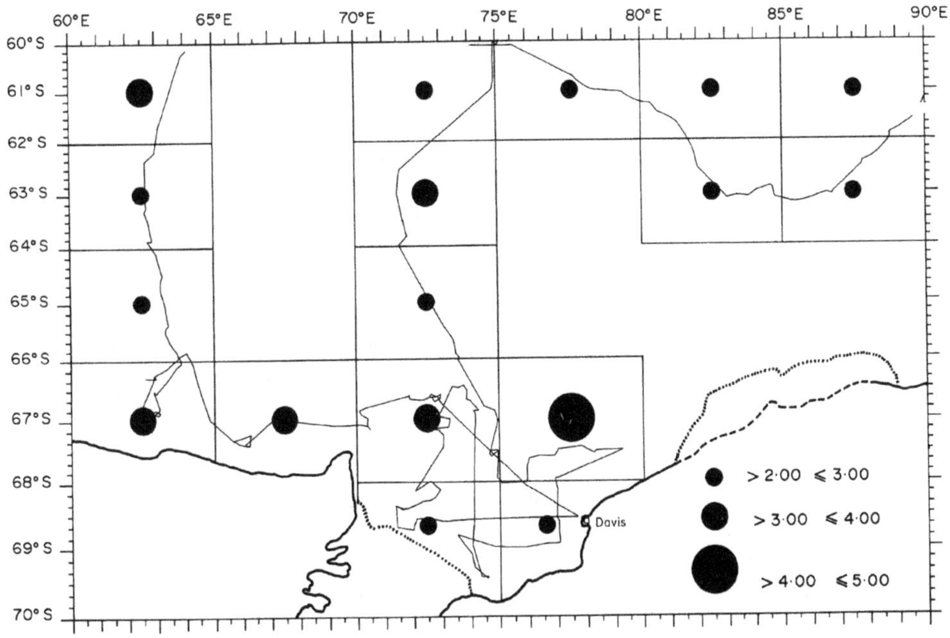

Fig. 3. Mean species diversity of birds seen in Prydz Bay.

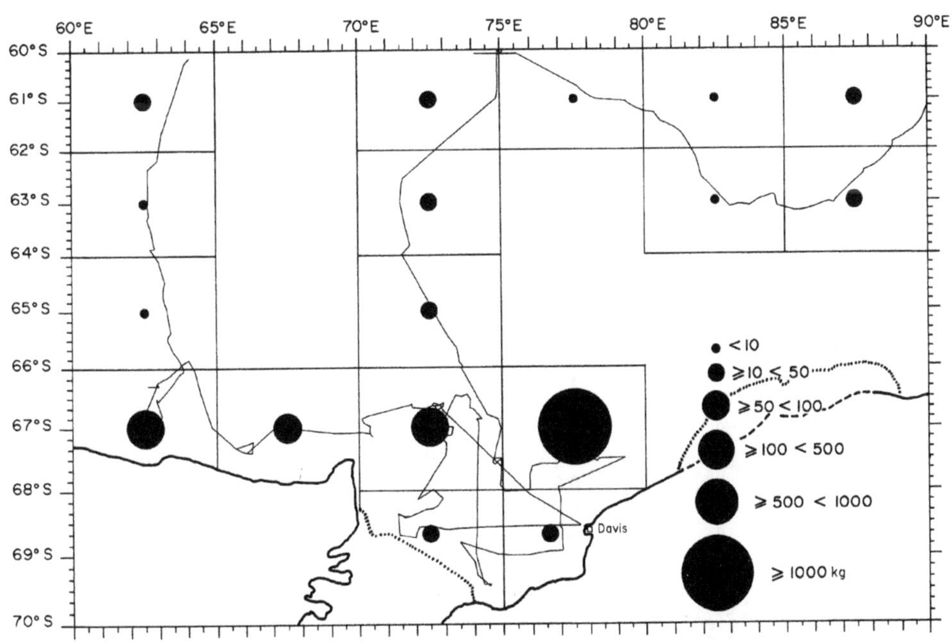

Fig. 4. Mean biomass (kg) of all birds seen in Prydz Bay.

Ice, weather and sea conditions

Sea ice extended to 58 °S and covered 100% of the study area in mid-November 1982 but by mid-December it had receded to south of 65 °S and covered only 25% of the study area (Northern Ice Limit, 1982). The mean total abundance and species diversity were greatest at the edge of the pack-ice (0 – 2 Oktas) while mean biomass was greatest when pack-ice was densest (5 – 8 Oktas). An Okta is one unit on a scale of one to eight. Table 2 shows that prions, wandering albatross, white-chinned petrels and light-mantled albatross were rarely seen in ice conditions greater than 0 – 2 Oktas. Conversely, Adélie and emperor penguins were rarely seen in ice conditions less than this. South polar skuas and southern giant-petrels were rarely seen away from pack-ice. Cape petrels, antarctic petrels and snow petrels were regularly sighted over all ice conditions though numbers were highest at the edge of the pack-ice (0 – 2 Oktas). Southern fulmars occurred in greatest numbers in open water and at the edge of the pack-ice.

Over the study period wind speeds ranged from 0 knots to 27 knots, wave heights 0 m to 6 m, air temperature −7.0 °C to +11.0 °C, air pressure 950 mm Hg to 1 025 mm Hg and sea temperatures −2.0 °C to +3.0 °C (Table 3). There were no significant differences in the total abundance of all birds observed or in the abundance of any of the six most numerous species over the range of wind speeds, water temperatures or air pressures experienced. Total abundance increased significantly with air temperature ($H = 12.57$, $P < 0.05$) and decreased with wave heights less than 0.1 m ($H = 8.6$, $P < 0.05$).

Diurnal rhythms

Sea birds, like many other birds, have been shown to exhibit diurnal rhythms of activity (Cullen, 1954; Moss, 1965; Muller-Schwarze, 1969; Yeates, 1971;

Table 2. Average number per observation (rounded to one decimal place) of individuals of each species seen under various ice conditions in Prydz Bay.

Common name	Bergs			Pack-ice		
	No ice	< 10 bergs	> 10 bergs	0 – 2 Oktas	3 – 5 Oktas	5 – 8 Oktas
Emperor penguin	0	0	1.2	0.3	1.5	4.7
Adélie penguin	2.7	1.0	0.4	6.7	3.7	11.5
Wandering albatross	0	0	0	0	0	0
Light-mantled albatross	0	0.3	0	0	0	0
Southern giant-petrel	0	0	0	0	0	0.4
Southern fulmar	15.7	0.4	0.8	10.4	0.6	0.4
Antarctic petrel	14.9	34.2	3.7	50.9	4.2	10.0
Cape petrel	0.7	0.9	0.5	5.0	0.3	0.2
Snow petrel	1.6	1.3	3.3	19.3	2.2	2.3
Prion sp.	0.1	0.3	0	0	0	0
White-chinned petrel	0.2	5.3	0	0.1	0	0
Shearwater sp.	0	0	0	0	0	0
Wilson's storm-petrel	0.6	0.6	1.0	0.5	1.2	1.2
South polar skua	0	0	0	0	0	0.2
Arctic/antarctic tern	0	0	0	0	0	0
No. of observations made for each ice condition	47	30	37	77	49	125

0 indicates <0.05.

Table 3. Range of physical conditions experienced during observations and their correlation with total bird abundance. H is the statistic used in the Kruskal-Wallis test.

Parameter	Range	H	P	
Wind speed	0,27 knots	5.98	0.425	ns
Wave height	0,6 metres	8.60	0.013	*
Air temperature	−7, +11 °C	12.57	0.027	*
Sea temperature	−2.0, +3.0 °C	5.80	0.215	ns
Air pressure	950, 1025 mm Hg	3.65	0.056	ns
Ice cover	0, 8 oktas	12.79	0.025	*

P = probability. ns = not significant, * = significant at the 5% level.

Paulin & Sagar, 1977). This study was unable to find any significant change in the total number of birds observed at sea between 04:00 h and 23:59 h local time. The period between 00:00 h and 03:59 h was not included in the analysis as less than five observations were made in this interval. Similar analyses for counts of antarctic petrels, southern fulmars, snow petrels, Adélie penguins, cape petrels and Wilson's storm-petrels also gave non-significant results.

Species associations

The birds that commonly occurred together fall into three broadly associated groups. These are Prion sp. and white-chinned petrels; cape petrels, southern fulmars, antarctic petrels, snow petrels and Wilson's storm-petrels; and emperor and Adélie penguins and south polar skuas. The only close associations appear to be between antarctic and snow petrels and between white-chinned petrels and Prion sp.

Discussion

The biomass of seabirds in the Prydz Bay region may be at least equal to that of the world's densest concentrations of seabirds such as those associated with upwellings off the coasts of south west Africa (Benguela), south east Arabia and Peru (Table 4).

The greatest total abundance and species diversity of birds in Prydz Bay occurred in the grid Block between latitudes 66° and 68° South and longitudes 75° and 80° East. This area is on the edge of the continental shelf where the depth drops to approximately 3 000 m, the greatest depth for all blocks surveyed and where the warmest, least saline water in Prydz Bay is found (Smith *et al.*, 1982). Therefore it seems reasonable to suggest that birds concentrate there for food that may be associated with any local front or upwelling that may occur in this grid. Ainley & Jacobs (1981) have suggested that birds aggregate along the continental margin because euphausiids often concentrate in this region (see also Marr, 1962).

The distribution and apparent concentrations of species can be related to ice conditions, weather and water temperature (Manikowski, 1971; Brown, 1980; Ainley & Jacobs, 1981).

Ice conditions were also found to influence the distribution, abundance and diversity of birds in Prydz Bay. Species were most abundant at the edge of the pack-ice (0–2 Oktas), while biomass was greatest in thick ice (5–8 Oktas). The peak in biomass was due to the high penguin numbers in the thick ice. Cline *et al.* (1969) and Routh (1949) reported similar findings for species abundance and biomass in ice, but they gave no explanations as to why such concentrations exist. Brown (1980) proposed that birds concentrate at the edge of the pack-ice due

Table 4. Comparison of the numbers and biomass of birds seen per 10 min observation in Prydz Bay and other 'sea bird rich' areas.

Area	Number per observation	Biomass (kg) per observation	Source
Senegal	5.1	6.1	Brown, 1979
Peru	95.8	99.2	Brown, 1979
Benguela	100+	100+	Summerhayes *et al.*, 1974
S.E. Arabia	25+	8+	Bailey, 1966
Prydz Bay	45.5	120.5	This study

to the increased availability of food.

The total abundance of birds at sea was found to increase significantly with air temperature and decrease with wave heights below 0.1 m. Spurr (1978) showed an inverse relationship between the numbers of incubating Adélie penguins at a breeding colony and air temperature, and as it might be expected that other incubating species will have similar attendance patterns, it is not surprising that fewer birds of all species are seen at sea during their breeding seasons when temperatures are low. However, this study was not able to detect any significant changes in the abundance of any particular species with temperature.

Significantly fewer birds were seen in very calm sea conditions. This may be because birds are less likely to forage over water where they have a greater chance of being detected and avoided by their prey (Dunn, 1975). Although total abundance did not change significantly over the range of wave heights greater than 0.1 m, the numbers actually feeding could have decreased as wave heights increased since birds may avoid feeding in areas where they have difficulty in visually detecting food.

Analysis of species association revealed that birds were seen together in three main groupings: prions and white-chinned petrels; fulmarine petrels, and the penguins and skuas. Whilst the strongest associations of birds in the first two groups were possibly influenced by ice conditions, i.e. prions and white-chinned petrels were rarely found away from the ice (Table 2), the basic reason for the association was probably the similarity in diet and foraging technique of the species involved. There is, as yet, little quantitative information on these activities of antarctic seabirds.

The third grouping, which consisted of emperor penguins, Adélie penguins and south polar skuas existed to some extent as an artifact of the survey method, as penguins were seen when they were on pack-ice but not when they were in the water. Nonetheless, within these limits, an association existed. South polar skuas were possibly found in association with the penguins because skuas commonly prey on injured, sick or young penguins (Young, 1963; Trillmich, 1978; Trivelpiece & Volkman, 1982).

Acknowledgements

I would like to thank G. W. Johnstone, K. R. Kerry, J. M. Cullen, I. Norman, E. J. Woehler, H. R. Burton and others for their constructive comments on the manuscript and D. Ratkowsky for help and advice with statistical matters. J. Cox drew the figures and Y. Hess typed the manuscript.

References

Ainley, D. G. & S. S. Jacobs, 1981. Seabird affinities for ocean and ice boundaries in the Antarctic. Deep Sea Res. 28: 1173–1185.

Bailey, R. S., 1966. The sea birds of the south coast of Arabia. Ibis 108: 224–264.

Bailey, R. S. & W. R. P. Bourne, 1972. Notes on sea-birds 36. Counting at sea. Ardea 60: 124–127.

BIOMASS Working Party on Bird Ecology, 1982. Recording observations of birds at sea. BIOMASS handbook No. 18. Cambridge, 14 pp.

Brown, R. G. B., 1979. Sea birds of the Senegal upwelling and adjacent waters. Ibis 121: 283–292.

Brown, R. G. B., 1980. Seabirds as marine animals. In J. Burger, B. L. Olla & H. E. Winn (eds), Behaviour of Marine Animals, 4. Plenum Press, N.Y.: 1–34.

Clifford, H. T. & W. Stephenson, 1975. An introduction to numerical classification. Academic Press N.Y., 229 pp.

Cline, D. R., D. B. Siniff & A. W. Erickson, 1969. Summer birds of the pack ice in the Weddell Sea, Antarctica. Auk 86: 701–706.

Cooke, F. & E. L. Mills, 1972. Summer distribution of pelagic birds off the coast of Argentina. Ibis 114: 245–251.

Cullen, J. M., 1954. The diurnal rhythm of birds in the Artic summer. Ibis 96: 31–46.

Dixon, T. J., 1977. The distance at which sitting birds can be seen at sea. Ibis 119: 372–375.

Dunn, E. K., 1975. Changes in the fishing abilities of terns associated with windspeed and sea surface conditions. Nature 244: 520–521.

Gill, F. B., 1967. Observations on the pelagic distributions of seabirds in the western Indian Ocean. Proc. U.S. natn. Mus. 123: 1–33.

Griffiths, A. M., W. R. Siegfried & R. W. Abrams, 1982. Ecological structure of a pelagic seabird community in the southern ocean. Polar Biol. 1: 39–46.

Johnstone, G. W. & K. Kerry, 1976. Ornithological observations in the Australian sector of the Southern Ocean. Proc. 16th Inter. Orn. Congress, 1974. Aust. Acad. Sci., Canberra: 725–738.

Manikowski, S., 1971. The influence of meteorological factors on the behaviour of sea birds. Acta zool. cracoviensia 16: 581–657.

Marr, J. W. S., 1962. The natural history and geography of the Antarctic Krill. Discovery Reports 32: 33–464.

Montague, T. L., J. M. Cullen & K. Fitzherbert, 1986. The diet of the short-tailed Shearwater during its breeding season. Emu. 86: 207–213.

Moss, R., 1965. Diurnal rhythms of fulmars (Fulmarus glacialis) in the Arctic autumn. Ibis 107: 533–535.

Muller-Schwarze, D., 1969. Daily pattern of activity of the Adélie penguin during the polar day in Antarctica. Umschau 66: 603–604.

Northern Ice Limit, 1982. Prepared by U.S. Navy-NOAA Joint Ice Center. Svitland, 52 pp.

Paulin, C. P. & P. M. Sagar, 1977. A diurnal rhythm of activity by the Adélie penguin. Notornis 24: 158–160.

Powers, K. D., 1982. A comparison of two methods of counting birds at sea. J. Field Orn. 53(3): 202–304.

Routh, M., 1949. Ornithological observations in the antarctic seas 1946, 1947. Ibis 91: 577–606.

Smith, N. R., Z. Dong, K. R. Kerry & S. Wright, 1982. Water masses and circulation in the region of Prydz Bay, Antarctica. Deep Sea Res. 31: 1121–1147.

Sokal, R. R. & F. J. Rohlf, 1969. Biometry. W. H. Freeman, San Francisco, 776 pp.

Spurr, E., 1978. Diurnal activity of Adélie penguin Pygoscelis adéliae at Cape Bird. Ibis 120: 147–152.

Stonehouse, B., 1975. The biology of penguins. Macmillan Press, N.Y., 555 pp.

Summerhayes, C. P., P. K. Hofmayer & R. H. Roiux, 1974. Seabirds off the S.W. coast of Africa. Ostrich 45: 83–109.

Trillmich, F., 1978. Feeding territories and breeding success of south polar skuas. Auk 95: 23–33.

Trivelpiece, W. & N. J. Volkman, 1982. Feeding strategies of sympatric south polar Catharacta maccormicki and brown skuas C. lonnbergi. Ibis: 124: 50–54.

Yeates, G. W., 1971. Diurnal activity of the Adélie penguin (Pygoscelis adeliae) at Cape Royds, Antarctica. J. nat. Hist., Lond. 5: 103–112.

Young, E. C. Feeding habits of the south polar skua Catharacta maccormicki. Ibis 105: 301–318.

Hydrobiologia 165: 239–253 (1988)
J. M. Ferris, H. R. Burton, G. W. Johnstone & I. A. E. Bayly (eds) Biology of the Vestfold Hills, Antarctica
Kluwer Academic Publishers

The breeding season diet of Adélie penguins at the Vestfold Hills, East Antarctica

R. A. Puddicombe[1,2] & G. W. Johnstone[1]
[1]*Antarctic Division, Department of Science, Kingston, Tasmania, 7050, Australia;* [2]*Present address: Australian National Parks and Wildlife Service, G.P.O. Box 636, Canberra, A.C.T., 2601, Australia*

Key words: Antarctica, Adélie, penguin, diet, krill, fish

Abstract

The stomach contents of adult Adélie penguins (*Pygoscelis adéliae*) were sampled at Magnetic Island, 4.5 km offshore from Davis Station, Antarctica. An intensive sampling program was conducted during the posthatching (mid-December to late February) part of the 1982–1983 breeding season, and during the prehatching (mid-October to mid-December) part of the 1983–1984 breeding season. Combining the samples from both seasons, weekly sampling covered almost the entire 4.5 months the penguins spent ashore at the breeding colonies and resulted in a total of 574 samples.

Samples obtained prior to hatching were very small and consisted mostly of amphipods (14 species represented). During the chick rearing period, the weight of samples increased steadily. *Euphausia crystallorophias* was an important food item throughout the sampling period, particularly from hatching onwards. *Euphasia superba* also occurred regularly. After a brief peak in late December, it was usually present but it was seldom the major component. Fish of ten species, predominantly *Pleuragramma antarcticum*, (up to 195 mm in length), became a significant component when the chicks formed crêches in early January. They became the principal dietary component at fledging in February.

Introduction

The antarctic marine ecosystem has experienced considerable change over the last 80 a with a massive reduction in the number of baleen whales and the more recent development and expansion of the fishery for the antarctic krill, *Euphausia superba* (Dana) (Laws, 1977).

The continually increasing interest in, and exploitation of, the natural resources of the antarctic continent and its surrounding waters by a number of nations requires a comprehensive knowledge of the antarctic ecosystem if that ecosystem is to be effectively managed and conserved. For this reason, the BIOMASS Working Party on Bird Ecology in 1978 instigated the International Survey of Antarctic Seabirds (ISAS) which aimed principally at krill-eating Adélie *Pygoscelis adéliae* (Hombron & Jacquinot) and royal/macaroni *Eudyptes chrysolophus* (Brandt) penguins (Siegfried & Croxall, 1983). These authors noted that penguins 'constitute a major class of predators potentially or actually competing with man and certain other animals, such as whales, for food resources' and quoted estimates of direct and indirect annual krill consumption by antarctic seabirds of up to 115×10^6 t, with penguins the major consumers. This figure is slightly greater than the consumption by cetaceans and two-thirds of the consumption by seals calculated by similar methods (Croxall, 1984).

Siegfried & Croxall (1983) stressed the need for 'more refined assessments of the impacts of seabirds on the main potentially exploitable resources (krill, squid and fish) of the Southern Ocean', and emphasized the need for local studies of diet, population size and breeding dynamics of selected species to investigate to what extent these species could be used to indicate changes in the abundance of marine resources, especially those resulting from commercial exploitation.

Prydz Bay in East Antarctica is at the centre of a nutrient rich oceanic gyre circulating clockwise, south of 63 °S and between 58 °E and 90 °E (Smith *et al.*, 1984). The krill population, thought to be contained by the action of this gyre, at one time supported a large whale population which has been drastically over-harvested and reduced this century. More recently, a significant krill fishery has developed in this same region.

The purpose of this study was to investigate the diet of Adélie penguins in the Prydz Bay region during the breeding season, and in particular, to determine the importance of *E. superba* in their diet. Previously reported data and the results of this study

Table 1. Adélie penguin diet reported by geographical location.

Location	Reference	Foods reported
Scotia Arc/Antarctic Peninsula		
South Orkney Islands	Sladen (1958)	Euphausiids
	White & Conroy (1975)	100% euphausiids
	Lishman (1985)	Almost entirely *E. superba* plus small amounts of fish and amphipods
South Shetland Islands	Volkman *et al.* (1980)	99% euphausiids
Graham Land	Sladen (1958)	Euphausiids
Palmer Land	Eklund (1945)	Nototheniid fish and *E. superba*
Weddell Sea	Bellingshausen (1945)	Almost all euphausiids
Ross Sea		
Ross Sea	Falla (1937)	Primarily euphausiids
Cape Crozier (Ross I.)	Emison (1968)	60% *E. crystallorophias*, 39% fish, 2% amphipods
Cape Bird (Ross I.)	Paulin (1975)	46% *E. crystallorophias*, 44% fish, 10% amphipods
Cape Royds (Ross I.)	Wilson (1907)	Small fish, cephalopods, euphausiids, crustaceans
Cape Hallett	Logan (in litt.)	97.3% euphausiids, 1.7% fish, 0.9% amphipods, 0.1% other crustaceans
	Kinsky & Ensor (in litt.)	Primarily *E. superba* and *E. crystallorophias*, with fish increasing in importance later in the breeding season
East Antarctica		
Cape Adare	Wilson (1907)	Almost all *E. superba*
	Levick (1915)	Euphausiids
Wilkes Land	Penney (1968)	Euphausiids
	Cowan (1983)	*E. crystallorophias*, *E. superba*, fish, amphipods
Vestfold Hills	This study	42% *E. crystallorophias*, 32% fish, 24% *E. superba*, 2% amphipods

are summarized in Table 1. The majority of these data have been collected in the South Orkney Islands, Antarctic Peninsula and Ross Sea areas. The only information available from the remaining 200° of longitude is from general observations made by Wilson (1907) and Levick (1915) at Cape Adare, and Penney (1968) in Wilkes Land.

This study is the first major examination of Adélie penguin diet from Eastern Antarctica and is the first study anywhere to collect Adélie penguin food samples covering an almost entire (albeit composite) breeding season. This work also satisfies Siegfried & Croxall's (1983) recommendation regarding the need to have information on diet from local areas in order that the impact of commercial harvests on prey species in adjacent waters can be assessed.

Study site

Samples were collected at Magnetic Island (68°30′S, 77°50′E), 4.5 km north-west of Davis Station (Fig. 1) in the Vestfold Hills, East Antarctica. In this region Adélie penguins arrive to breed in early to mid-October, eggs are laid in the first half of November, chicks hatch from mid-December and fledge in February. The colony disperses in March with the last birds seen in mid- to late March.

The total number of Adélie penguins breeding along the Vestfold Hills coastline and on at least 17 offshore islands (Fig. 1 & Horne, 1983) has been estimated at 130000 pairs (Johnstone *et al.*, 1973). Horne (1983) presented estimates of the number of adult Adélie penguins breeding at Magnetic Island in the order of 4000 to 9000 pairs. In the present authors' opinion the number is considerably higher,

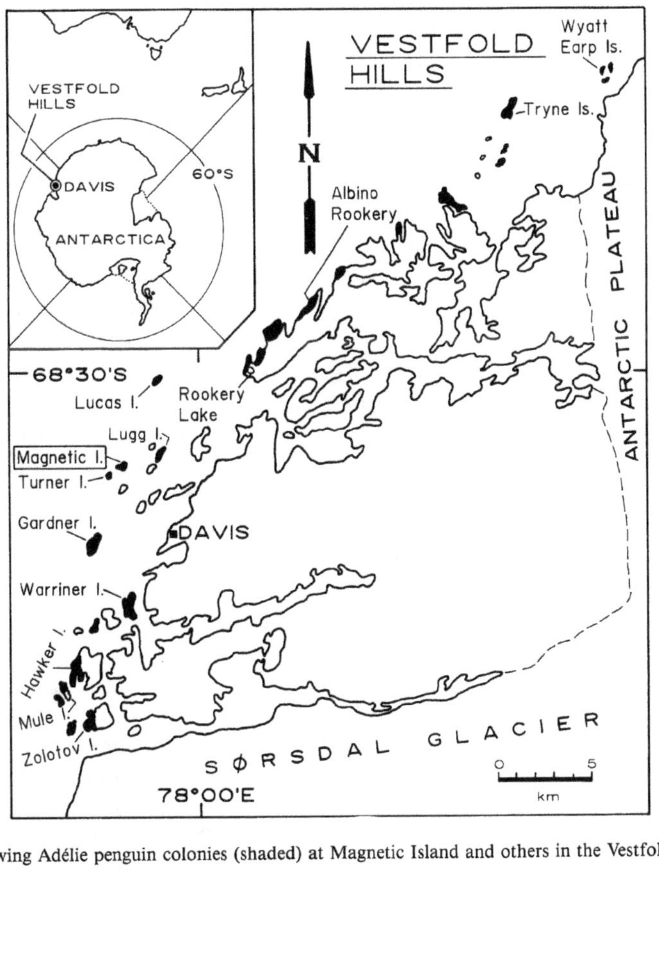

Fig. 1. Locality map showing Adélie penguin colonies (shaded) at Magnetic Island and others in the Vestfold Hills.

probably in the order of 15 000 or more breeding pairs.

Magnetic Island is surrounded by a shallow shelf up to 20 m deep with deeper channels running through it to about 1 km offshore. Further offshore, particularly to the south, water depth increases rapidly to over 100 m and remains at a depth of around 300 m – 750 m for some 200 km offshore (R. Williams, pers. commun.).

Material and methods

Sampling regime

Samples were collected approximately weekly from mid-December 1982 to the end of February 1983 (451 samples) and from mid-October to mid-December 1983 (123 samples). Thus most of the 4.5 months the Adélie penguins spent ashore was sampled.

Samples were only collected from clean adult penguins as they arrived at the colony. Birds were caught either by hand or with a hand-held net. Few samples were obtained during the incubation phase as few mobile birds were encountered and, of those sampled, most were completely empty or contained only stones. After hatching, traffic of birds between the ice edge and the colony increased greatly. Sampled birds were not sexed or aged, but most birds were presumed to have been breeding adults.

The method of water-offloading described by Randall & Davidson (1981) and modified by Wilson (1984) was used. Approximately 800 ml of warm fresh water was poured into a graduated funnel (connected to the same tube and inserted in the same way) until water began to run out of the bird's mouth. The tube was then quickly removed, one finger held in the angle of the mouth while ensuring that this did not obstruct emesis of food, and the penguin inverted over a bucket, held by the feet with one hand and by the head with the other and squeezed dorsoventrally between the operator's knees. In this position the introduced water lay on top of the food mass and when pressure was applied the stomach contents were effectively flushed out. Warm water was used not only to prevent freezing (ambient temperatures during the sampling period ranged from −27 °C to +8 °C), but also to prevent undesirable lowering of the penguins' body temperature. Each bird was flushed with water only once.

Most samples were taken between 09:00 h and 18:00 h local time and all were taken between 08:00 h and 24:00 h. After collection, samples were drained through a fine (0.5 mm) mesh sieve and preserved in 70% ethyl alcohol.

Sample analysis

Samples were sorted by placing the entire contents into trays and removing amphipods and all minor components such as stones. This usually left one or more of the following components; *Euphausia crystallorophias* (Holt & Tattersall), *E. superba* and fish. These were sorted by removing the least common of the components from the remainder. Once separated, each of the components was drained, counted, weighed wet, preserved and labelled.

Individual food items in most samples were sufficiently intact to be identified and counted. This was not so for fish material, particularly the remains of large fish which usually consisted of pieces of flesh and skeletal material. Some otoliths were also present. Euphausiid material was usually relatively undigested, making separation into species and counting of individuals simple.

In 23 well-digested samples, numbers of euphausiids could be estimated by counts of eyes but species could not be assigned with any certainty. In these samples, the total number of euphausiids and total weight were obtained and divided equally between *E. crystallorophias* and *E. superba*. Comparing samples divided in this way to other samples collected at the same times indicates that such a division may introduce a minor bias in favour of *E. superba* in some of the nine weeks in which such samples occurred. Subsampling for the purposes of counting was performed with samples over 600 ml.

Measurements of carapace lengths from mid-line posterior dorsal to anterior tip of rostrum (standard 4; Mauchline, 1980) were taken of 262 *E. superba* and 343 *E. crystallorophias*. Krill were selected for measurement by randomly selecting three diet samples per week. Each of these samples was placed in a sorting tray divided by a 10 × 10 grid and then, where possible, 10 specimens each of *E. superba* and

E. crystallorophias were selected from random grid references. Body lengths (mm) of *E. superba* were derived using the formula

$$X = 2.857\ Y + 2.63$$

(where X = body length, Y = carapace length) (Murano *et al.*, 1979). Body lengths (mm) of *E. crystallorophias* were derived using the formula

$$X = 1.512\ Y + 13.28$$

(P. Thomas, pers. commun.).

Results

Sample weight

The weights of samples changed markedly in mid-December concurrent with hatching (Fig. 2), reflecting the sudden increase of food brought ashore by

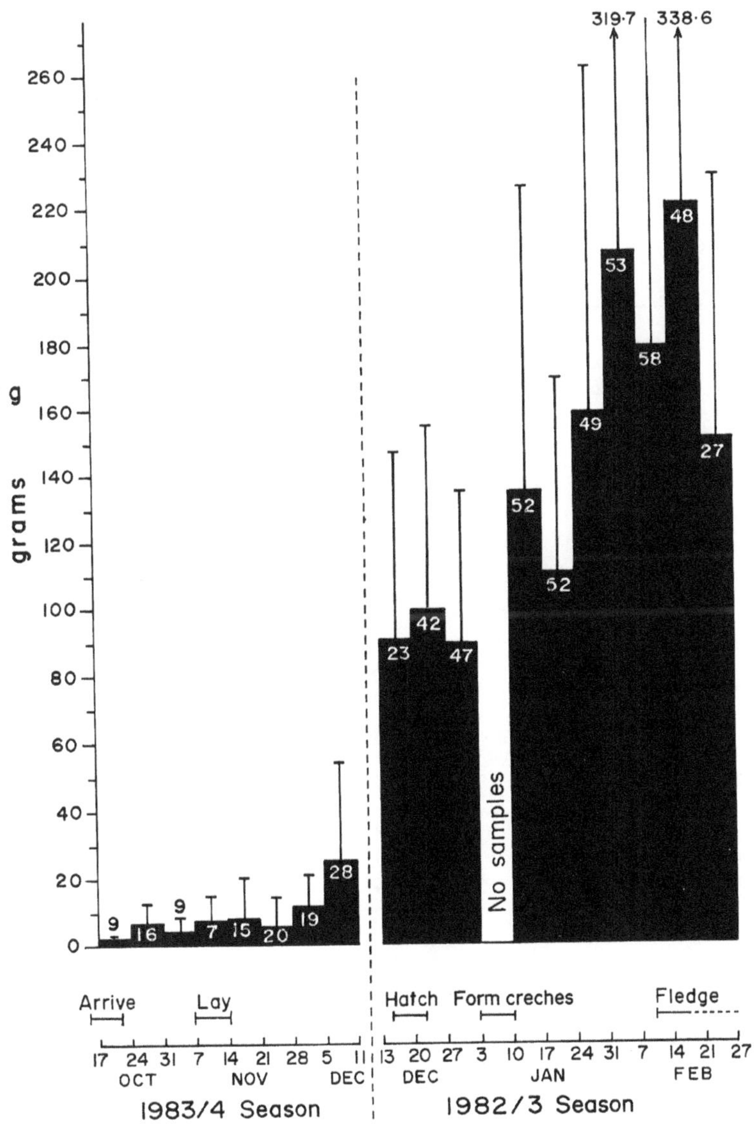

Fig. 2. Number, mean weight and standard deviation of stomach content samples collected from Adélie penguins.

adults when feeding chicks. From chick hatching to fledging, the mean weight of samples increased from approximately 90 g to 220 g, reflecting the chicks' increasing food requirements and capacity with increasing body size. The largest sample, obtained on 14 February 1983, weighed 589.4 g representing 10%–15% of the adult bird's body weight.

Sample contents

Items found in the samples are listed in Table 2. They included four major categories of prey species. In order of decreasing weight these were *E. crystallorophias*, fish (10 species), *E. superba* and amphipods (14 species). Greater specific detail and frequency of occurrence is given for fish in Table 3 and amphipods in Table 4. There were in addition, many minor items, including two parasitic worms, a cestode (probably *Diphyllobothrium* cf *quadratum*) and a nematode, *Contracaecum* sp. These were thought to be fish parasites ingested by the penguins when feeding on fish (R. J. G. Lester, pers. commun.). Fish material was invariably present in any sample containing either of these worms.

Table 2. Frequency of occurrence and total weight of food and 'non-food' items in 574 Adélie penguin stomach samples.

	Frequency	Total weight of item in all samples (g)
Food Items		
E. crystallorophias	435	28469.6
Fish (10 species)	421	21633.3
E. superba	377	16449.2
Amphipod (14 species)	472	1259.9
Squid	5	9.3
Mysid	4	4.8
Pteropod	30	4.5[a]
Holothurian	1	2.7
Isopod (*Valvifera* sp.)	2	1.2
Coelenterate	1	1.0
Prawn	1	0.9
Polychaete (*Tomopteris* sp.)	4	0.9
Gastropod	3	0.2
'Non-Food' Items		
Stone/Shell	151	286.9
Seaweed	6	3.4
Egg shell	5	1.5
Echinoderm shell	2	0.2
Hydrozoan – sessile	5	0.1
Sponge spicule	2	0.1
Unidentified	10	7.7
Fish Parasites		
Cestodes (?*Diphyllobothrium* sp.)	173	[b]
Nematodes (*Contracaecum* sp.)	146	[b]
Copepods (*Clavellodes* sp.)	1	[b]

[a] Thought to derive from fish stomach contents.
[b] Weights not recorded.

Table 3. Frequency of occurrence of fish identified in 574 samples from Adélie penguin stomachs.

Taxa	Frequency
Family Nototheniidae	
Pleuragramma antarcticum Boulenger	208
Trematomus sp.	93
Pagothenia borchgrevinki (Boulenger)	14
Trematomus cf *hansoni*	7
Nototheniidae indet.	5
Family Channichthyidae	
Chionodraco sp.	66
Channichthyidae indet.	24
Pagetopsis macropterus (Boulenger)	8
Family Bathydraconidae	
Gymnodraco acuticeps Boulenger	5
Family Myctophidae	
Myctophidae indet.	1

Frequency of occurrence of major food items

The percentage frequency of occurrence of the major food categories are presented in Fig. 3 and Table 5. Prior to hatching, amphipods were the most frequently encountered food category occurring in 94% of samples, *E. crystallorophias* occurred quite commonly (45%) while *E. superba* (14%) and fish (20%) were relatively infrequent. After hatching, the occurrence of amphipods fell but the other major categories increased. Once the chicks formed crêches the frequency of occurrence of all major categories was around 80%, with fish increasing to almost 100%.

Table 4. Frequency of occurrence of amphipods identified in 574 samples from Adélie penguin stomachs.

Taxa	Frequency
Hyperiidea	
Family Hyperiidae	
Hyperia macrocephala (Dana)	234
Parathemisto gaudichaudii (Guer)	24
Hyperoche sp.	1
Family Vibiliidae	
Cyllopus magellanicus Dana	223
Gammaridea	
Family Lysianassidae	
Orchomene rossi (Walker)	189
Cheirimedon fougneri Walker	129
Uristes murrayi (Walker)	25
Waldeckia obesa (Chevreux)	24
Lysianassidae indet.	1
Family Eusiridae	
Eusirus sp. nov.	133
Eusirus perdentatus Chevreux	36
Paramoera walkeri Stebbing	11
Family Paramphithoidae	
Epimeriella macronyx Walker	53
Epimeriella walkeri K. H. Barnard	1

Temporal change in dietary composition

Table 5 shows the average weight of samples for pre- and posthatching periods. *E. crystallorophias* was the most important food item overall in the posthatching period. It constituted approximately 42% by weight of the diet. Next was fish (32%), followed by *E. superba* (24%) and amphipods (1%). The remaining assorted food and 'non-food' items

Table 5. Frequency of occurrence (%) and mean weight (g) for major food components from Adélie penguin stomach samples, for pre- and posthatching periods.

	E. crystallorophias	Fish	*E. superba*	Amphipods	Other	Total
Prehatching %	45.5	20.3	13.8	94.3	0.0	-
g	5.03	0.36	0.60	4.31	0.63	10.93
Posthatching %	84.0	87.8	79.8	78.9	0.0	-
g	61.75	47.87	36.31	1.62	0.57	148.12

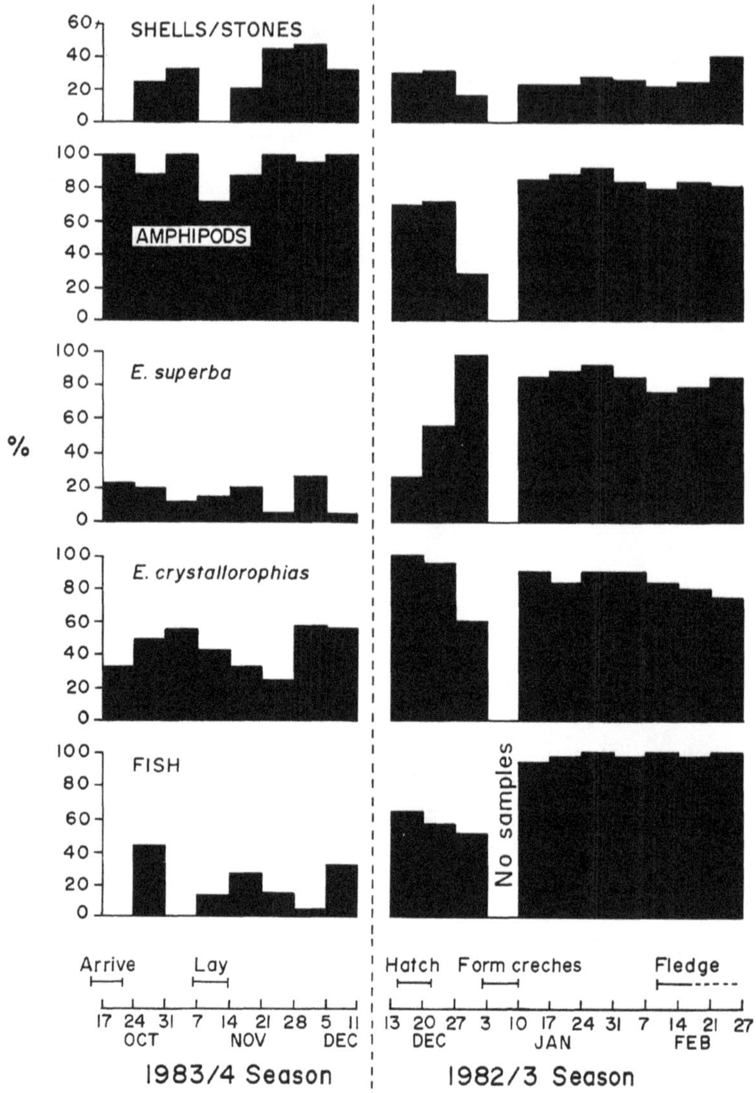

Fig. 3. Percentage frequency of occurrence of major dietary components from Adélie penguins stomach contents.

represented less than 0.5%. The relative proportions of items in the samples, however, varied considerably throughout the season (Figs. 3 & 4).

Prior to hatching, *E. crystallorophias* and amphipods constituted the bulk of the stomach contents with *E. superba* and fish present only occasionally and in very small quantities. However, during the incubation phase, only very small samples were obtained and many birds sampled had bile-stained stomach contents and faeces. Presumably they had digested most of their food before returning to the colony (Randall & Davidson, 1981).

After the chicks hatched, *E. crystallorophias* was still the predominant food item, with fish and *E. superba* becoming increasingly important. From early January when chicks were forming crêches and their food demands were increasing, the reliance by adults on fish increased whilst the intake of *E. crystallorophias* and *E. superba* followed no clear pattern (Fig. 4).

Relative importance of E. crystallorophias *and* E. superba

Throughout most of the breeding season *E. superba* was of secondary importance to *E. crystallorophias* both numerically and by weight. In only three of 18 weeks (late December 1982 and February 1983) did

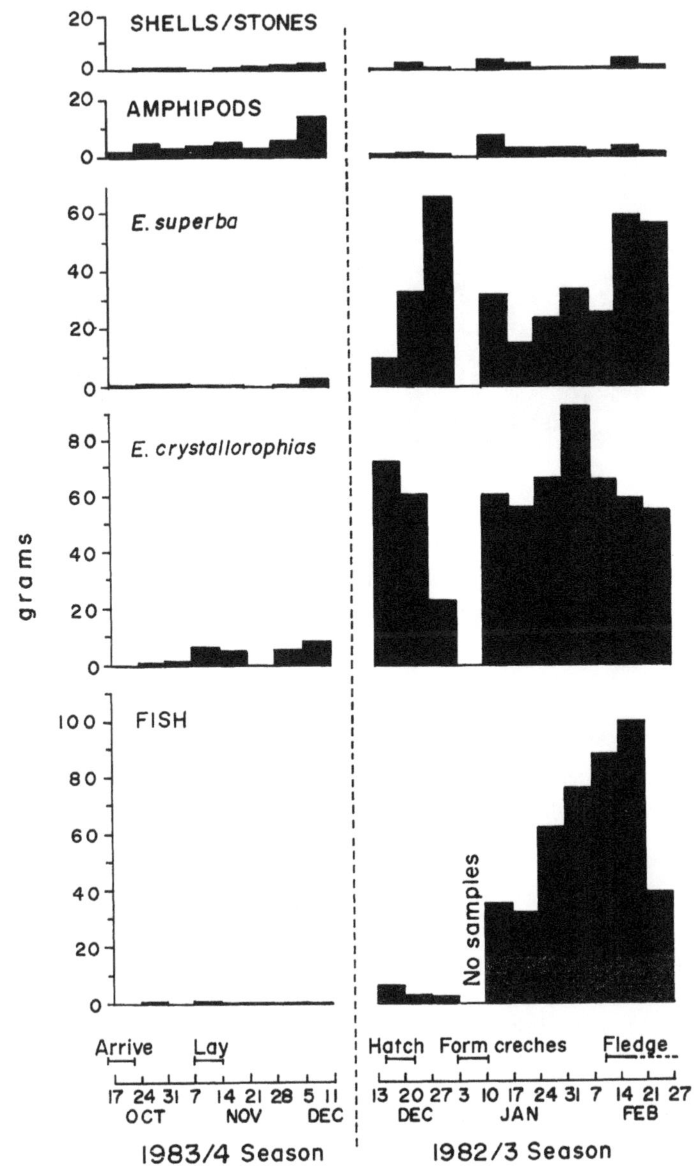

Fig. 4. Mean weight of major food components of Adélie penguin stomach contents.

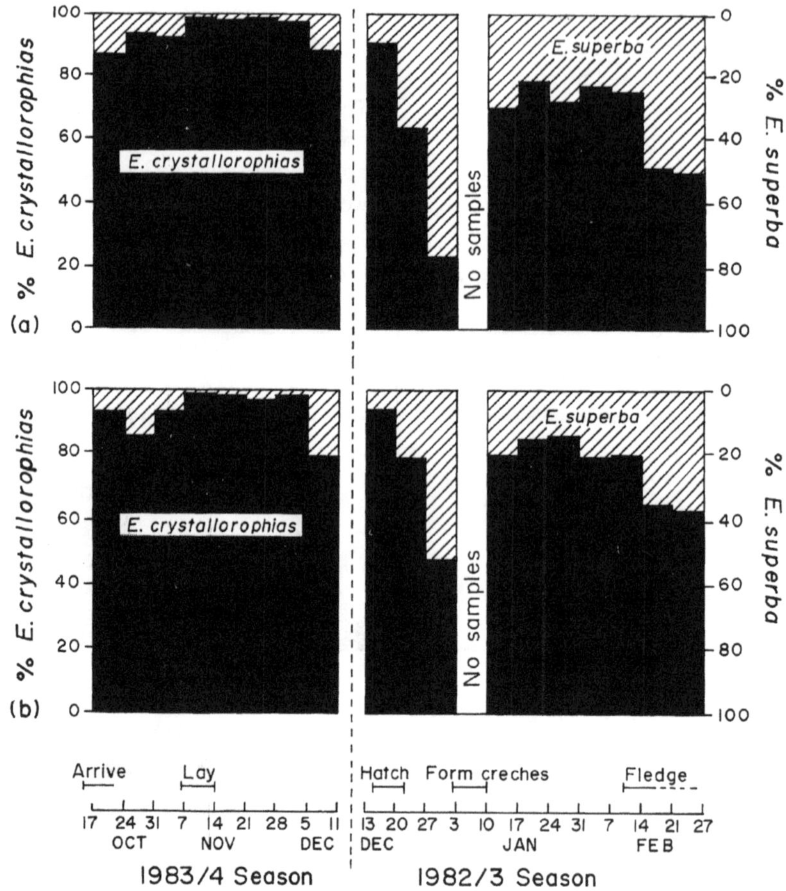

Fig. 5. Mean percentage composition by weight (a) and by numbers (b) of *E. crystallorophias* and *E. superba* from Adélie penguin stomach content samples.

samples produce a larger percentage frequency of occurrence of *E. superba* (Fig. 3).

Size of euphausiids

The carapace lengths of the 262 specimens of *E. superba* measured ranged from 6.2 mm to 19.0 mm (mean ± S.D., 13.25 ± 2.77 mm), representing total body lengths ranging from 20.34 mm to 56.91 mm (mean ± S.D., 40.63 ± 7.85 mm).

The carapace lengths of the 343 specimens of *E. crystallorophias* ranged from 5.2 mm to 13.8 mm (mean ± S.D., 9.32 ± 1.21 mm), representing total body lengths ranging from 21.14 mm to 34.15 mm (mean ± S.D., 27.36 ± 1.83 mm).

Relative importance of euphausiids and fish

The weight of both euphausiid species combined was consistently greater than the weight of fish in combined weekly samples throughout the breeding season.

Discussion

Sampling method

This study is based on a large number of samples collected over two breeding seasons, which, when combined, provide data for all phases of the breeding season.

The interpretation of our results depends on:
1) the extent to which the sampling method obtained representative samples of the penguins' stomach contents;
2) the extent to which the samples represented the diet of the local population;
3) the degree of annual variation in diet from year to year.

1) Penguins were not killed to examine whether complete stomach samples were obtained. This decision was made both for humane reasons and to minimize the adverse effects of this study on the Adélie penguin population of Magnetic Island.

An indication of the completeness of the samples is sample size. The mean posthatching sample weight was 148 g, with a range of 0.3 g – 589.4 g. Johnstone (unpubl. data) used an emetic to collect Adélie penguin stomach contents on Magnetic Island in January 1982; the eight samples varied in weight from 310 g to 693 g (mean 457 g). In another diet study of Adélie penguins Logan (in litt.) reported samples up to 840 g, using the same sampling technique as this study. Complete stomach samples obtained by Kinsky & Ensor (in litt.) ranged up to 1 000 g. Volkman *et al.* (1980) reported that the total stomach contents (of killed birds) weighed 350 g on average in the period following the hatching of chicks. Evidently, our sampling technique did not obtain complete samples in all cases.

Samples obtained throughout the chick raising period increased steadily in size. Similar trends have been reported by Hunter (1983) for giant petrels *Macronectes* sp., Lishman (1985) for Adélie and chinstrap penguins *Pygoscelis antarctica* (Forster), Croxall & Furse (1980) for chinstrap penguins and Williams (1981) for rockhopper *Eudyptes chrysocome* Forster and macaroni penguins *Eudyptes chrysolophus*. More recent sampling at Magnetic Island (N. Gales, pers. commun.) has shown that multiple flushing is essential to obtain complete samples. This is consistent with other more recent work on penguins (Hindell, in press; Gales, 1987).

2) The extent to which the samples accurately represented the overall diet of Adélies in the region therefore cannot be assessed directly. However, a preliminary study in the 1981 – 1982 season included samples from the Wyatt Earp Islands and the Rauer Islands, some 30 km to the north-east and south-west respectively of Magnetic Island. Samples from these sites were not apparently different (Johnstone, unpubl. data). Sampling in subsequent seasons at these locations has confirmed that the diet varies little within this 60 km stretch of coast (Green & Johnstone, in press; E. J. Woehler, unpubl. data).

3) Logan (in litt.) compared the results of her study of Adélie diet at Cape Hallett in January 1983 with those of the 1965 – 1966 study by Kinsky & Ensor (in litt.), and found considerable differences in sample size and composition particularly with regard to the occurrence of fish in the diet. Caution therefore should be used in interpreting our data which were derived from two consecutive summers.

Food species and relative importance of dietary items

The area of Prydz Bay adjacent to the Vestfold Hills bears similarities to the Ross Sea region in that both lie over the continental shelf and so are relatively shallow. The neritic *E. crystallorophias* is the predominant euphausiid occurring in the Ross Sea and is a major component of Adélie penguin diet in both areas (Table 1).

The pelagic shoaling *P. antarcticum* formed the bulk of the 'fish' category. *P. antarcticum* occurs almost entirely over the continental shelf (Williams, 1985a). The other species of fish recorded are benthic and commonly occur in shallow water (Table 3).

Emison (1968) found that stomach contents composed primarily of fish tended to be heavier than those composed primarily of crustaceans so it could be advantageous energetically for Adélie penguins to take the larger fish rather than the small euphausiids. The fact that *P. antarcticum* itself feeds on copepods and *E. crystallorophias* (Williams,

1985b; Gorelova & Gerasimchuk, 1981) suggests that these species may be routinely associated, making such a choice possible.

Likewise, the small but consistent amphipod component (Table 4) comprised both pelagic species such as *Hyperia macrocephala* (Dana) and *Parathemisto gaudichaudii* (Guer), and also inshore or demersal species commonly found in shallow water, such as *Orchomene rossi* (Walker), *Waldeckia obesa* (Chevreux), *Paramoera walkeri* Stebbing and *epimeriella macronyx* (Walker).

Squid remains were found in only five samples. This may be because squid in this region are too large to be easily taken by Adélie penguins or are rare in the surface waters where Adélie penguins feed. However, there are many possible reasons for the consistent omission of squid from the diet of Adélie penguins in this and other studies. Apart from their 'general availability' to Adélie penguins, their food value could be a factor. Clarke & Prince (1980) reported that the most obvious difference between krill, fish and squid meals fed to albatross chicks was the very low calcium content of squid meals. Gailey-Phipps & Sladen (1982) noted that eggshell formation imposes a heavy calcium demand on the fasting female (ashore for approximately 17 days) while calcium was also found to be an important dietary component for the healthy development of growing captive chicks (Beldegreen & Asper, 1980).

Gailey-Phipps & Sladen (1982) also noted that fat-based metabolism is important for penguins during both moult and breeding while Penney (1978) stated that stunting of captive bred Adélie chicks was overcome through the addition of *inter alia*, corn oil to replace fish oil in their diet.

Krill and fish satisfy both of these dietary requirements while squid does not. It may be that adult Adélie penguins may not solely feed, in an opportunistic manner, on all available organisms of an appropriate size range, but may selectively choose prey which firstly is available and secondly is capable of supplying the nutritional demands of the adults and growing chicks.

Comparisons with other studies

Our results closely resemble those from studies in the northern Ross Sea region (Table 1). Kinsky & Ensor (in litt.) found both *E. superba* and *E. crystallorophias* present in the diet with a trend of increasing reliance on fish later in the breeding season. Their study site at Cape Hallett was situated less than 200 km from the edge of the continental shelf, as is Magnetic Island. Our results are also similar to those of Emison (1968). He found very few *E. superba* in material collected at Ross Island, some 500 km from the edge of the continental shelf. Ainley *et al.* (1983) found that the only euphausiid in three Adélie penguins collected over the continental shelf of the Ross Sea was *E. crystallorophias*, but two collected over the continental slope contained almost entirely *E. superba*. Other seabirds in the same region showed a similar pattern (Ainley *et al.*, 1984). *E. superba* would appear to be more significant in the diet of Adélie penguins in the Scotia Arc/Antarctic Peninsula area (Table 1).

Feeding/digestion times

An observation made during this study, and also reported by Sladen (1958), was that the stomachs of most birds entering the colony during the egg laying and incubation periods were empty or at best contained only a very small amount of food (Fig. 2). Most of the 127 stomachs examined at this time were bile-stained which, according to Sladen (1958), indicates that the birds had not fed for at least three days. Another possibility alluded to by Rand (1960) is that the penguins may be able to voluntarily speed up or delay digestion rates.

The empty birds returning to the rookery, particularly in the incubating phase, present a puzzle in that, although they had to cover a considerable distance across the sea-ice to get to Magnetic Island, at speeds reported to vary between 3 km h^{-1} and 10 km h^{-1} (Emlen & Penney, 1964), the trip would nonetheless only take a maximum of 5 h non-stop travel (and even less time later in the season when the fast ice has broken out) so that they should still have some food present in their stomachs if they had fed immediately before returning. Jackass penguins *Spheniscus demersus* (Linné) require at least 10 h to digest 50 g of anchovies: they require several hours more to digest squid (Wilson *et al.*, 1985). The results from Magnetic Island would suggest that the birds had swum some considerable distance from their feeding grounds to the ice edge or had rested between feeding and return to the colony.

Diet composition and quality

Previous authors (Rand, 1960; Stonehouse, 1967; Gailey-Phipps & Sladen, 1982) have commented that penguins as a group tend to be unspecialized feeders, preying opportunistically on whatever food source is available. Our results show considerable change in dietary composition over the breeding season with the clearest trend being the pronounced swing to fish late in the season. This may reflect an increased availability of fish in late summer but other explanations are possible. Our data do not allow much directed discussion of them. However, the annual cycle of breeding and moult should reasonably be expected to strongly influence the food consumption of penguins.

Fish and krill reportedly have similar calorific contents with figures of 0.95 kcal g^{-1} wet weight for fish and 1.04 kcal g^{-1} wet weight for *E. superba* (Croxall & Prince, 1982). Volkman & Trivelpiece (1980) reported calorific contents by dry weight as 5.2 kcal g^{-1} for fish and 4.7 kcal g^{-1} for krill. Although the lipid and calorific content of both fish and krill can vary considerably depending on species, sex, age, reproductive condition and season (Clarke & Prince, 1980; Clarke, 1984), there is probably little difference between the relative calorific content per gram of equivalent sex and age groups of *E. superba* and *E. crystallorophias.*

Conclusions and future work

If the diet of Adélie penguins in the Prydz Bay region is to be used to interpret the effects of harvesting pressures on marine organisms in that area, more information is required. This information must be gained by consistent methodology and the multiple flushings technique is recommended.

Firstly, the annual variability of the diet must be determined. Secondly, dietary information from the Vestfold Hills must be compared with similar information from elsewhere in the Prydz Bay region so that the representative character of the Vestfold Hills data may be assessed. Adélie penguin dietary samples have now also been collected from the Rauer Islands (E. J. Woehler, unpubl. data) and will provide some comparisons. Thirdly, the winter diet of Adélie penguins is still unknown. Samples collected during this time would provide information on diet prior to breeding, including the contribution of *E. superba.* However, no presently available logistic means appear capable of providing these samples. Not only should the diet of Adélie penguins be monitored annually in the Prydz Bay area but also the size of the Adélie penguin breeding population in the Vestfold Hills and surroundings should be determined.

The results of this study indicate that the diet of the Adélie penguin, as assessed during the breeding season, is likely to have only very limited usefulness as an indicator of the abundance of *E. superba* stocks in Prydz Bay. However, the winter availability of *E. superba* in Prydz Bay may be a limiting factor determining Adélie penguin breeding numbers and success. At this stage it is premature to judge the usefulness of the Adélie penguin population of the Vestfold Hills as an indicator of the status of *E. superba* in Prydz Bay.

Acknowledgements

This study was funded by the Antarctic Division, Department of Science. The field work was made possible by the willing assistance of the men of Davis Station in 1983. Particular thanks also go to R. Williams, J. Lowry and R. Lester for identifying fish, amphipods and fish parasites respectively. Thanks go to K. Green for measuring the krill and providing this information for use in this article. We also thank E. J. Woehler, H. R. Burton and two anonymous referees for their comments and revisions of this paper.

References

Ainley, D. G., R. E. LeResche & W. J. L. Sladen, 1983. Breeding biology of the Adélie Penguin. Univ. Calif. Press, Los Angeles, 240 pp.

Ainley, D. G., E. F. O'Connor & R. J. Boekelheide, 1984. The marine ecology of birds in the Ross Sea, Antarctica. Ornithological Monographs No. 32. American Ornithologists' Union, Wash. D.C., 97 pp.

Beldegreen, R. A. & E. A. Asper, 1980. Breeding and husbandry of Adélie penguins at Sea World of Florida. Int. Zoo. Yearbook 20: 197–200.

Bellingshausen, F. G., 1945. The voyage of Captain Bellingshausen to the antarctic seas 1819–1821. Vol. 1. Translated and edited by F. Debenham. Hakluyt Society, London, 259 pp.

Clarke, A., 1984. The lipid content and composition of some antarctic macrozooplankton. Br. Antarct. Surv. Bull. 63: 57–70.

Clarke, A. & P. A. Prince, 1980. Chemical composition and calorific value of food fed to Mollymauk chicks Diomedea melanophris and D. chrysostoma at Bird Island, South Georgia. Ibis 122: 488–494.

Cowan, A. N., 1983. A modified penguin stomach tube. Corella 7(3): 59–61.

Croxall, J. P., 1984. Seabirds. In R. M. Laws (ed.), Antarctic Ecology, Vol. 2. Academic Press, Lond.: 533–619.

Croxall, J. P. & J. R. Furse, 1989. Food of Chinstrap penguins, Pygoscelis antarctica and Macaroni penguins Eudyptes chrysolophus at Elephant Island, South Shetland Islands. Ibis 122: 237–245.

Croxall, J. P. & P. A. Prince, 1982. A preliminary assessment of the impact of seabirds on marine resources at South Georgia. Comm. Natn. Fr. Rech. Antarct. 51: 501–509.

Eklund, C. R. 1945. Condensed ornithology report, East Base, Palmer Land, Proc. Am. Phil. Soc. 89: 299–304.

Emison, W. B., 1968. Feeding preferences of the Adélie penguin at Cape Crozier, Ross Island. In O. L. Austin (ed.). Antarctic Bird Studies, Antarctic Research Series. American Geophysical Union, Wash. D.C. 12: 191–212.

Emlen, J. T. & R. L. Penney, 1964. Distance navigation in the Adélie penguin. Ibis 106: 417–431.

Falla, R. A., 1937. Birds. B.A.N.Z. Antarct. Research Expedition 1929–1931. Report, Series B, 2: 1–304.

Gailey-Phipps, J. J. & W. J. L. Sladen, 1982. Survey on nutrition of penguins. JAVMA 181 (11): 1305–1309.

Gales, R. P., 1987. Validation of the stomach flushing technique for obtaining stomach contents of penguins. Ibis. 129: 335–343.

Gorelova, T. A. & V. V. Gerasimchuk, 1981. Data on nutrition and daily consumption of juvenile Pleuragramma antarcticum Boulenger. In N. V. Parin (ed.), Fishes of the Open Ocean. Acad. Sci. USSR, Moscow: 103–109.

Green, K. & G. W. Johnstone, in press. Changes in the diet of Adélie penguins breeding in East Antarctica. Aust. Wildl. Res.

Hindell, M. A., in press. The diet of the King Penguin, Aptenodytes patagonicus, at Macquarie Island. Ibis.

Horne, R. S. C., 1983. The distribution of penguin breeding colonies on the Australian Antarctic Territory, Heard Island, the MacDonald Islands, and Macquarie Island. ANARE Res. Notes 9. Antarct. Div., Dept. Sci., Hobart, 82 pp.

Hunter, S., 1983. The food and feeding ecology of the Giant petrels Macronectes halli and M. giganteus at South Georgia. J. Zool., Lond. 200: 521–538.

Johnstone, G. W., D. J. Lugg & D. A. Brown, 1973. The biology of the Vestfold Hills, Antarctica. ANARE Sci. Rep. Publ. 123. Antarct. Div., Dept. Sci., Melbourne, 62 pp.

Laws, R. M., 1977. The significance of vertebrates in the antarctic marine ecosystem. In G. A. Llano (ed.), Adaptations within Antarctic Ecosystems. Smithsonian Institution, Wash. D.C.: 411–438.

Levick, G. M., 1915. Natural history of the Adélie penguin. British Antarctic ('Terra Nova') Expedition, 1910. Natural History Report. Zoology 1(2): 55–84.

Lishman, G. S., 1985. The food and feeding ecology of Adélie penguins (Pygoscelis adeliae) and Chinstrap penguins (P. antarctica) at Signy Island, South Orkney Islands. J. Zool. Lond. A205: 245–263.

Mauchline, J., 1980. Measurements of body length of Euphausia superba (Dana). BIOMASS handbook series 4: 1–9.

Murano, M., S. Seuawa & M. Kata, 1979. Moult and growth of the Antarctic Krill in laboratory. Trans. Tokyo Univ. Fish 3: 99–106.

Paulin, C. D., 1975. Feeding of the Adélie penguin Pygoscelis adeliae. Mauri Ora 3: 27–30.

Penney, R. L., 1968. Territorial and social behaviour in the Adélie penguin. In O. L. Austin (ed.), Antarctic Bird Studies, Antarctic Research Series. American Geophysical Union, Washington, D.C. 12: 83–131.

Penney, R. L., 1978. Breeding Adélie penguins Pygoscelis adeliae in captivity. Int. Zoo. Yearbook 18: 13–21.

Rand, R. W., 1960. The biology of guano-producing sea-birds. Department of Commerce and Industry Division of Fisheries, Union of South Africa. Investig. Rep. No. 41. March: 2–25.

Randall, R. M. & I. S. Davidson, 1981. Device for obtaining food samples from stomachs of Jackass penguins. S. Afr. J. Wildl. Res. 11: 121–125.

Siegfried, W. R. & J. P. Croxall, 1983. Progress and prospects of ornithological research within BIOMASS. In T. Nemoto & T. Matsuda (eds), Proceedings of the BIOMASS Colloquium in 1982. National Institute of Polar Research, Tokyo.

Sladen, W. J. L., 1958. The Pygoscelid penguins. Falkland Islands Dependencies Survey. Sci. Rep. No. 17. H.M. Stationary Office, London: 97 pp.

Smith, N. R., Z. Dong, K. R. Kerry & S. Wright, 1984. Water masses and circulation in the region of Prydz Bay, Antarctica. Deep Sea Res. 31: 1121–1147.

Stonehouse, B., 1967. The general biology and thermal balances of penguins. Advances in Ecological Research 4: 131–196.

Volkman, N. J., P. Presler & W. Trivelpiece, 1980. Diets of Pygoscelid penguins at King George Island, Antarctica. Condor 82: 373–378.

Volkman, N. J. & W. Trivelpiece, 1980. Growth in pygoscelid penguin chicks. J. Zool., Lond. 191: 521–530.

White, M. G. & J. W. H. Conroy, 1975. Aspects of competition between Pygoscelid penguins at Signy Island, South Orkney Islands. Ibis 117: 371–373.

Williams, A. J., 1981. Chick feeding rates of Macaroni and Rockhopper penguins at Marion Island. Ostrich 53: 129–134.

Williams, R., 1985a. The potential impact of a krill fishery upon pelagic fish in the Prydz Bay area of Antarctica. Polar Biol. 5: 1–4.

Williams, R., 1985b. Trophic relationships between pelagic fish and Euphausiids in antarctic waters. In W. R. Siegfried, P. R. Condy & R. M. Laws (eds), Antarctic Nutrient Cycles and Food Webs, Springer-Verlag, Berlin: 452–459.

Wilson, E. A., 1907. Aves. Brit. Natl Antarctic Expedition Rept, 1901–1904. Nat. Hist. 2: 1–121.

Wilson, R. P., 1984. An improved stomach pump for penguins and other seabirds. J. Field. Orn. 55: 109–112.

Wilson, R. P., G. D. La Cock, M. P. Wilson & F. Mollagee, 1985. Differential digestion of fish and squid in Jackass penguins Spheniscus demersus. Ornis Scandinavica 16: 77–79.

Hydrobiologia 165: 255–261 (1988)
J. M. Ferris, H. R. Burton, G. W. Johnstone & I. A. E. Bayly (eds) Biology of the Vestfold Hills, Antarctica
Kluwer Academic Publishers

Predicting the swimming and diving behaviour of penguins from muscle biochemistry

J. Baldwin
Department of Zoology, Monash University, Clayton, Victoria, 3168, Australia

Key words: penguin, diving, muscle, aerobic, anaerobic

Abstract

Energy metabolism in the pectoralis and supracoracoideus muscles of seven species of penguins was investigated by determining muscle fibre diameter, myoglobin content, pH buffering capacity and the distribution and properties of lactate dehydrogenase isoenzymes.

The penguins can be arranged as follows in order of increasing anaerobic capabilities of the muscles: little < rockhopper and royal < gentoo < Adelie, emperor and king.

As a good correlation exists between muscle biochemistry and known diving behaviour of emperor, king, gentoo and little penguins, predictions can be made about the behaviour of species for which only the biochemical data are available.

Introduction and methods

Penguins (Family Spheniscidae) are able to dive deeper and remain submerged for longer than any other birds. Feeding exclusively on marine animals, the choice of food items will be related to both depth of occurrence of the prey and the swimming and diving ability of the penguin. Thus information on the relative swimming and diving behaviour of different penguin species, apart from being important for interpreting adaptations associated with diving, also is useful in a more general ecological context.

Until recently, data on the natural diving behaviour of penguins have been largely anecdotal or based upon circumstantial evidence. However, the use of suitable depth recorders that can be attached to free ranging birds promises to rectify this situation (Kooyman *et al.*, 1982; Lishman & Croxall, 1983).

Another approach that can be used to compare the swimming and diving behaviour of different penguin species is to examine specific biochemical and physiological adaptations associated with these activities. By 'calibrating' these adaptations with reference to penguins for which swimming and diving behaviour are well documented, it should be possible to make useful predictions about other species for which direct behavioural data are not available.

I have used this second approach in examining the biochemical organization of muscles, used to power swimming, for seven species of penguins ranging from the largest to the smallest. The selection of biochemical parameters was based on the relationship between the metabolic pathway used to provide ATP for muscle contraction and oxygen availability.

During short bursts of rapid locomotion, or when cut off from environmental oxygen, oxygen tensions may become limiting and ATP is generated by substrate level phosphorylations coupled to the catabolism of carbohydrates following activation of the glycolytic pathway. Although the rate of energy production may be high, anaerobic glycolysis is limited in duration by the size of muscle glycogen stores, and the ability to tolerate proton accumulation. Adaptations associated with heavy reliance on anaerobic muscle work include a high muscle pH buffering capacity (Castelline & Somero, 1981), lactate dehydrogenase isoenzymes that function preferentially as pyruvate reductases (Everse & Kaplan, 1973) and a predominance of large diameter white muscle fibres (George & Berger, 1966).

Under conditions where muscle oxygen tensions remain high, for example during sustained locomotion at submaximal speeds or dives of short duration, ATP is obtained by oxidative phosphorylation following the complete oxidation of fuels, usually lipids, through the Krebs cycle. In this way, limited muscle glycogen stores are conserved for use during temporary hypoxia at the expense of the generally more extensive lipid reserves that can be used only aerobically. Adaptations associated with aerobic muscle work include high myoglobin concentrations (Pages & Planas, 1983), lactate dehydrogenase isoenzymes functioning predominantly as lactate oxidases (Everse & Kaplan, 1973), and increased proportions of the smaller red muscle fibres (George & Berger, 1966).

While other parameters, particularly the maximum activities of enzymes in aerobic and anaerobic pathways, are useful for assessing the relative contributions of aerobic and anaerobic metabolism in muscle, problems associated with collecting penguins limited the study to properties that could be determined accurately in tissues fixed for microscopy or frozen at $-20\,°C$ for several weeks before analysis.

Results and discussion

Experimental animals

All penguins used in the study were adults collected in the field. King *Aptenodytes patagonicus* Miller, royal *Eudyptes schlegeli* Finsch, rockhopper *Eudyptes chrysocome* (Forster), and gentoo *Pygoscelis papua* (Forster) penguins were collected on subantarctic Macquarie Island. Little penguins *Eudyptula minor* (Forster) were obtained from Phillip Island, Victoria, Australia. Adelie penguins *Pygoscelis adelie* (Hombron & Jacquinot) came from the Tryne Island colony offshore from the Vestfold Hills, Antarctica, and emperor penguins *Aptenodytes forsteri* Gray were collected at Amanda Bay, in the nearest colony to the Vestfold Hills.

Structure, function and biochemical properties of penguin swimming muscles

Swimming penguins use the wings for propulsion, with the webbed feet acting only as rudders. The pectoralis and supracoracoideus are the major muscles powering wing movements. Both arise from the ventral surface of the sternum, with the pectoralis acting to depress the humerus and wing, and the supracoracoideus elevating the humerus and wing. In most flying birds the major power stroke providing both lift and forward propulsion is produced by the pectoralis muscle during the wing downstroke. However, the supracoracoideus muscle also may be used for a powered upstroke in birds specialised for hovering (hummingbirds) and steep take-off (quail), and this is accompanied by enlargement of the supracoracoideus relative to the pectoralis. As swimming penguins approximate neutral buoyancy, muscle work can be directed towards producing forward propulsion rather than lift. Propulsion is obtained from both the upstroke and downstroke of the wingbeat cycle, these strokes being of similar duration and magnitude (Clark & Bemis, 1979). The importance of the supracoracoideus muscle in powering the upstroke is reflected in pectoralis/supracoracoideus weight ratios that are low compared with those of other water birds that use the feet rather than the wings when swimming and diving (Table 1).

Muscle fibres of the avian pectoralis and supracoracoideus are classified into three types: red, white and intermediate. Muscles from different species vary in the relative proportions of these fibre types. The highly aerobic red fibres are characterized by their smaller diameters in mixed fibre muscles, the presence of large numbers of mitochondria and associated oxidative enzymes, and high lipid con-

Table 1. Relative weights of bird pectoralis and supracoracoideus muscles. Penguin data are from Baldwin *et al.* (1984), and data on other birds are from Hartman (1961).

Species	Pectoralis : supracoracoideus
King penguin	2.1
Royal penguin	2.0
Rockhopper penguin	2.2
Gentoo penguin	1.9
Little penguin	2.1
Brown Pelican *Pelecanus occidentalis* Linnaeus	5.8
Least Grebe *Podiceps dominicus* (Linnaeus)	7.0
Lesser Scaup *Aythya affinis* (Eyton)	7.6
Double-crested Cormorant *Phalacrocorax auritus* (Lesson)	10.9
Anhinga *Anhinga anhinga* (Linnaeus)	10.9

tent. Anaerobic white fibres are of large diameter, with fewer mitochondria, higher activities of glycolytic enzymes, and high glycogen and low lipid content. The properties of the intermediate fibres fall between those of the red and white fibres (George & Berger, 1966).

The mean diameters of muscle fibres in the pectoralis and supracoracoideus muscles of five penguin species are given in Table 2. In each muscle the population of fibres is of fairly uniform cross sectional area. The values for all the penguin muscles fall within the range reported for red fibres of other birds (22.5 to 62.5 μm), although there is considerable overlap with the smaller of the intermediate and white fibres (George & Berger, 1966). As the smallest diameter red fibres occur in the most highly aerobic flight muscles, differences in fibre diameter between penguin species presumably reflect increasing dependence on anaerobic muscle work as fibre diameter increases. Under dark field illumination the little penguin muscle fibres show abundant birefringent inclusions identical in size and distribution with lipid droplets identified in sections stained with oil red O (Mill & Baldwin, 1983). These inclusions are rare in muscle fibres of the king penguin and of intermediate abundance in the muscle fibres of rockhopper, royal and gentoo penguins. Electron micrographs of the little penguin swimming muscles also show numerous large lipid droplets and many large mitochondria with well developed cristae (Mill & Baldwin, 1983). Unfortunately comparable electron microscopy has not been carried out with the other penguin species.

Table 2. Diameter of muscle fibres in penguin pectoralis and supracoracoideus muscles. Values are means with standard deviations in brackets for 353 fibres. Data from Baldwin *et al.* (1984).

Species	Fibre diameter (μm)	
	Pectoralis	Supracoracoideus
King	50.9 (7.6)	60.4 (8.2)
Royal	44.5 (8.4)	42.9 (5.4)
Rockhopper	43.6 (6.6)	43.8 (5.9)
Gentoo	43.2 (6.9)	55.9 (6.4)
Little	36.9 (6.0)	46.2 (6.8)

Muscle myoglobin can serve both as a means of facilitating rapid oxygen transfer from the blood to the muscle cell mitochondria, and as an oxygen store. Thus high myoglobin content can be interpreted as an adaptation to high levels of aerobic metabolism or to prolonged diving (Castellini & Somero, 1981; Pages & Planas, 1983). The large differences between the myoglobin content of the pectoralis muscles of active aerobic flyers and diving penguins (Table 3) clearly imply an adaptation for oxygen storage rather than enhanced oxygen transfer in the latter.

There is a strong correlation between the capacity of skeletal muscle homogenates to buffer pH, and the extent to which the muscle functions anaerobically during locomotion or experiences prolonged hypoxia during diving. A high pH buffering capacity is interpreted as an adaptation for counteracting changes in hydrogen ion concentration associated with anaerobic muscle work (Castellini & Somero, 1981; Hochachka & Mommsen, 1983). The pH buffering capacities of the penguin pectoralis and supracoracoideus muscles are given in Table 4.

Muscle lactate dehydrogenase enzymes are

Table 3. Myoglobin concentration in the pectoralis and supracoracoideus muscles of penguins, and the pectoralis muscles of active aerobic flying birds. Penguin data from Baldwin *et al.* (1984), and data on other birds from Pages & Planas (1983).

Species	Myoglobin ($g\ 100\ g^{-1}$ wet wt muscle)	
	Pectoralis	Supracoracoideus
King	4.3	4.6
Royal	4.7	4.5
Rockhopper	3.7	3.8
Gentoo	4.3	4.3
Little	2.8	2.8
Emperor	2.4	2.4
Adelie	3.6	3.6
Manx Shearwater *Puffinus puffinus* (Brünnich)	0.6	
Black-headed seagull *Larus ridibundus* Linnaeus	0.6	
European Starling *Sturnus vulgaris* Linnaeus	0.2	
Domestic pigeon *Columba livia* Gmelin	0.5	

tetramers composed of two subunit types, M and H. Predominantly aerobic muscles contain isoenzymes with a high proportion of H subunits. These isoenzymes display substrate inhibition at high pyruvate concentrations, and function effectively as lactate oxidases during aerobic muscle work. The proportion of M subunits increases with dependence of the muscle on anaerobic glycolysis. Isoenzymes with a high proportion of M subunits are less sensitive to pyruvate inhibition, and function as pyruvate reductases for the regeneration of cytoplasmic NAD^+ during anaerobic muscle work (Everse & Kaplan, 1973). Wilson *et al.* (1963) showed that the ratio of lactate dehydrogenase activity at low (0.33 mM) and high (10 mM) pyruvate concentration provided an index of the relative contributions of aerobic and anaerobic metabolism in pectoral muscle. Birds with high pyruvate inhibition ratios were sustained flyers utilizing basically aerobic metabolism, while low values occurred in anaerobically powered burst flyers. The relative proportions of M subunits and pyruvate inhibition ratios for lactate dehydrogenases in the pectoralis and supracoracoideus muscles of penguins, and in the pectoralis muscle of selected flying birds, are listed in Table 5.

Table 4. pH buffering capacities of penguin pectoralis and supracoracoideus muscles. Data from Baldwin *et al.* (1984).

Species	Buffering capacity (Slykes)	
	Pectoralis	Supracoracoideus
King	67	67
Royal	59	59
Rockhopper	58	62
Gentoo	56	60
Little	49	53
Emperor	70	68
Adelie	67	68

Extrapolating from the biochemical properties of penguin swimming muscles to swimming and diving behaviour

The biochemical data reveal marked differences between the penguin species in the anaerobic capabilities of the muscles used to power swimming.

The king, emperor and Adelie penguins consistently show highest pH buffering capacities, lowest pyruvate inhibition ratios for lactate dehydrogenase and high propertions of lactate dehydrogenase M subunits. In addition, muscle fibres of the king penguin pectoralis and supracoracoideus are of larger diameter and contain fewer lipid droplets than muscle fibres of royal, rockhopper, gentoo or little penguins.

The anaerobic capabilities of the gentoo swimming muscles fall between those of the king, emperor and Adelie group, and those of the rockhopper and royal penguins. Fibre diameters in the supracoracoideus muscle are second only to those of the king penguin, pH buffering capacities of both pectoralis and supracoracoideus overlap the range observed for the king penguin, the percentage of lactate dehydrogenase M subunits is higher than in the Adelie and emperor, and lactate dehydrogenase inhi-

Table 5. The relative proportion of lactate dehydrogenase M subunits and lactate dehydrogenase pyruvate inhibition ratios for penguin pectoralis and supracoracoideus muscles, and for pectoralis muscles of selected flying birds. Penguin data from Baldwin *et al.* (1984), and data on other birds from Wilson *et al.* (1963).

Species	M subunits (%)		Pyruvate inhibition ratio	
	pectoralis	supracoracoideus	pectoralis	supracoracoideus
King	97.7	98.0	0.7	0.9
Royal	89.9	86.6	1.1	1.1
Rockhopper	90.5	85.4	1.1	1.2
Gentoo	96.0	97.8	1.0	1.0
Little	67.1	58.2	1.3	1.4
Emperor	95.6	93.6	0.6	0.6
Adelie	94.2	94.1	0.7	0.8
Wilson's Storm petrel *Oceanites oceanicus* (Kuhl)	<10		3.8	
Herring Gull *Larus argentatus* Pontopidan			1.8	
Pintail *Anas acuta* Linnaeus			1.6	
Great Blue Heron *Ardea herodias* Linnaeus			1.3	
Common Pheasant *Phasianus colchicus* Linnaeus	100		0.7	

bition ratios are lower than for the rockhopper, royal and little penguins.

The least anaerobic muscles are those of the little penguin. Both the pectoralis and supracoracoideus have abundant oil droplets, the pectoralis has the smallest diameter muscle fibres, and both muscles display the lowest pH buffering capacities, the lowest percentage of lactate dehydrogenase M subunits and highest pyruvate inhibition ratios.

Table 6. Maximum speeds of penguins swimming in the Penguinarium at the Detroit Zoo (from Clark & Bemis, 1979).

Species	Maximum speed (m s^{-1})
Emperor	2.26
King	3.35
Adelie	1.98
Rockhopper	2.18
Little	1.72

As the potential anaerobic capabilities of the swimming muscles are presumed to reflect adaptations associated with bursts of rapid swimming and dives of maximal duration one can pose the questions: (1) how does the ordering of the penguin muscles in terms of anaerobic capabilities agree with actual data on swimming and diving behaviour? (2) What predictions can be made about swimming and diving behaviour in species for which biochemical data, but not behavioural data, are available.

Accurate data on the swimming behaviour of penguins in nature are very limited, and even less information is available on the maximum speeds achieved during short bursts. The highest speed reported by Kooyman *et al.* (1971) for an emperor penguin swimming between two holes in the antarctic ice was 2.67 m s^{-1}. Maximum speeds observed by Clark & Bemis (1979) for emperor, king, Adelie, rockhopper and little penguins swimming in the Penguinarium at the Detroit Zoo are given in Table 6. However, in these cases it is not certain that the penguins were displaying maximum speeds that would require anaerobic muscle work.

More information is available on depths and duration of diving by non-captive penguins. It is generally agreed that the majority of routine natural dives are aerobic with the birds surfacing before oxygen stores are exhausted. As these oxygen stores can be replenished rapidly compared with the time taken to recover from prolonged anaerobic diving, this strategy allows a greater percentage of time to be spent feeding under water (see Butler & Jones, 1982). The very high myoglobin concentrations in the penguin pectoralis and supracoracoideus muscles are an adaptation for extending aerobic dive times. In mammals oxygen stores are linearly related to body mass, and oxygen consumption scales as (body mass)$^{0.75}$. Therefore, all other things being equal, aerobic dive times should be proportional to (body mass)$^{0.25}$ (Scholander, 1940; Calder, 1969; Butler & Jones, 1982). If the same scaling factors hold for penguins, longer maximum aerobic dive times would be expected in the larger species.

Anaerobic capabilities, while not used routinely, do set the maximum duration and depth of dives. They also serve as a safety factor, for example when time must be spent locating breathing holes when swimming under ice, and thereby increase significantly the depth at which prey can be pursued.

Swimming muscles of the two largest penguins, the emperor and king, appear best adapted for prolonged anaerobic work. These penguins are also the most highly accomplished divers. Free ranging king penguins are known to dive routinely beyond 50 m and to reach depths of 240 m (Kooyman *et al.*, 1982), while emperor penguins are reported to remain submerged for as long as 18 min and to dive as deep as 265 m (Kooyman *et al.*, 1971).

The best data on the diving behaviour of little penguins comes from a study of free ranging Bass Strait birds equipped with capillary pressure gauges that give a single maximum depth record (Montague, 1983). Of a total of 30 values obtained over several days, the greatest depth recorded was 60 m, with 25 recordings of maximum depths less than 40 m. Observations of inshore feeding groups showed dive durations of 10 to 15 seconds. As the food items (mainly small fish and cephalopods) utilized by little penguins usually occur in large schools near the surface there may be little need for prolonged deep diving to a point where oxygen stores are exhausted. Such diving behaviour is in keeping with the very limited anaerobic capabilities of the swimming muscles.

The anaerobic capabilities of the gentoo pectoralis and supracoracoideus lie between those of the highly anaerobic emperor and king penguin, and the basically aerobic little penguin muscles. Kooyman (1975) recorded a maximum dive duration of 2 min for gentoos observed feeding in shallow water. The deepest dive reported for a gentoo penguin is of a bird caught in a net at 100 m while feeding on the bottom (Conroy & Twelves, 1972). Free ranging gentoos equipped with capillary depth gauges showed maximum diving depths ranging from 3.5 m to 70 m, with a mean maximum depth of 22.3 m obtained for 19 birds (Adams & Brown, 1983).

So in answer to the first question, there is a good correlation between the anaerobic capabilities for the penguin pectoralis and supracoracoideus muscles and diving behaviour. Using this relationship between muscle biochemistry and diving behaviour established for the emperor, king, gentoo and little penguins, predictions now can be made about the behaviour of other species for which only biochemical data are available. The seven species of penguins examined in this study can be arranged as follows in order of increasing anaerobic capacity of the muscles used to power swimming: little < royal and rockhopper < gentoo < Adelie, emperor and king. This ranking implies that, relative to the gentoo, the Adelie makes greater use of anaerobic metabolism and may undertake longer and deeper dives, while the diving capabilities of the rockhopper and royal penguins lie somewhere between those of the little and gentoo penguins.

References

Adams, N. J. & C. R. Brown, 1983. Diving depths of the gentoo penguin (Pygoscelis papua). Condor 85: 503–504.

Baldwin, J., J.-P. Jardel, T. Montague & R. Tomkin, 1984. Energy metabolism in penguin swimming muscles. Mol. Physiol. 6: 33–42.

Butler, P. J. & D. R. Jones, 1982. The comparative physiology of diving in vertebrates. In D. Lowenstein (ed.), Advances in Comparative Physiology and Biochemistry, 8. Academic Press, N.Y.: 180–364.

Calder, W. A., 1969. Temperature relations and underwater endurance of the smallest homeothermic diver, the water shrew. Comp. Biochem. Physiol. 30: 1075–1082.

Castellini, M. A. & G. N. Somero, 1981. Buffering capacity of vertebrate muscle: correlations with potentials for anaerobic function. J. Comp. Physiol. B 143: 191–198.

Clark, B. D. & W. Bemis, 1979. Kinematics of swimming of penguins at the Detroit Zoo. J. Zool., Lond. 188: 411–428.

Conroy, J. W. H. & E. L. Twelves, 1972. Diving depths of the gentoo penguin Pygoscelis papua and blue-eyed shag Phalacrocorax atriceps from the South Orkney Islands. Bull. br. antarct. Surv. 30: 106–108.

Everse, J. & N. O. Kaplan, 1973. Lactate dehydrogenases. Structure and function. Adv. Enzymol. 37: 61–133.

George, J. C. & A. J. Berger, 1966. Avian Myology. Academic Press, N.Y., 660 pp.

Hartman, F. A., 1961. Locomotor mechanisms of birds. Smithsonian miscellaneous collections 143: 1–91.

Hochachka, P. W. & T. P. Mommsen, 1983. Protons and anaerobiosis. Science 219: 1391–1397.

Kooyman, G. L., 1975. The physiology of diving in penguins. In B. Stonehouse (ed.). The biology of penguins. Macmillan, Lond.: 115–137.

Kooyman, G. L., R. W. Davis, J. P. Croxall & D. P. Costa, 1982. Diving depths and energy requirements of king penguins. Science 217: 726–727.

Kooyman, G. L., C. M. Drabek, R. Elsner & W. B. Campbell, 1971. Diving behaviour of the emperor penguin Aptenodytes forsteri. Auk 88: 775–795.

Lishman, G. S. & J. P. Croxall, 1983. Diving depths of the chinstrap penguin Pygoscelis antarctica. Bull. br. antarct. Surv. 61: 21–25.

Mill, G. K. & J. Baldwin, 1983. Biochemical correlates of swimming and diving behaviour in the little penguin Eudyptula minor. Physiol. Zool. 56: 242–254.

Montague, T. L., 1983. The food and feeding ecology of the little penguin Eudyptula minor at Phillip Island, Victoria, Australia. M.Sc. thesis, Monash University, Australia.

Pages, T. & J. Planas, 1983. Muscle myoglobin and flying habits in birds. Comp. Biochem. Physiol. 74A: 289–294.

Scholander, P. F., 1940. Experimental investigation on the respiratory function in diving mammals and birds. Hvalradets Skrifter Norske Videnskaps-Akademy, Oslo 22: 1–131.

Wilson, A. C., R. D. Cahn & N. O. Kaplan, 1963. Functions of the two forms of lactic dehydrogenase in the breast muscle of birds. Nature 197: 331–334.

Hydrobiologia 165: 263–269 (1988)
J. M. Ferris, H. R. Burton, G. W. Johnstone & I. A. E. Bayly (eds) Biology of the Vestfold Hills, Antarctica
Kluwer Academic Publishers

Studies of viruses in penguins in the Vestfold Hills

I. R. Morgan & H. A. Westbury
'Attwood' Veterinary Research Laboratory, Department of Agriculture and Rural Affairs Westmeadows, Victoria, 3047, Australia

Key words: viruses, penguins, paramyxovirus, serology survey, Antarctica

Abstract

Blood and cloacal swabs were collected from 1 073 Adelie penguins (*Pygoscelis adeliae*) in 13 separate colonies near Davis in the Vestfold Hills, Antarctica. Three paramyxoviruses, distinct from Newcastle disease virus, were isolated from cloacal swabs obtained from penguins at three widely separated colonies. These isolates, together with others obtained from penguins in other areas of Antarctica, suggest that such viruses are widespread in antarctic penguins. Their significance in the ecology of antarctic penguins has not been determined. No evidence of infection with Newcastle disease virus or avian influenza virus was obtained during the survey.

Introduction

Previous studies on Macquarie Island (54°31′S, 158°58′E) (Morgan *et al.*, 1978, 1981) and in the Antarctic at Casey (66°17′S, 110°32′E) (Morgan & Westbury, 1981) have shown that subantarctic and antarctic penguins may become infected with a variety of avian viruses including Newcastle disease virus, influenza viruses and paramyxoviruses. This paper reports the results of sampling from penguins in the vicinity of the Australian National Antarctic Research Expedition (ANARE) station at Davis (68°35′S, 77°58′E). The aim of this survey was to obtain more information on the level and extent of virus infections in antarctic penguins. A further aim was to examine whether there appeared to be any association between the presence of virus infections in penguins and the proximity of human habitation.

Materials and methods

Blood and cloacal samples were collected from Adelie penguins *Pygoscelis adeliae* (Hombron & Jacquinot) at several colonies near Davis between 6 December 1981 and 17 January 1982. Birds were captured from the margins of breeding colonies with the aid of a long handled net. Care was taken not to disturb birds seated on eggs or tending chicks.

Blood samples

Venepuncture of the brachial vein was performed in the mid-third of the humeral part of the flipper. The blood was collected in evacuated glass tubes (Venoject Evacuated Blood Collecting Tubes, Terumo Corporation, Japan) using a system of double ended needles and tube holders.

To reduce the effects of the cold temperatures on clot retraction, the samples were held in the inner layers of the operator's clothing until clot retraction had commenced. The samples were then transferred to foam plastic racks for transport to Davis where they were held for approximately 24 h at 20 °C before centrifugation. Serum was aspirated from the clots and stored in individual sterile plastic vials at −20 °C at the station and during transport to Australia.

Cloacal samples

Cottonwool tipped swabs were used to collect cloacal samples. The swabs were moistened with the transport medium, Brain Heart Infusion Broth (Oxoid Limited, Hampshire, England) containing 1.5 mg ml^{-1} of penicillin and 5 mg ml^{-1} of streptomycin, prior to collection of the sample. Swabs were stored in individual tubes containing 3 ml of the transport medium. This was stored at $-20\,°C$ on return to the station, usually within 8 h of collection of the swab.

Virus isolation

On return to Australia, the cloacal samples were thawed and 0.1 ml aliquots inoculated into the amnio-allantoic sac of 8 d old embryonated eggs. These were incubated for 5 d and the amnio-allantoic fluids harvested from dead embryos from day two of incubation and from all remaining embryos at day five of incubation. Before harvesting of the amnio-allantoic fluid, eggs were chilled at 4 °C for 2–3 h. The amnio-allantoic fluids were tested for the presence of haemagglutinins. Any fluids showing the presence of haemagglutinating activity were tested, by the haemagglutination-inhibition (HI) test, for the presence of Newcastle disease virus (NDV) using antiserum prepared against the V4 strain of NDV (Simmons, 1967).

Haemagglutination test

The HI test was undertaken according to the technique of Allan & Gough (1974) in V-shaped microtitre trays (Cooke Engineering Company, Alexandria, Virginia) using 10 haemagglutinating units of virus, twofold serial dilutions of serum inactivated at 45 °C for 30 min and a 0.75% chicken erythrocyte suspension. The tests used the V4 strain of NDV (Simmons, 1967), avian influenza virus A/fowl/Victoria/1975 (H7, N6) (Turner, 1976) and avian paramyxovirus strains designated APMV-1M and APMV-3M isolated from penguins on Macquarie Island (Morgan *et al.,* 1981). All penguin sera were absorbed with chicken red blood cells before use in the HI tests.

Results

Sample collection

One thousand and seventy-three (1 073) Adelie penguins were sampled at 13 sites. It proved difficult to collect blood samples from almost one half (622/1 030) of the adult Adelie penguins sampled. In most unsuccessful cases venepuncture produced a minimal blood flow. These birds may have had a reduced peripheral blood flow due to the cold conditions. This hypothesis is supported by the observation that venepuncture was most successful on sunny days. No attempt was made to collect blood samples from Adelie penguin chicks due to their small size.

Figure 1 and Table 1 provide details of the sampling sites and numbers of birds sampled.

Most Adelie penguins were sampled between 6 December 1981 and 31 December 1981. The first eggs hatched in most colonies on or about 14 December 1981, so that sampling was conducted over the later part of the incubation period and the early part of chick rearing.

Table 1. Details of penguins sampled at Davis 1981/82.

Location	Adults		Chicks
	No. swabs	No. blood samples	No. swabs
Wyatt Earp Islands	80	64	–
Tryne Island	80	48	–
Un-named Island Sth of Tryne I.	80	37	–
Albino Rookery	80	48	–
Rookery Lake	120	90	–
Lucas Island	80	51	–
Lugg Island	60	34	–
Magnetic Island	87	54	40
Turner Island	80	43	–
Gardner Island	63	45	2
Warriner Island	60	23	–
Hawker Island	80	46	1
Zolotov Island	80	39	–

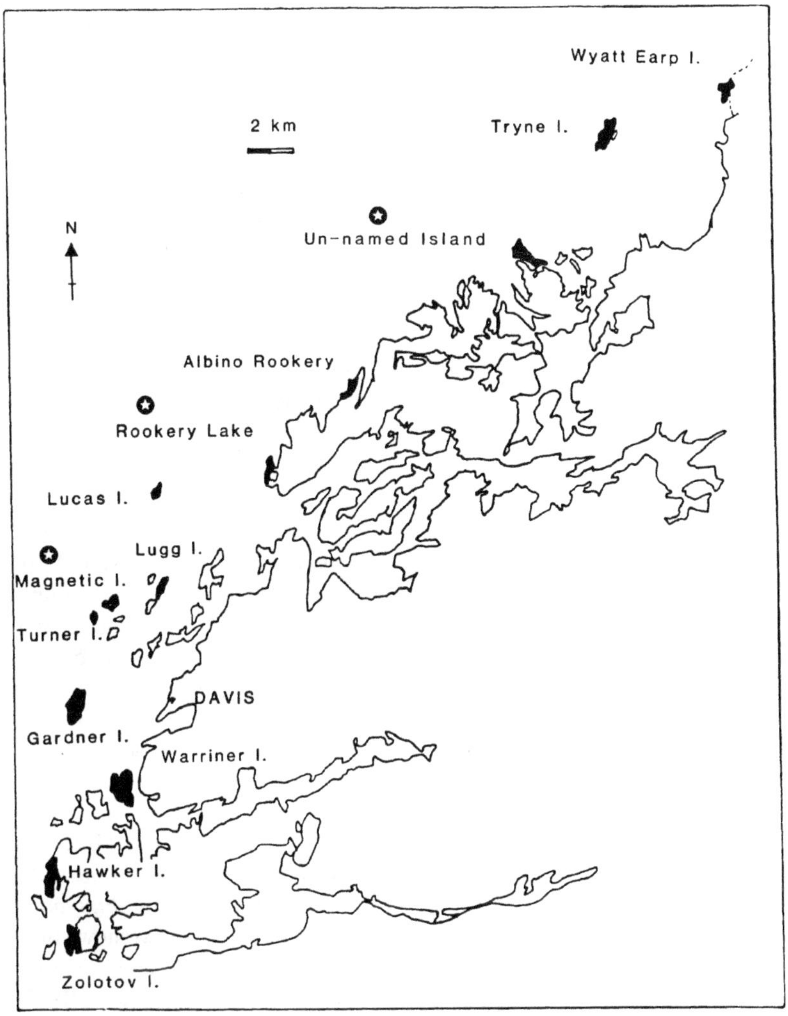

Fig. 1. Location of penguin colonies sampled near Davis. ✪ paramyxovirus isolated from penguins.

No unusual mortality of chicks, or adults, was observed in any of the colonies sampled.

Virus isolation

Three virus isolates were obtained from penguin faecal swabs collected at three widely separated sites at Magnetic Island, Rookery Lake and an island between Long Peninsula and Tryne Island (see Fig. 1). These sites are 4.5, 10 and 23 km respectively from Davis.

The virus isolates were shown not to be NDV or Influenza A viruses by the use of the HI test. On the basis of their morphology under the electron microscope they were classified as paramyxoviruses. The isolates have been concentrated and sent to the NDV Reference Centre at the Ministry of Agriculture, Fisheries and Food Central Veterinary Laboratory, Weybridge, England. They will be tested, against paramyxoviruses previously isolated from subantarctic and antarctic penguins, to determine if they belong to either of the two groups so far identified.

Serology

None of the sera showed evidence of antibody to NDV, Influenza A virus or the paramyxoviruses APMV-1M and APMV-3M.

Discussion

The isolation of paramyxoviruses from Adelie penguins in the Davis region has demonstrated that these viruses are widespread in Antarctica. They have now been isolated at Casey (Morgan & Westbury, 1981) and at Cape Bird (77°13′S, 166°28′E) near McMurdo Sound (Dr F. J. Austin – pers. commun.). The latter virus isolate, from an Adelie penguin, is serologically similar to the two paramyxoviruses isolated from Adelie penguins at Casey. These, in turn are serologically similar to the paramyxoviruses APMV-1M and APMV-5M isolated from royal penguins *Eudyptes schlegeli* Finsch on Macquarie Island (Morgan *et al.,* 1978, 1981). Paramyxoviruses have therefore been isolated from penguins in widely separated areas of the Antarctic and subantarctic. The pathogenicity of the paramyxoviruses isolated from penguins is very low in domestic chickens (W. Allan, pers. commun.), but their effect on penguins is unknown.

In recent years numerous paramyxoviruses, distinct from NDV, have been isolated from birds in a wide variety of environments (Alexander, 1980). Most of these have been isolated from apparently healthy birds and transmission experiments, in general, have shown most of them to be non-pathogenic. Therefore, it is most probable that the paramyxoviruses isolated from penguins produce asymptomatic infections and have little effect on them. However, there is some evidence to suggest that there is some synergy between avian paramyxoviruses and other micro-organisms (Lang *et al.,* 1975). If paramyxoviruses infect penguins already infected with other agents, there is a possibility that a more serious disease syndrome could result.

Infection with some paramyxoviruses isolated from wild birds has been shown to protect domestic chickens from the effects of virulent NDV (Alexander *et al.,* 1979). If the penguin paramyxoviruses exert similar effects they may have a beneficial role in penguins by producing an immunity to any NDV introduced into antarctic populations. Further comment on the significance of the penguin paramyxoviruses awaits the results of work being carried out in England. Quarantine requirements have precluded further work on these viruses in Australia.

No serological evidence of infection with NDV, Influenza A virus or penguin paramyxoviruses was obtained. This is in contrast to the results obtained at Casey (Morgan & Westbury, 1981) and a limited survey of 28 sera from Bechervaise Island and Verner Island near Mawson (67°37′S, 62°52′E) in 1981 (I. Morgan – Department of Agriculture internal report, 1981). The latter sera demonstrated the presence of antibody to NDV in three of 17 Adelie penguins from Bechervaise Island. In addition, antibody to APMV-1M paramyxovirus was demonstrated in nine of 17 Adelie penguins from the same island, while two of 11 birds from Verner Island also showed high titres to the same virus. In a later survey 56 Adelie penguin serum samples from Bechervaise Island showed no evidence of antibody to NDV or APMV-1M paramyxovirus.

At Casey, 74 of 285 sera showed evidence of antibody to APMV-3M paramyxovirus, while serum antibodies to avian influenza virus were detected in six of the 285 sera.

Experience with natural infections in water birds (Spalatin & Hanson, 1975) and with experimental infections of little penguins *Eudyptula minor* (Forster) (Morgan *et al.,* 1978) has shown that antibody is only transiently present in serum after infection with NDV. Similarly, Morgan *et al.* (1978) showed that antibody to APMV-1M paramyxovirus in little penguins was present for only a short time after infection. In both infections antibody appeared to be detectable in serum for only one to three months after infection.

It is possible that this survey collected samples at a time when few birds were infected with viruses. Sampling later in the breeding season may have revealed the presence of antibody to penguin paramyxoviruses or NDV.

In this regard it is interesting to note American experiences at the U.S. base at McMurdo Sound. A highly pathogenic NDV was isolated from Adelie penguins following their transport to America from Antarctica (Pierson & Pfow, 1975). These birds had no contact with other birds from the time of their collection, during transport and after their arrival in the United States. It was concluded that the birds had become infected with NDV in Antarctica. Subsequently, the U.S. Department of Agriculture collected serum and cloacal samples from Adelie penguins at McMurdo Sound but failed to isolate NDV or find NDV antibody in the birds sampled (R. P. Hanson, pers. commun.).

The absence of antibody to APMV-1M or APMV-3M paramyxoviruses in the penguin colonies where virus isolates were obtained suggests two possibilities. The first is that the paramyxovirus isolates from Davis are different to the paramyxoviruses previously isolated from the Antarctic and subantarctic. In this case, antibody to the Davis isolates would not be detected using APMV-1M or APMV-3M paramyxoviruses in HI tests. The second hypothesis is that the penguin paramyxoviruses spread very slowly between birds and that only a few birds had been infected at the time of sampling. Their presence may not have been detected at the level of sampling conducted. With a transient antibody response, previously infected birds would show no evidence of antibody in their serum. That some time factor may be involved in the detection of serum antibody in penguins from colonies where virus infections have occurred is supported by observations on previous surveys.

In the Macquarie Island survey (Morgan *et al.*, 1978, 1981) there was a delay between the isolation of paramyxovirus and the detection of antibody to that virus in penguins from the same colony. Antibodies to APMV-1M paramyxovirus were not detected in royal penguins until 18 d after its isolation from a royal penguin at that site. Antibodies to both NDV and paramyxovirus APMV-1M were not detected until the penguins in the breeding colonies had two to three week old chicks.

Penguin sampling at Casey and Mawson, where a number of penguins with viral antibodies were detected, was carried out at the end of the breeding season. In the Casey sampling, 39% of adult penguins had paramyxovirus antibodies in their serum. These levels suggest that the sampling at Davis should have detected some birds with viral antibody.

The above evidence, together with the observations of the present survey, suggests that there may be some factor which limits the spread of virus infections until chicks are hatched. This may mean that penguins are more susceptible at this time to viral infections. If this were the case it would be important to limit any opportunity for viral contamination of penguin colonies at this time of the year.

Acknowledgements

The Antarctic Division of the Department of Science provided transport and logistic support and without this aid the project could not have been attempted. The support and willing assistance in the field given by Dr Gavin Johnstone of the Antarctic Division is a pleasure to record, as is the field assistance provided by various members of the 1981 ANARE.

References

Alexander, D. J., 1980. Avian paramyxoviruses. Vet. Bull. 50: 737–752.

Alexander, D. J., N. J. Chettle & G. Parsons, 1979. Resistance of chickens to challenge with the virulent Herts 33 strain of Newcastle disease virus induced by prior infection with serologically distinct avian paramyxoviruses. Res. vet. Sci. 26: 198–201.

Allan, W. H. & R. E. Gough, 1974. A standard haemagglutination inhibition test for Newcastle disease. 1. A comparison of macro and micro methods. Vet. Rec. 95: 120–123.

Lang, G., A. Gagnon & J. Howell, 1975. The occurrence of paramyxovirus Yucaipa in Canadian poultry. Can. vet. J. 16: 233–237.

Morgan, I. R. & H. A. Westbury, 1981. Virological studies of Adelie penguins (Pygoscelis adeliae) in Antarctica. Avian Dis. 25: 1019–1026.

Morgan, I. R., I. W. Caple, H. A. Westbury & J. Campbell, 1978. Disease investigations of penguins and elephant seals on Maquarie Island. Res. Project Ser. 47. Department of Agriculture, Victoria, Australia.

Morgan, I. R., H. A. Westbury, I. W. Caple & J. Campbell, 1981. A survey of virus infection in sub-antarctic penguins on Macquarie Island, Southern Ocean. Aust. vet. J. 57: 333–335.

Pierson, G. P. & C. J. Pfow, 1975. Newcastle disease surveillance in the United States. J. am. vet. med. Ass. 167: 801–803.

Simmons, G. C., 1967. The isolation of Newcastle disease virus in Queensland. Aust. vet. J. 43: 29–30.

Spalatin, J. & R. P. Hanson, 1975. Epizootiology of Newcastle disease in waterfowl. Avian Dis. 19: 573–582.

Turner, A. J., 1976. The isolation of fowl plague virus in Victoria. Aust. vet. J. 52: 384.

Hydrobiologia 165: 269–277 (1988)
J. M. Ferris, H. R. Burton, G. W. Johnstone & I. A. E. Bayly (eds) Biology of the Vestfold Hills, Antarctica
Kluwer Academic Publishers

Marking and monitoring studies of the Kerguelen stock of southern elephant seals *Mirounga leonina* and their bearing on biological research in the Vestfold Hills

M. N. Bester
Mammal Research Institute, University of Pretoria, Pretoria 0002, South Africa

Key words: marking, monitoring, elephant seals, Kerguelen Province

Abstract

Southern elephant seals *Mirounga leonina* breed and moult on many subantarctic islands during the austral spring and summer. In the Kerguelen Province the subpopulations of *M. leonina* at Kerguelen (49°21′S, 70°12′E), Marion (46°54′S, 37°45′E) and Possession (46°25′S, 51°45′E) Islands have declined since 1970 and their present status at Heard Island (53°05′S, 73°30′E) is unknown. Population studies during their terrestrial phase have failed to explain the declines. Long distance movements of individuals between the subpopulations in question and also the Vestfold Hills, Antarctica (68°35′S, 77°58′E) have been recorded. The availability of food resources, competition with rapidly increasing fur seal populations and competition with fishing fleets have all been implicated in their decline. These explanations assume that communal feeding grounds are utilized. As they are predators entirely dependent on marine feeding, a study of their spatial and temporal distribution during their pelagic existence is of the utmost importance. Parameters describing growth, reproduction rates, population dynamics, and feeding ecology of the subpopulations in the Kerguelen Province may furthermore serve as indices of change within the marine ecosystem. The presence of a relatively large and predominantly male nonbreeding population of *M. leonina* at the Vestfold Hills in Antarctica which originates from the Kerguelen/Heard Island group, and which shows annual return, should be included in the marking and monitoring studies of the Kerguelen stock of southern elephant seals. Studies here, including an update of the size and social structure of the Heard island subpopulation, may elucidate the observed decline of, in particular, the adult bull component of the breeding stock.

Introduction

The southern elephant seal *Mirounga leonina* Linnaeus breeds on islands on both sides of the Antarctic Convergence and comprises three stocks – these being referred to as the South Georgia, Macquarie and Kerguelen stocks (Laws, 1960). The last includes not only elephant seals of the Kerguelen archipelago, but also those found at Heard, Marion and Prince Edward Islands, Iles Crozet, and Amsterdam and St. Paul Islands (Fig. 1).

Two pelagic (feeding) and two terrestrial (fasting) phases can be distinguished in the annual cycle of at least the adult breeding elephant seals. The breeding season lasts from August to November in the Kerguelen Province (Angot, 1954; Carrick & Ingham, 1960; Condy, 1979). Bulls haul out first in August and pregnant cows in early September, reaching a peak by mid-October. Cows give birth to a single pup approximately 6 d after hauling out and are mated about 18 d after giving birth; most matings occur between October and early November. Cows return to sea approximately three weeks after birth when the pups are weaned. Cows are gregarious and form

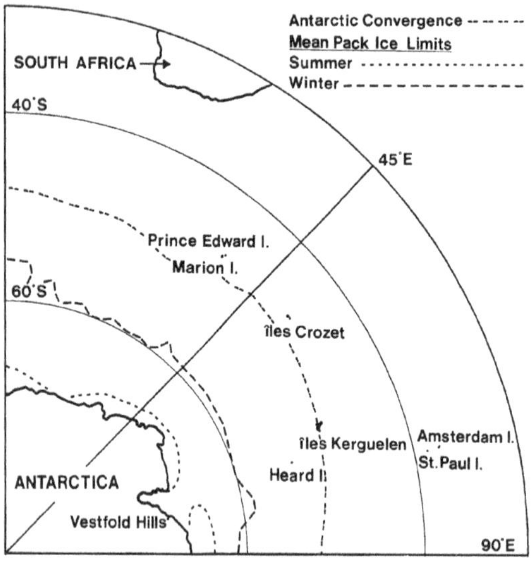

Fig. 1. Distribution of the Kerguelen stock of southern elephant seals.

groups during the breeding season, which attract large bulls (beachmasters) to take up station among them. Groups containing more than 50–60 cows include an additional bull(s) (assistant beachmaster(s)), with challengers in close proximity, usually gathered on beaches with flat, even profiles and easily accessible from the sea.

Weaned pups complete their first moult and remain on the beaches fasting for about a month. By mid-November some local dispersal has occurred and all depart to sea by the second week in January (Condy, 1979; Panagis, 1981; Lenglart & Bester, 1982). Commencing in early November and overlapping slightly with the presence of the last breeding bulls, subadults of both sexes haul out and move inland to moult. Yearlings also haul out to moult, generally remaining on the beaches.

Adults return to moult later, cows reaching peak numbers in January and bulls in February/March. Individuals moult within approximately 30 d and from May to August very few elephant seals occur on the islands; presumably at this time they are feeding at sea.

The present paper outlines and motivates the proposed marking and monitoring studies of southern elephant seals within the Kerguelen Province. Results of ongoing components of these studies are detailed and recommendations for biological research in the Vestfold Hills are made.

The marking and monitoring program

The timing of the haul out cycle is constant from year to year. For example, for island populations within the Kerguelen Province the maximum number of cows breeding is ashore between 15 and 18 October (Angot, 1954; Barrat & Mougin, 1978; Carrick *et al.*, 1962b; Condy, 1979; Van Aarde, 1980a), with marked annual synchrony at particular islands (Barrat & Mougin, 1978; Pascal, 1979; Skinner & Van Aarde, 1983). This allows scheduling of different tasks within a monitoring program, which may include censusing, tagging and resighting of all age and sex classes, at times of expected (predictable) maximum haul out and onshore attendance, and before dispersal.

An account of the seasonal cycle shows that at any time an unknown proportion of the population is at sea. Counts of pups provide the normal index of seal population size (Eberhardt *et al.*, 1979) and these can be combined with data on the age structure of the population to estimate the total population size (Laws, 1960; McCann, 1985). Curves that predict total pup production have been developed. They depend on the relationship between pup number and the number of cows ashore, cows and weaned pups ashore, and cows, weaned pups and dead pups ashore (Pascal, 1979; Van Aarde, 1980a; McCann, 1985). The relationship between census date and the proportion of cows ashore is particularly useful when it takes observers extended periods to visit the entire census area.

The use of mark/recapture methods in estimating pup production on the main breeding beaches at Kerguelen, within the constraints imposed by the validity of model assumptions, may be feasible and would certainly reduce the number of people required for a full census (13 days for two people) at the peak of pupping. Underyearlings (weaned, fully moulted pups) tagged in one locality randomly mix with their year class within a well-defined dispersion area by late November, and before they become completely pelagic (Lenglart & Bester, 1982).

To derive total population size, from its relationship with annual pup production for a particular population, requires the construction of life tables using information on survival, fecundity and sex ratio. This has been done for the South Georgia elephant seal population (Laws, 1960; McCann, 1981, 1985). It depends on a sufficiently large and random sample (150 according to Caughley, 1977) for analysis of the age structure of the population, and on sightings of known-aged (marked) animals. Where culling is ruled out, as in small populations, resighting of tagged seals and the building up of case histories of individuals through regular intensive searches, particularly during the breeding season, are necessary. This should yield information on female reproductive rates, age of recruitment to the breeding population, and longevity. This necessitates an intensive tagging/resighting effort which will be affected by the degree of female and male site fidelity, emigration and tag loss, all of which must be evaluated. Elephant seals have been tagged regularly since 1973 at Marion Island using monel metal tags (Condy & Bester, 1975), and since 1983 using plastic rototags. At Iles Kerguelen, seals have been tagged sporadically since 1975 (Van Aarde & Pascal, 1980; Lenglart & Bester, 1982), despite attempts to do so annually. Only limited tagging has been carried out at Ile de la Possession (Iles Crozet).

As effective monitoring studies should be well designed, carefully controlled and periodically repeated with specific objectives (Laws, 1984), it has been decided to investigate the demography of elephant seals of the Kerguelen Province on a regional level, together with more detailed studies of their diets, the availability of prey and individual movement within the region as a whole (Bester & Jouventin, 1984). The rationale for this lies in the disquieting population trends of southern elephant seals in this region as set out below.

Population trends of southern elephant seals

The number of breeding elephant seals have been declining since 1970 at the Prince Edward Islands (Marion Island), Iles Crozet (Ile de la Possession) and Iles Kerguelen (Peninsule Courbet) (Barrat & Mougin, 1978; Condy, 1978; Van Aarde, 1980a). These decreases are all of the same order varying from 4.6% a^{-1} (Kerguelen; 1970–1977; 2 data points) to 8.0% a^{-1} (Marion; 1973–1982; 9 data points) for adult cows (Van Aarde, 1980a; Skinner & Van Aarde, 1983). The reason for this decline is obscure and it is possible that factors implicated in these trends may vary with subpopulations.

Terrestrial phase

The Marion Island population has declined stepwise since 1973 (Skinner & Van Aarde, 1983) and breeding seals occur at very low densities on land, ruling out any intraspecific density dependent effects during their terrestrial phase. Here the decrease in cow numbers and decline in the annual pup crop are ascribed to a possible malfunction of a mechanism ensuring pregnancy due to a scarcity of bulls (Skinner & Van Aarde, 1983). At Iles Kerguelen, on the other hand, the population decline since 1970 may be part of a long-term fluctuation in the breeding population size related to density. An analysis based on a subdivision of the beaches on the Peninsule Courbet into five distinct regions, for which comparable data on breeding cow numbers during various years since 1952 were available, showed that the degree of long-term fluctuations in cow numbers increased with density (Van Aarde, 1980a). Slight increases in cow density on densely inhabited coastlines resulted in large increases in harem sizes and a decrease in the number of harems (Van Aarde, 1980b). Here a density dependent mechanism has been identified which acts primarily through pup mortality as a function of the density of bulls active in and around harems, bull numbers being positively correlated with cow numbers (Bester & Lenglart, 1982).

Both the Kerguelen and Marion elephant seal breeding populations have, however, shown a change in the adult sex ratio in favour of cows, which suggests a selective removal of bulls during their pelagic phase (Van Aarde, 1980a; Bester & Lenglart, 1982; Skinner & Van Aarde, 1983). Factors responsible have not been identified. Populations of subantarctic fur seals *Arctocephalus tropicalis* (Gray) and/or antarctic fur seals *Arctocephalus gazella* (Peters), sharing the island habitats with the elephant seals during the austral summer, do not compete with them on land as they are spatially and temporally separated during their breeding season (Ling, 1969; Bester, 1982).

Pelagic phase

Killer whales *Orcinus orca* Linnaeus are abundant around Marion Island and Ile de la Possession and show a regular annual visitation cycle closely related to the seasonality in occurrence of elephant seals (Condy *et al.*, 1978). They may have contributed to the decline of elephant seals (Barrat & Mougin, 1978; Condy, 1978). This appears not to be the case at Iles Kerguelen where killer whales were seldom seen (Van Aarde, 1980a), and the magnitude of the decline of elephant seals suggests that other factors, for example the availability of food, are involved (Condy, 1978; Van Aarde, 1980a).

The diet and foraging behaviour of elephant seals, both geographically and with respect to the position in the water column, are poorly known. Laws (1977, 1984) and McCann (1985) concluded that elephant seals consume approximately 75% squid and 25% fish, feeding mainly on fish in inshore waters and on cephalopods elsewhere (Laws, 1956, 1984). The large, rapidly expanding *A. tropicalis* population at Marion Island (Kerley, 1983), which feeds on fish, cephalopods and euphausiids (Rand, 1956; Condy, 1981), may be competing with the declining elephant seal population, especially with the newly independent underyearling elephant seals in the local oceanic zone. Underyearling elephant seals appear to spend most of their summer reasonably close to the island after attaining nutritional independence and are probably in competition with the fur seals, many of which make short feeding trips throughout the summer (Condy, 1981). At Kerguelen, however, the fur seal *A. gazella* population is unlikely to compete; it is small (Bester, 1981) and may well utilize a different food resource; for example at South Georgia it feeds largely on krill (Doidge & Croxall, 1985). On the other hand, whereas no commercial exploitation of fish takes place around the Prince Edward Islands and Iles Crozet, commercial catches of fish on the Kerguelen shelf area may have contributed to the downward trend in elephant seal numbers since 1970 (Van Aarde, 1980a; Pascal, 1982) through a reduction in food availability.

In looking for some common factors responsible for the observed declines in the elephant seal subpopulations in the Kerguelen Province, Condy (1984) suggested two possibilities: first, that the factor(s) responsible for the decline is(are) acting simultaneously, and possibly independently, at or near each of the breeding islands when the subpopulations are gathered during spring and summer; and second, that the factor(s) is(are) operating at some other location, such as a common feeding ground, where the subpopulations may mingle in winter.

Indices of change within the marine ecosystem

The general model for self-regulation of populations in long-lived species proposes that it is comprised of stages of regulation with immature mortality rates as the first stage, followed by age of first reproduction and then reproductive rates of adult females. The next and last stage, as populations reach maximum sustainable levels, is assumed to involve changes in adult survival rates. The simplest and most direct mechanism leading to this sequence of events is the over-utilization of some essential resource, food being the best known (Eberhardt, 1977). Reduced food availability would lead to increased juvenile mortality. Growth rates are reduced in the survivors which reach maturity appreciably later, resulting in delayed age of first breeding (Eberhardt, 1977). As reproductive success in adult females is closely correlated with physical condition, reproductive success is very likely to decline just before conditions reach the point where adult survival rate declines in long-lived species with low reproductive rates (Siniff *et al.*, 1978). Laws (1984) furthermore concluded that the principal causative factor for faster growth-rates and associated precocious sexual maturity in the South Georgia elephant seal population as compared to the Macquarie Island population (Carrick *et al.*, 1962a; Bryden, 1968) was greater food availability.

The initial standing stock of fish for the Kerguelen Shelf area (50000 km^2) was estimated to be in excess of 230000 t and with a maximum sustainable yield (MSY) of 80000 t a^{-1} (Everson, 1977), compared with a recently suggested standing stock of only 130000 t and a MSY of about 20000 t a^{-1} (Hureau, 1979). Both estimates are far below the 394000 t of fish the Kerguelen subpopulation of elephant seals would consume annually using a population estimate of 157000 (derived from Van Aarde, 1980a) and biomass and food consumption estimates presented by McCann (1985) if they fed exclusively on the shelf area. McCann (1985) considers that these estimates of fish stocks need revision, and that it would therefore seem premature to attribute the declines in elephant seal numbers at Iles Kerguelen to overfishing in the vicinity (Pascal, 1982; Van Aarde, 1980a). Cow numbers over the entire period tended to increase, and bull numbers to decrease (Van Aarde, 1980a). Since the difference in body size between the sexes is so large, it seems likely that they are not necessarily affected in the same way by environmental changes, and that they possibly exploit different food resources (McCann, 1985).

It can be hypothesized that large sized elephant seals, i.e. older subadult and adult bulls, in this extremely sexually dimorphic species range over and feed in marine areas segregated from adult cows and younger (smaller) age classes of both sexes and may therefore be affected differently by environmental conditions. This postulate is important in view of the proposed selective removal of bulls during their aquatic phase, the overall decline in breeding population sizes (Barrat & Mougin, 1978; Van Aarde, 1980a; Bester & Lenglart, 1982; Skinner & Van Aarde, 1983), and the self-regulating population model advocating security of the fully adult population as being the key to persistence (Eberhardt, 1977; Siniff *et al.*, 1978).

Although elephant seals move widely about the oceans, as evidenced by their visits to the continents abutting on the Southern Ocean, and penetrate both tropical and antarctic waters (King, 1983), Table 1 suggests a tendency for small immatures of both sexes (< 3 a of age) to disperse widely from their natal island(s) within the Kerguelen Province in the vicinity of the Antarctic Convergence. Subadult and adult bulls of 5 a and older, and therefore at or after the growth spurt associated with puberty in the male (Ling & Bryden, 1981) show a southward displacement into colder water as demonstrated by their presence in the Vestfold Hills, 2000 km distant from Iles Kerguelen which is the most northerly known source of southern elephant seals. At least one other extremely dimorphic seal species breeding in high latitudes, the northern fur seal *Callorhinus ursinus* (Linnaeus), which has a short lactation period releasing adult post-partum females from maternal responsibilities relatively early (±4 months) is thought to show sexual segregation during the pelagic phase outside the reproductive season, as adult males remain in high latitude (colder) waters (Gentry, 1981). The migration of different classes within baleen whale species, by sex, age and reproductive status, is staggered in relation to energy needs. Larger and older individuals tend to reach higher latitudes than smaller and younger ones, and the extent to which different migrating species penetrate the colder waters is also correlated with their body size (Laws, 1977).

A study of the trophodynamics, i.e. food consumption in quality and quantity, the feeding behaviour, energetic and biochemical considerations, and the role of elephant seals in the Southern Ocean ecosystem, the implications of this for growth and survival, and the use of growth rates and reproductive parameters as indicators of changes in the ecosystem, are therefore indicated. These are fields of research with most application in management and conservation, in applying the principles embodied in the Convention for the Conservation of Antarctic Seals and the Convention on the Conservation of Antarctic Marine Living Resources (*vide* Laws, 1984: 705).

The Vestfold Hills in perspective

The Vestfold Hills summer nonbreeding population of southern elephant seals is almost totally male with Class III animals (estimated at 6–10 a old) and to a lesser extent Class IV animals (8–12+ a old), in the majority (Tierney, 1977). He suggested that the physiological need to moult in a terrestrial site,

Table 1. Resightings of southern elephant seals tagged at localities within the Kerguelen Province.

Locality marked	Resight Date/Locality	[a]Age	Sex	Displace-ment (km)	Source
Heard Island	Nov. 1953 Marion I.	2	♀	2740	Carrick & Ingham (1960)
(53°05′S, 73°30′E)	Nov. 1953 Marion I.	2	♂	2740	Carrick & Ingham (1960)
	Aug. 1957 Vestfold H.	5	♂	1600	Ingham (1960)
	Nov. 1957 Vestfold H.	6	♂	1600	Ingham (1960)
	– 1960 Kerguelen I.	6/7	♂	440	Bajard (1962)
Iles Kerguelen	Apr. 1980 Marion I.	2/3	♀	2460	MRI Records
(49°21′S, 70°12′E)	Feb. 1983 Heard I.	6	♂	440	Tremont & Vining (1983)
	Mar. 1980 Heard I.	3	♂	440	Burton (1985)
	Mar. 1980 Heard I.	3	–	440	Burton (1985)
	Mar. 1980 Vestfold H.	3	♂	2000	Burton (1985)
	Feb. 1984 Vestfold H.	8	♂	2000	H. R. Burton (in litt.)
Marion Island	– 1978. Possession I.	4	♀	1140	Mus. natn. Hist. Nat., Paris
(46°54′S, 37°45′E)					
Ile de la Possession	Nov. 1980 Marion I.	–	♂	1140	MRI Records
(46°25′S, 51°45′E)	Nov. 1980 } Marion I. Jan. 1982 }	1 } 2 }	♀	1140	MRI Records
	Jan. 1981 Marion I.	1	♂	1140	MRI Records
	Jan. – Mar. 1981 Marion I.	1	♂	1140	MRI Records
	Mar. – May. 1981 Marion I.	1	♀	1140	MRI Records
	Mar. 1981 Marion I.	1	♂	1140	MRI Records
	Apr. 1981 Marion I.	1	♂	1140	MRI Records
Vestfold Hills	Oct. – Nov. 1974 Kerguelen Is.	[b]	♂♂ (×5)[c]	2000	Burton (1985)
(68°38′S, 77°58′E)	Nov. 1975	[b]	♂♂ (×9)[c]	2000	Burton (1985)
	– 1976	[b]	♂♂ (×6)[c]	2000	Van Aarde & Pascal (1980)

[a] Age in years to closest birthdate: assumed to be in October of each year.
[b] Subadults.
[c] Number of animals sighted.

the availability of suitable beaches, and the relative ease with which southern elephant seals can approach the Vestfold Hills from the Southern Ocean, explain the presence of the elephant seals. Significant numbers begin to arrive during February when the fast ice has always broken out and the pack ice is minimal. The very low numbers of cows, most of which were assumed to be mature (Tierney, 1977) may result from most of them being faithful to their natal island for the moult haul out.

At Macquarie Island the moulting season of cows lasts from mid-December to early March, with a peak in late January similar to the cycle at Iles Kerguelen and Heard Island (Angot, 1954; Carrick *et al.*, 1962a; Ling & Bryden, 1981). Most of the moulting season of cows therefore occurs before the favourable period of access to the Vestfold Hills; even if they had been at sea in the Prydz Bay area they could not have gained access to the shore. Similarly, the moulting season of Class III bulls at Macquarie Island lasts from late November to late March, with a modal date of commencement in mid-January (Ling & Bryden, 1981), so that they also should be under-represented at the Vestfold Hills. These bulls may, however, not have been present at their natal islands during the previous breeding season as social maturity is only attained at 10–11 a (South Georgia, 1978) or 14 a (Macquarie Island, 1950s) according to McCann (1981) and Carrick *et al.* (1962b). Consequently, they may have had the opportunity to range further at sea, as they are large enough to travel far from their island of origin (Tierney, 1977); and they may therefore settle near the limits of the geographical range of the species (stock) to form nonbreeding groups at sites such as the Vestfold Hills (Carrick & Ingham, 1960). Mature, pregnant cows show strong fidelity to their birthsites, whilst immature cows are largely at sea, during the breeding season (Nicholls, 1970). This would limit the time available for dispersal between weaning their pups and the moult haul out. In this respect, the relatively few cows present, most of which were judged to be mature (Tierney, 1977), may not have been recruited to the breeding population as some may not pup for the first time until 7 a of age (Carrick *et al.*, 1962b) or may have missed a pregnancy.

A nondirectional dispersal of elephant seals throughout the circumpolar feeding zone during winter, and concentration on the breeding and moulting sites in summer were assumed by Carrick *et al.* (1962b). The recorded annual return of some individuals to the Vestfold Hills (Tierney, 1977) and movement (an assumed return) of others from there to Iles Kerguelen (Table 1), however, suggest a two-way directional migration for at least part of the population. A convenient navigational cue for these elephant seals may be the Kerguelen/Gaussberg Ridge, an extensive system of submarine plateaux and banks forming an almost continuous shallow water connection 200 m – 600 m deep between Iles Kerguelen, Heard Island and the Antarctic Continent (Williams, 1983). Tierney (1977) postulated that the elephant seals follow the Kerguelen Rise to the pack ice edge, then feed along this to the Vestfold Hills where they moult.

The presence of an unknown number of large, predominantly socially immature males feeding in the Prydz Bay area is suspected from their occurrence ashore at the Vestfold Hills. They presumably take advantage of the greater food abundance during the austral summer (Foxton, 1956; Laws, 1977) which would explain their penetration of polar waters. Burton (1985) furthermore believes that this preponderance of immature males migrating to beaches far distant from their birth and potential breeding sites, suggests food limitation near, or overcrowding on, these sites. Since this group largely represents possible recruits to the adult bull breeding population of the Kerguelen and Heard Islands, and possibly the Kerguelen Province as a whole where they seem to be in difficulty, monitoring of the Vestfold Hills nonbreeding population is indicated. The prospect of annual observations of elephant seals at two sites approximately 2000 km apart along a possible migration route extending from the Antarctic Convergence to the Antarctic Continent, suggests the deployment and subsequent recovery of sophisticated instrumentation such as Time/Depth Recorders (TDR's) and Platform Transmitter Terminals (PTT's) for use in satellite tracking of individuals. In conjunction with mass flipper tagging this will provide information on the spatial and temporal distribution of southern elephant seals during their little known aquatic phase. Together with a study on the trophodynamics of the elephant seals with its implications for growth and survival as has been detailed in the present paper, it would allow an evaluation of possible resource competition with man in the region as a whole.

Acknowledgements

I am indebted to the South African Department of Transport for providing logistical and financial support on the advice of the South African Scientific Committee for Antarctic Research (SASCAR) who supported the research on the islands. Terres Australes et Antarctiques Francaises (TAAF) provided the logistics during the collaborative programme at Iles Kerguelen, and SASCAR provided funds to attend the present symposium, for which I am grateful.

References

Angot, M., 1954. Observations sur les mammifères marins de l'Archipel de Kerguelen – avec une étude détaillée de l'éléphant de mer, Mirounga leonina (L.). Mammalia 18: 1–111.

Bajard, P., 1962. L'éléphant de mer (Mirounga leonina). Biologie-Exploitation du troupeau hôte de l'Archipel de Kerguelen. Thèse, Faculté de Médecine et de Pharmacie de Lyon, 57 pp.

Barrat, A. & J. L. Mougin, 1978. L'éléphant de mer Mirounga leonina de L'Ile de la Possession, Archipel Crozet (46°25′S, 51°45′E). Mammalia 42: 143–173.

Bester, M. N., 1981. Fur seals Arctocephalus gazella and leopard seals Hydrurga leptonyx at the Courbet Peninsula, Kerguelen. S. afr. J. antarct. Res. 10/11: 35–37.

Bester, M. N., 1982. The effect of the subantarctic environment on aspects of the terrestrial phase of fur seal populations. Com. nat. fr. Rech. antarct. 51: 469–476.

Bester, M. N. & P. Jouventin, 1984. Rationale and strategy for a collaborative research programme between SASCAR and TAAF on pinnipeds inhabiting south Indian Ocean islands (the Kerguelen province). S. afr. J. Sci. 80: 32–34.

Bester, M. N. & P-Y. Lenglart, 1982. An analysis of the southern elephant seal Mirounga leonina breeding population at Kerguelen. S. afr. J. antarct. Res. 12: 11–16.

Bryden, M. M., 1968. Control of growth in two populations of elephant seals. Nature 217: 1106–1108.

Burton, H. R., 1985. Tagging studies of male southern elephant seals (Mirounga leonina L.) in the Vestfold Hills area, Antarctica, and some aspects of their behaviour. In J. K. Ling & M. M. Bryden (eds), Studies of Sea Mammals in South latitudes. S. aust. Mus., Adelaide: 19–30.

Carrick, R. & S. E. Ingham, 1960. Ecological studies of the southern elephant seal, Mirounga leonina (L.), at Macquarie Island and Heard Island. Mammalia 24: 325–342.

Carrick, R., S. E. Csordas & S. E. Ingham, 1962a. Studies on the southern elephant seal, Mirounga leonina (L.) 4. Breeding and development. C.S.I.R.O. Wildl. Res. 7: 161–197.

Carrick, R., S. E. Csordas, S. E. Ingham & K. Keith, 1962b. Studies on the southern elephant seal, Mirounga leonina (L.) 3. The annual cycle in relation to age and sex. C.S.I.R.O. Wildl. Res. 7: 119–160.

Caughley, G., 1977. Analysis of vertebrate populations. J. Wiley & Sons, N.Y., 234 pp.

Condy, P. R., 1978. The distribution and abundance of southern elephant seals Mirounga leonina (Linn.) on the Prince Edward Islands. S. afr. J. antarct. Res. 8: 42–48.

Condy, P. R., 1979. Annual cycle of the southern elephant seal Mirounga leonina (Linn.) at Marion Island. S. afr. J. Zool. 14: 95–102.

Condy, P. R., 1981. Annual food consumption and seasonal fluctuations in biomass of seals at Marion Island. Mammalia 45: 15–22.

Condy, P. R., 1984. The population of the southern elephant seal, Mirounga leonina, at Marion Island, 1973–1983. S. afr. J. Sci. 80: 26–27.

Condy, P. R. & M. N. Bester, 1975. Notes on the tagging of seals on Marion and Gough Islands. S. afr. J. antarct. Res. 5: 45–47.

Condy, P. R., R. J. Van Aarde & M. N. Bester, 1978. The seasonal occurrence and behaviour of killer whales Orcinus orca, at Marion Island. J. Zool., Lond. 184: 449–464.

Doidge, D. W. & J. P. Croxall, 1985. Diet and energy budget of the Antarctic fur seal Arctocephalus gazella. In W. R. Siegfried, P. R. Condy & R. M. Laws (eds), Nutrient cycling and food webs in the Antarctic. Springer-Verlag, Berlin: 543–550.

Eberhardt, L. L., 1977. Optimal policies for conservation of large mammals, with special reference to marine ecosystems. Envir. Cons. 4: 205–212.

Eberhardt, L. L., D. G. Chapman & J. R. Gilbert, 1979. A review of marine mammal census methods. Wildl. Monogr. 63, 46 pp.

Everson, I., 1977. The living resources of the Southern Ocean. FAO United Nations Southern Ocean Fisheries Programme, GLO/SO/77/1, 156 pp.

Foxton, P., 1956. The distribution of the standing crop of zooplankton in the Southern Ocean. Discovery Rep. 18: 191–236.

Gentry, R. L., 1981. Northern fur seal Callorhinus ursinus (Linnaeus, 1758). In S. H. Ridgway & R. J. Harrison (eds), Handbook of Marine Mammals. Academic Press, Lond. 1: 143–160.

Hureau, J. C., 1979. La faune ichtyologique du secteur indien de l'Océan Antarctique et estimation du stock de poissons autour des Iles Kerguelen. Mém. Mus. natn. Hist. Nat., Paris; N.S., Série C, 43: 235–247.

Ingham, S. E., 1960. The status of seals (Pinnipedia) at Australian antarctic stations. Mammalia 24: 422–430.

Kerley, G. I. H., 1983. Relative population sizes and trends, and hybridization of fur seals Arctocephalus tropicalis and A. gazella at the Prince Edward Islands, Southern Ocean. S. afr. J. Zool. 18: 388–392.

King, J. E., 1983. Seals of the world. Br. Mus. (Nat. Hist.), Lond., 240 pp.

Laws, R. M., 1960. The southern elephant seal (Mirounga leonina Linn.) at South Georgia. Norsk Hvalfangst-Tidende 10 & 11: 466–476; 520–542.

Laws, R. M., 1956. The elephant seal (Mirounga leonina Linn.). II. General, social and reproductive behaviour. Sci. Rep. Falkland Is. Dep. Surv. 15: 1–88.

Laws, R. M., 1977. Seals and whales of the Southern Ocean. Phil. Trans. r. Soc., Lond. B 279: 81–96.

Laws, R. M., 1984. Seals. In R. M. Laws (ed.), Antarctic Ecology. Academic Press, Lond. 2: 621–715.

Lenglart, P-Y. & M. N. Bester, 1982. Post-weaning dispersion of southern elephant seal Mirounga leonina underyearlings at Kerguelen. Rev. Ecol. (Terre Vie) 36: 175–186.

Ling, J. K., 1969. A review of the ecological factors affecting the annual cycle in island populations of seals. Pacific Sci. 23: 399–413.

Ling, J. K. & M. M. Bryden, 1981. Southern elephant seal Mirounga leonina Linnaeus, 1758. In S. H. Ridgway & R. J. Harrison (eds), Handbook of Marine Mammals. Academic Press, Lond. 2: 297–327.

McCann, T. S., 1981. The social organization and behaviour of the southern elephant seal, Mirounga leonina (L.). Ph.D. Thesis, University of Lond., 163 pp.

McCann, T. S., 1985. Size, status and demography of southern elephant seal Mirounga leonina populations. In J. K. Ling & M. M. Bryden (eds), Studies of Sea Mammals in South latitudes. S. aust. Mus., Adelaide: 1–17.

Nicholls, D. G., 1970. Dispersal and dispersion in relation to the birthsite of the southern elephant seal, Mirounga leonina (L.), of Macquarie Island. Mammalia 34: 598–616.

Panagis, K., 1981. Local movement of southern elephant seal pups Mirounga leonina (Linn.) at Marion Island. S. afr. J. antarct. Res. 10/11: 28–31.

Pascal, M., 1979. Essai de dénombrement de la population d'éléphants de mer (Mirounga leonina L.) des Iles Kerguelen (49°S, 69°E). Mammalia 43: 147–159.

Pascal, M., 1982. Evolution numérique de la population d'éléphants de mer (Mirounga leonina L.) de l'Archipel des Kerguelen au cours des 30 dernières années. Com. nat. fr. Rech. Antarctiques 51: 119–135.

Rand, R. W., 1956. Notes on the Marion Island fur seal. Proc. zool. Soc. Lond. 126: 65–82.

Siniff, D. B., I. Stirling & L. L. Eberhardt, 1978. Problems in the conservation of polar marine mammals. In M. A. McWhinnie (ed.), Polar Research, To the Present and the Future. Westview Press, Inc. U.S.A.: 161–174.

Skinner, J. D. & R. J. Van Aarde, 1983. Observations on the trend of the breeding population of southern elephant seals, Mirounga leonina, at Marion Island. J. appl. Ecol. 20: 707–712.

Tierney, T. J., 1977. The southern elephant seal, Mirounga leonina (L.), in the Vestfold Hills, Antarctica. Aust. Wildl. Res. 4: 13–24.

Tremont, S. & R. Vining, 1983. Elephant seal observations. In R. Vining (ed.), Heard Island Expedition 1983, Scientific Reports. Garvan Institute of Medical Research. St. Vincent's Hospital, Sydney: 25.

Van Aarde, R. J., 1980a. Fluctuations in the population of southern elephant seals Mirounga leonina at Kerguelen Island. S. afr. J. Zool. 15: 99–106.

Van Aarde, R. J., 1980b. Harem structure of the southern elephant seal Mirounga leonina at Kerguelen Island. Rev. Ecol. (Terre Vie) 34: 31–44.

Van Aarde, R. J. & M. Pascal, 1980. Marking southern elephant seals at Iles Kerguelen. Polar Rec. 20: 62–65.

Williams, R., 1983. The inshore fishes of Heard and McDonald Islands, Southern Indian Ocean. J. Fish Biol. 23: 283–292.

Hydrobiologia 165: 279–284 (1988)
J. M. Ferris, H. R. Burton, G. W. Johnstone & I. A. E. Bayly (eds) Biology of the Vestfold Hills, Antarctica
Kluwer Academic Publishers

Variations in underwater vocalizations of Weddell seals (*Leptonychotes weddelli*) at the Vestfold Hills as a measure of breeding population discreteness

J. A. Thomas[1,4], R. A. Puddicombe[2,5], M. George & D. Lewis[3]
[1]*Hubbs Marine Research Institute, 1700 South Shores Road, San Diego, California, 92109, USA;* [2]*Antarctic Division, Department of Science, Kingston, Tasmania, 7050, Australia;* [3]*Oceanic Research Foundation, P.O. Box 247, Windsor, N.S.W., 2756, Australia;* [4]*Present address: Naval Ocean Systems Center, P.O. Box 997, Kailua, Hawaii, 96734, USA;* [5]*Present address: Whales and Marine Section, Australian National Parks and Wildlife Service, G.P.O. Box 636, Canberra, A.C.T., 2601, Australia*

Key words: Weddell seal, vocalization, Vestfold Hills, Antarctica

Abstract

Weddell seal vocalizations from Davis Station showed similarities to those from McMurdo Sound and Palmer Peninsula, but none were identical. One vocalization, DD1, was unique to Davis Station. At all sites trills, or territorial defense calls, were the most common and had more types than other calls. Chugs, an aggressive sound, were common at all sites. Weddell seal calls from Davis Station showed similarities to sounds from McMurdo by sharing the use of prefixes and suffixes. Sounds from Davis Station shared the use of both ascending and descending trills and whistles with calls from Palmer Peninsula.

Commonalities in underwater vocalizations may indicate the degree of mixing between breeding populations, in which case, Weddell seals in the Davis area probably are from a breeding population distinct from those at either McMurdo Sound or Palmer Peninsula.

Introduction

The vocal nature of Weddell seals (*Leptonychotes weddelli*) has intrigued researchers throughout the history of antarctic exploration. Early accounts of explorers mention seals making 'musical noise' (Wilson, 1905). Lindsey (1937) made phonograph recordings of Weddell seals at the Bay of Whales. Schevill & Watkins (1971) and Ray & Schevill (1967) made numerous recordings of underwater sounds in McMurdo Sound. Thomas & Kuechle (1982) quantitatively examined the acoustic characteristics of Weddell seal sounds in McMurdo Sound and classified them into discrete categories. These underwater sounds consisted of numerous frequency-modulated sweeps, chirps, pulses, and clicks. The functions of Weddell seal vocalizations were determined using playback studies (Thomas *et al.*, 1983) at McMurdo Sound. In the same period, Ian Stirling recorded the seals' calls near the Palmer Peninsula. Upon examination, distinct geographic variations in vocalizations between Palmer Peninsula and McMurdo Sound were documented (Thomas & Stirling, 1983). In addition, unique mirror-image vocalization patterns and calls in 'trios' at Palmer Peninsula were documented.

We predicted that variations in the seals' repertoire would reflect the extent of geographic isolation between breeding populations. Therefore, the amount of variability in vocalizations around the continent should reflect the degree of mixing between breeding colonies and would allow us to identify discrete populations. The Australian station at Davis in East Prydz Bay lies approximately between

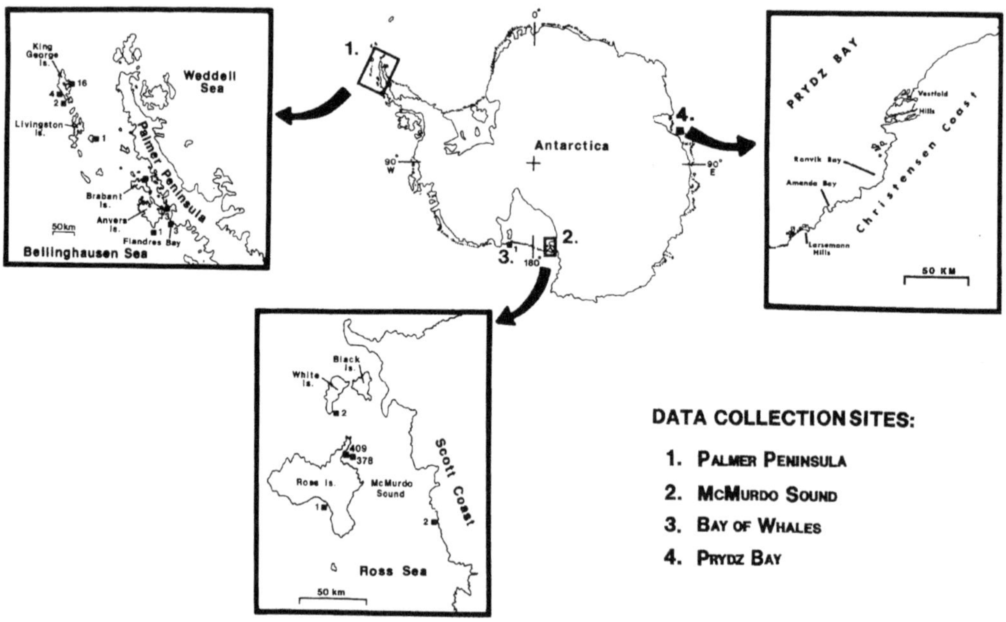

Fig. 1. Map showing the recording sites at four regions of the Antarctic; Palmer Peninsula, McMurdo Sound, Bay of Whales and Prydz Bay.

McMurdo Sound and Palmer Peninsula and makes an ideal site for sampling Weddell seal vocalizations and testing our hypothesis (Fig. 1).

Materials and methods

Breeding colonies of Weddell seals, comprising some 3000 individuals, inhabit the near vicinity of Davis Station (Lugg, 1966). Two research teams conducted studies in East Prydz Bay during 1983 and recorded underwater vocalizations of Weddell seals using similar sampling schemes and equipment to those used by Thomas & Stirling (1983). Rhys Puddicombe made monthly recordings near Davis Station from May to September and fortnightly recordings from October to December 1983 (144 samples). The recording equipment was a Marantz C207-LP cassette recorder, coupled with an ITC 6050C hydrophone (system frequency response linear to 20 kHz). The recording method was to erect a tent near a natural breathing hole in the ice, lower the hydrophone into the hole, and record. During each recording session, a 24 h cycle was sampled using 10 min recordings every 2 h. Recordings were made at five sites (Brookes Hut, mouth of Long Fjord, opposite Brookes Hut, Gardner Island, and Tryne Fjord) in and off the Vestfold Hills.

At approximately the same time, Mimi George and David Lewis made underwater recordings of Weddell seals along the coast extending from 30 km to 100 km south-west of Davis Station. Recordings were made using a Sony cassette recorder and Yak-Yak hydrophone (system frequency response linear to 15 kHz) with a 40 dB preamplifier. One hundred and ten 15 min recordings were made in an opportunistic manner during the months of March 1983 to February 1984. On three occasions, sampling was done every 2 h for a 24 h period at the same location. Sites included colonies at: Cape Drakon, Rauer Islands, Brattstrandoyana Island, Hovdoyana Islands off Amanda Bay, and Larseman Hills.

In the laboratory, Jeanette Thomas made sonograms of sounds using a Unigon SC104A real-time analyzer and a Tekronix hard-copy unit. Frequency and time characteristics of vocalizations were compared. Calls were classified into corresponding categories reported from McMurdo Sound and Palmer

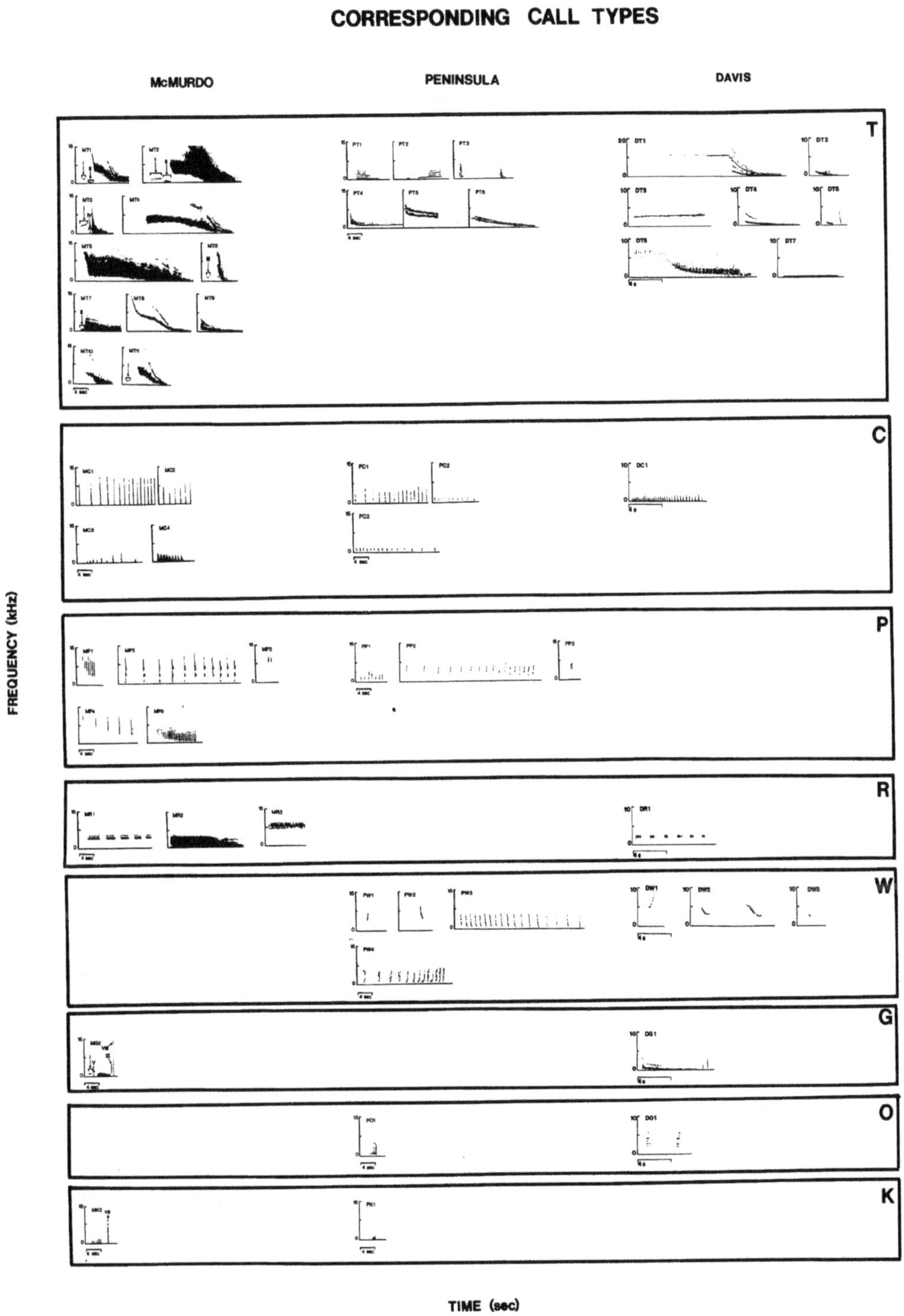

Fig. 2. Sonograms of Weddell seal vocalizations recorded at McMurdo Sound, Palmer Peninsula and Davis Station. These vocalizations have a corresponding form at another site.

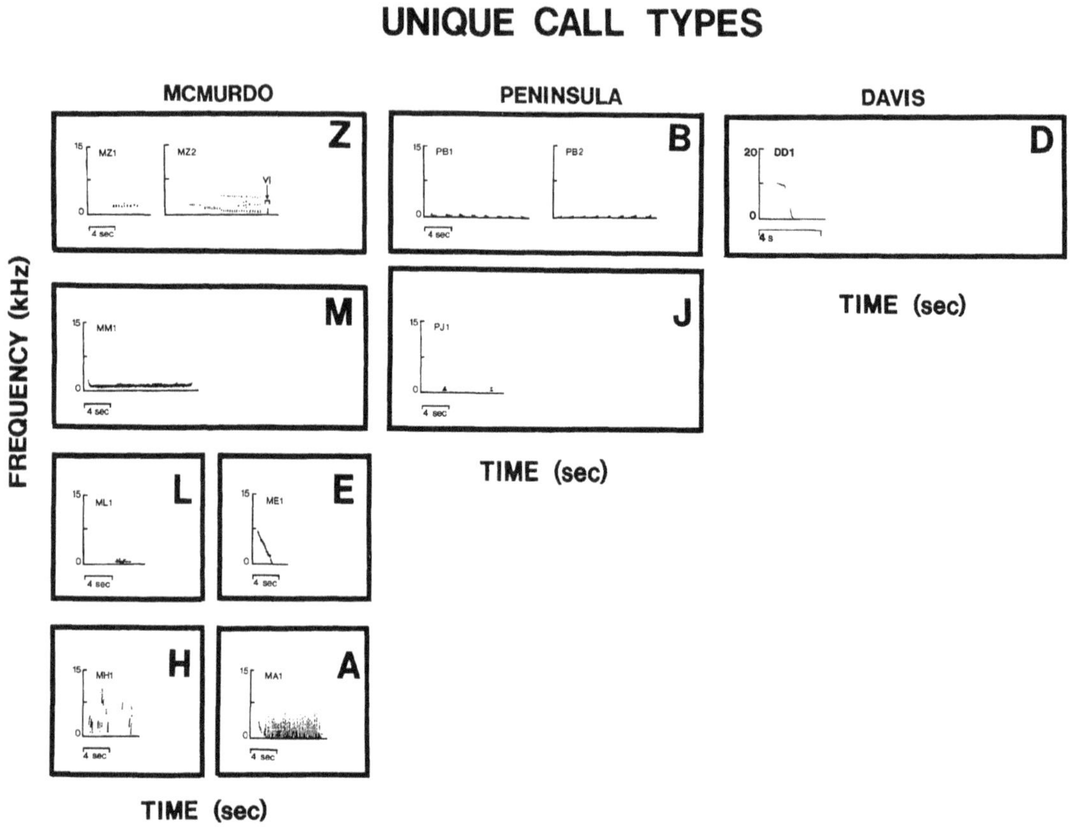

Fig. 3. Sonograms of Weddell seal vocalizations that are unique to only one of the three sites.

Peninsula (Fig. 2) or were assigned to a new category if no similar sounds had been reported (Fig. 3). A comparison of the vocalization patterns from the Vestfold Hills with those recorded at McMurdo Sound and Palmer Peninsula enabled us to determine the extent of similarity.

Results and discussion

Vocalizations of Weddell seals were classified using a three-place code developed by Thomas & Kuechle (1982). The first letter represents the site, the second letter the major call category, and the number the type within that call category. Figures 2 & 3 illustrate the vocalizations present at the three sites. The major call categories from McMurdo Sound (M) include T, C, P, R, G, K, Z, M, E, L, A, H and from Palmer Peninsula (P) include T, C, P, K, W, B, J, O. Types within a category vary from a single type up to 11 variations (e.g. MT11). In keeping with this code, vocalizations recorded near Davis Station were designated with a D. Only one vocalization (DD1) did not fit into an existing major call category and was assigned a new letter (Fig. 3).

Table 1 illustrates the ways in which the calls recorded at Davis Station are similar to those from McMurdo Sound and Palmer Peninsula. The most common Weddell seal vocalization at all three sites was the T or trill vocalization. Thomas *et al.* (1983) reported that T sounds serve a territorial defense function and are therefore prevalent during the breeding season. DT3 and DT7 both ascend in frequency, which makes them more similar to trills at

Table 1. Davis Station compared with McMurdo Sound and Palmer Peninsula detailing the resemblance between Weddell seal vocalizations.

Davis Station	McMurdo Sound	Palmer Peninsula
DT1	initial sweep similar to MT1 & MT5; harmonics similar to MT7 & MT9	harmonics similar to PT1
DT2	sweep similar to MT9	-
DT3	-	ascending sweep similar to PT2
DT4	harmonic structure and sweep similar to MT9	sweep similar to PT4
DT5	prefixes similar to MT1, MT2, MT3, MT6, MT7 & MT11; sweep similar to MT6	-
DT6	sweep similar to MT2 & MT10	-
DT7	-	ascending sweep similar to PT2
DC1	retarding series similar to MC3; frequency similar to MC4	retarding series similar to PC3
DR1	frequency and accelerating series similar to MR1	-
DW1	-	single, ascending similar to PW1
DW2	-	series, descending similar to PW3
DW3	-	-
DG1	acoustic structure and suffix similar to MG2	-
DO1	-	harmonics, single, or series similar to PO1
DD1	-	-

Palmer Peninsula. All McMurdo trills have a descending frequency. The prefix in DT5 at Davis Station is similar to trill prefixes at McMurdo Sound (MT1, MT2, MT3, MT6, MT11), whereas such prefixes are absent at Palmer Peninsula. The use of a similar initial sweep-frequency occurs in some trills from both Davis (DT1) and McMurdo Stations (MT1, MT5).

The W, or whistle vocalization, is absent from the McMurdo repertoire, thus making all DW vocalizations more similar to those from Palmer Peninsula. Like the Palmer Peninsula whistles, those recorded at Davis Station can be either ascending or descending in frequency, single or in series. The function of this sound is unknown.

The G, or gutteral glug vocalization, is absent from the Palmer Peninsula repertoire. DG1 is very similar to MG2 both in acoustic structure and in the use of a two-part suffix. Thomas *et al.* (1983) reported this sound as a low intensity aggressive sound.

The DO1, or 'ooo' vocalization, is very similar to PO1 in frequency and harmonic structure. Similarly, both DO1 and PO1 are used as single sounds or in evenly repeated series. The 'ooo' vocalization is absent from McMurdo. The function of this sound is unknown.

The DC1, or chug vocalization, is similar to both MC3 and PC3 in having a retarding repetition rate; however, in terms of frequency structure DC1 is most similar to MC4. According to Thomas *et al.* (1983), chugs are a very common vocalization and are used in aggressive encounters. At both Palmer Peninsula and McMurdo Sound several types of chugs were present, but at Davis Station only a single type was encountered.

The DR1, or 'cricket call' vocalization, is very similar to MR1 in both frequency structure and the accelerating repetition rate. This sound is absent from the Palmer Peninsula repertoire. Thomas *et al.* (1983) assigned a non-communicative function to this sound because they could not elicit a response to playbacks of this vocalization.

DD1 occurs only at Davis Station. It has a descending frequency structure, but two separate components. The initial component is an almost pure-tone sweep and the ending component has harmonic structure. The function of this sound is unknown.

As with vocalizations at McMurdo Sound, mirror-image sounds (e.g. PT1 & PT2, PW1 & PW2, PW3 & PW4 and PB1 & PB2) and 'trio' vocalization patterns (PT4, PT5 & PT6) were absent in Weddell seal vocalizations at Davis Station. These vocalization patterns appear to be unique to Palmer Peninsula.

Sounds at Davis Station have similarities with both McMurdo Sound and Palmer Peninsula vocalizations. However, no vocalization at Davis Station is identical to Weddell seal sounds at either McMurdo Sound or Palmer Peninsula. This degree of difference indicates another geographic variant in their repertoire which probably reflects the existence of a distinct breeding population.

Conclusions

1. Vocalizations from Davis Station have some similarities to sounds from both McMurdo Sound and the Palmer Peninsula.
2. All major categories seem to be variable in acoustic structure at different sites.
3. Ascending vocalizations occur at both Davis Station and the Palmer Peninsula, but not at McMurdo Sound.
4. Dependent, auxiliary sounds (prefixes and suffixes), occur at Davis Station and McMurdo Sound, but not at Palmer Peninsula.
5. The predominant vocalization category at all sites is T (trills).
6. All sites have at least one unique major call category. DD1 category is unique to Davis Station.
7. The use of mirror-image vocalizations is unique to Palmer Peninsula.
8. The use of sounds in 'trios' is unique to Palmer Peninsula.
9. Vocalizations used in series can have an even, accelerating, or retarding repetition rate.
10. Underwater vocalizations are probably a good indication of the degree of mixing between breeding populations of Weddell seals around the antarctic continent.
11. Further sampling is needed to determine the minimum distance over which vocal variations can be detected.

Acknowledgements

Analysis of data was conducted at the bioacoustic laboratory at Hubbs Marine Research Institute. Data collection by Mr Puddicombe was supported by the Australian Antarctic Research Program and Davis Station. Funding for the expedition by Ms George and Dr Lewis was from public donation and sponsorship of the Oceanic Research Foundation.

References

Lindsey, A., 1937. The Weddell seal in the Bay of Whales, Antarctica. J. Mammal. 18: 127–144.

Lugg, D. J., 1966. Annual Cycle of the Weddell seal in the Vestfold Hills, Antarctica. J. Mammal. 47(2): 317–322.

Ray, C. & W. E. Schevill, 1967. Noisy underwater world of the Weddell seal. Animals 10: 109–113.

Schevill, W. E. & W. A. Watkins, 1971. Directionality of the sound beam in Leptonychotes weddelli (Mammalia: Pinnipedia). In W. Burt (ed.), Antarctic Pinnipedia. Antarct. Res. Ser. 18. Am. Geophys. Union, Natn. Acad. Sci. – Natn. Res. Coun., Wash. D.C.: 163–168.

Thomas, J. & V. Kuechle, 1982. Quantitative analysis of the underwater repertoire of the Weddell seal (Leptonychotes weddelli). J. Acoust. Soc. Am. 72: 1730–1738.

Thomas, J. & I. Stirling, 1983. Geographic variation in Weddell seal (Leptonychotes weddelli) vocalizations between Palmer Peninsula and McMurdo Sound, Antarctica. Can. J. Zool. 61: 2203–2210.

Thomas, J., K. Zinnel & L. Ferm, 1983. Investigation of Weddell seal (Leptonychotes weddelli) underwater calls using playback techniques. Can. J. Zool. 61: 1448–1456.

Wilson, E., 1905. On the whales, seals, and birds of the Ross Sea and South Victoria Land. In R. F. Scott (ed.), The Voyage of the Discovery 2. Smith, Elder & Co. Lond.: 269–294.

General index

Includes water bodies, geographical place names and areas, research and other institutes, families of living organisms. Genera and species are included in the species index.

Species index

GPSR Compliance
The European Union's (EU) General Product Safety Regulation (GPSR) is a set of rules that requires consumer products to be safe and our obligations to ensure this.

If you have any concerns about our products, you can contact us on

ProductSafety@springernature.com

In case Publisher is established outside the EU, the EU authorized representative is:

Springer Nature Customer Service Center GmbH
Europaplatz 3
69115 Heidelberg, Germany

www.ingramcontent.com/pod-product-compliance
Lightning Source LLC
Chambersburg PA
CBHW080933240926
51849CB00003BA/448
9789061936169